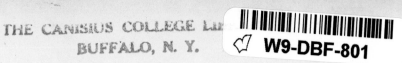
p9 193 line 8 — no need of dogma

196 — just before summary

469 — Bowman

Psychology
of
Adolescence

KARL C. GARRISON

PROFESSOR OF EDUCATION · UNIVERSITY OF GEORGIA

Psychology
of
Adolescence

FOURTH EDITION · NEW YORK · MCMLI

PRENTICE-HALL, INC.

PRENTICE-HALL PSYCHOLOGY SERIES · F. A. MOSS, EDITOR

To my son KARL

Preface

This fourth edition has been written to include the findings from selected recent studies of adolescents. The vast amount of recent published materials has made it necessary to reorganize the materials presented in the previous edition, and in some cases to substitute more recent and inclusive data for data secured from earlier sources. This edition has been enlarged by the addition of materials relative to *physiological changes during adolescence, adolescent peer relations, attitudes and beliefs,* and *moral and spiritual development.* In some areas the amount of material available is very extensive, with the result that many valuable materials had to be omitted; however, the studies reviewed and analyzed are representative, and have been chosen for their clarity, objectivity, and direct application.

The aim of this edition, like that of the three previous ones, is twofold. My experience has led me to believe, first, that its content and method will be welcomed by the many college students who are still in the later stages of adolescence. Those students are seeking information concerning a multitude of psychological problems, especially personality problems, and in writing I have kept these problems constantly in mind. The book is also designed to be of value to those entrusted with the care and guidance of adolescents. Parents—engrossed in domestic duties or in vocational and avocational pursuits—and even teachers too often forget the difficulties that beset youth. I hope that this book may afford both parents and teachers a more appreciative view of adolescents and a fuller recognition of the importance of their transition from childhood to adulthood.

The second aim of this book is to introduce the student to basic experimental studies, and thus lay the foundation for a critical appreciation of new studies that are constantly appearing, for the psychology of the various periods of human growth is at this time rapidly developing. The general student will find the facts actually given in detail in the text supplemented by specific references to sources in the bibliographies which follow the chapters. The more advanced or more alert student should find the sources named in the footnotes additionally helpful in his development of new techniques of study as well as an analytical view of new findings and principles in the field.

The original volume had its inception in my mind while I was an advanced student in genetic psychology at Peabody College. It was here that I first became familiar with Hall's writings and was impressed by the biological conception of individual development and the scientific study of the growing child. Throughout this edition, I have consistently clung to certain fundamental principles of growth and development that I formulated during my years of study and teaching courses in adolescent psychology. To a larger extent than ever before, the adolescent is coming to be regarded as a unified personality that can be neither catalogued by statistical procedures nor stereotyped by special tests. This does not mean that tests and statistics have no place in studying adolescents; they have a very important place. But the important thing is getting a more accurate picture of the adolescent growing and developing in accordance with his genetic constitution and the various environmental forces that have affected him from birth.

I have drawn heavily from recent scientific studies and from current source materials. Thus youth activity in this country, surveys of various aspects of the life of adolescents, clinical studies of adolescents, longitudinal studies, and representative research studies in related fields are reviewed, and are acknowledged by references. But, apart from these acknowledgments, it is difficult to give adequate credit to all the sources to which I am indebted. From correspondence and personal contacts as well as from published materials, I have secured valuable information and special help. I should like to express here my thanks to these associates as well as to the writers and publishers who have permitted the use of quotations and special data from certain studies. Special acknowledgment is due, moreover, to many students of adolescent psychology who have offered suggestions since the publication of the original edition in 1934, the revised edition in 1940, and the third edition in 1946.

Athens, Ga.   K. C. G.

Contents

PART I. INTRODUCTION

CHAPTER I. THE ADOLESCENT AGE

CHAPTER II. PROBLEMS OF ADOLESCENTS

PART II. GROWTH AND DEVELOPMENT DURING ADOLESCENCE

CHAPTER III. PHYSICAL AND MOTOR DEVELOPMENT

CHAPTER IV. PHYSIOLOGICAL CHANGES

CONTENTS

CHAPTER V. EMOTIONAL GROWTH

CHAPTER VI. MENTAL DEVELOPMENT

CHAPTER VII. ADOLESCENT INTERESTS

CHAPTER VIII. SOCIAL GROWTH AND DEVELOPMENT

CONTENTS

CHAPTER IX. GROWTH IN ATTITUDES AND RELIGIOUS BELIEFS

PART III. ADJUSTMENT OF ADOLESCENTS

CHAPTER X. THE ADOLESCENT AND HIS PEERS

CHAPTER XIII. THE ADOLESCENT IN THE COMMUNITY

CHAPTER XIV. THE ADOLESCENT PERSONALITY

CHAPTER XV. PERSONAL AND SOCIAL ADJUSTMENTS

CHAPTER XVI. JUVENILE DELINQUENCY

PART IV. THE GUIDANCE OF ADOLESCENTS

CHAPTER XVII. THE HYGIENE OF ADOLESCENCE: HEALTHFUL PERSONAL LIVING

CHAPTER XVIII. GUIDANCE: MORAL DEVELOPMENT AND CHARACTER FORMATION

CHAPTER XIX. EDUCATIONAL NEEDS OF THE ADOLESCENT

CHAPTER XX. VOCATIONAL CHOICE AND ADJUSTMENT

CHAPTER XXI. ADOLESCENCE AND DEMOCRACY

APPENDIXES

List of Illustrations

List of Tables

PART I

INTRODUCTION

I

The Adolescent Age

The meaning of adolescence. An examination of various definitions of adolescence reveals little difference of opinion regarding the physical facts that constitute the foundation for a general study of adolescence. Usually, adolescence is thought of as that period of life during which maturity is being attained; and especially is this true insofar as maturity relates to the development of the procreative powers of the individual. This period also marks a time in the individual's life when it is difficult to consider him either as a child or as an adult. Observations of, and experiences with, individuals during the "teen" period reveal that there is a fairly distinct time during which the individual cannot be treated as a child, and actually resents such treatment. Yet this same individual is by no means fully mature, and cannot be classed as an adult. During this transition from childhood to adulthood, therefore, the subject is referred to as an adolescent.

G. Stanley Hall[1] was the first to draw a vivid and striking picture of this stage of life, with all its specific characteristics, gradations, and peculiarities. His splendid portrayal of this period as the "storm and stress" time of life caught the attention of all who came into contact with his writings, which were, in fact, so impressive that they dominated the thinking of most American students of adolescent psychology for a number of years. Just three years before his death, he presented a rather clear description of the nature of the "flapper," which, he pointed out, has its beginning with the "teen" years.[2] He cites the definition of the term from the dictionary as: "[one] yet in the nest,

[1] G. Stanley Hall, *Adolescence* (2 vols.). New York: Appleton-Century-Crofts, Inc., 1904.
[2] G. Stanley Hall, "Flapper Americana Novisscina," *Atlantic Monthly,* 1922, Vol. 129, pp. 771–780.

and vainly attempting to fly while its wings have only pin feathers." His conception of adolescence as a transitional period has been described in a vivid and oftentimes exaggerated manner by various students of adolescent psychology.

Although puberty is as old as the human race, the concept of adolescence as a period of life including a number of years is largely a product of nineteenth- and twentieth-century culture. Many writers, following the pattern set forth by Hall, have characterized this age as one of "storm and stress"—indicating that most of the problems appearing among adolescents are inherent in the individuals themselves.

The many studies of human development conducted during the past two decades have shown that all development is a continuous process and that one period cannot be set apart as distinct and different from the preceding or following period. This concept will be emphasized throughout this and subsequent chapters. True, there are some profound changes occurring during adolescence that characterize this stage of life. Furthermore, the individual is reaching out into an enlarged world, and is expanding his mental and social outlook. This transition period is, therefore, characterized as one during which the individual is faced with many problems. Some of the major difficulties encountered will be considered in Chapter II as well as in subsequent chapters.

Physical symptoms of early adolescence. Studies of the physical growth of boys and girls show that there is an increased rate of growth in height just prior to the onset of pubescence. This is discussed at length in Chapter III. Since pubescence appears in girls earlier than in boys, the accelerated rate of growth in height occurs earlier among girls. This is quite noticeable when one observes a group of girls and boys in the seventh and eighth grades of our schools. These girls at the ages of 12, 13, and 14 will be on the average as tall as or taller than the boys of their age level.

There is also a pronounced increase in the rate of growth in weight just prior to the onset of pubescence. Some adolescents gain from 20 to 30 pounds during the year. The girls again pass the boys of their age level for weight during a two- or three-year period. Accompanying this increased growth, one finds important changes in body proportions. There is at first a rapid growth of the arms and legs, to be followed later by a more rapid growth of the trunk of the body. The hands, feet, and nose seem to play an important part in adolescent development. By the time the boy is 13 or 14 years of age his hands and feet have achieved a large percentage of their total development at maturity.

One of the earliest indications of the development of the girl during the pre-adolescent stage is the development of the breasts. The mammary nipple usually doesn't project above the level of the surrounding skin structures until the third year after birth. The nipple after this stage shows a slight elevation above the surrounding structures. There is no further pronounced change for the average girl until around the

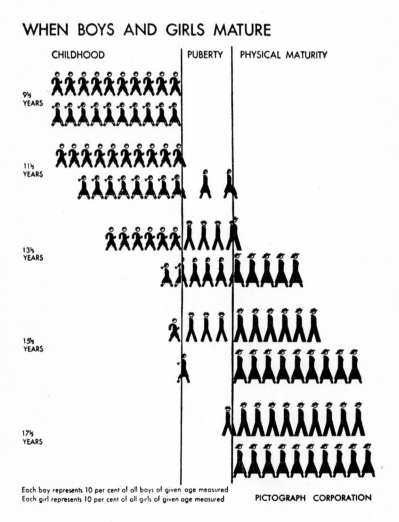

WHEN BOYS AND GIRLS MATURE

Each boy represents 10 per cent of all boys of given age measured
Each girl represents 10 per cent of all girls of given age measured

PICTOGRAPH CORPORATION

FIG. 1. *The Maturity of Boys and Girls*. (From Alice Keliher, *Life and Growth*. New York: D. Appleton-Century Co.)

tenth year, when the so-called "bud" stage appears. This is soon followed by the development of the "primary" breast, resulting mainly
from an increase in the fat surrounding and underlying the papilla
(nipple) and surrounding skin area.

Variation in pubescence. Studies that have been conducted relative
to the beginnings of puberty indicate that there is considerable variation in this phenomenon, and that variation between the sexes is especially pronounced. However, a more careful analysis reveals that a
great deal of overlapping of the sexes exists. Furthermore, the period
of pubescence is a distribution range, not an average chronological age,
since no single classification applies to a majority of a group of boys and
girls.

The study by Ramsey[3] furnishes one of the most complete analyses
available of the onset of pubescence among boys. Complete sex histories were obtained by means of personal interviews of 291 boys between the ages of 10 and 20 years. These boys were from the middle
or upper socio-economic strata of a midwestern city. The different
phases of the sexual development of the boys are shown in Table I.
These data indicate that not all of these sex characteristics appear in a
boy at the same time; however, the thirteen-year level seems to be the
modal period for the appearance of each of these characteristics, with
a distribution range for each characteristic from 10 to 16 years.

TABLE I

Percentage of Each Age Group, Showing Different Aspects of
Sexual Development (*After Ramsey*)

Age group	Ejaculation per cent	Voice change per cent	Nocturnal emission per cent	Pubic hair per cent
10	1.8	0.3	0.3	0.3
11	6.9	5.6	3.7	8.4
12	14.1	20.5	5.3	27.1
13	33.6	40.0	17.4	36.1
14	30.9	26.0	12.9	23.8
15	7.8	5.5	13.9	3.3
16	4.9	2.0	16.0	1.0

[3] G. V. Ramsey, "The Sexual Development of Boys," *American Journal of
Psychology*, 1943, Vol. 56, pp. 217–233.

In a study by Katherine Simmons[4] of a selected sampling of 200 girls whose parents were above average in both education and economic level and who were of Northern European stock, the average chronological age at pubescence was 12.56 years. The appearance of pubescence for more than two-thirds of the girls were between 11.5 years and 13.62 years. Terman and Baldwin[5] offer evidence indicating that children from the upper social strata generally mature a year or two in advance of those from lower strata; and along with this they found that superior children, as a group, matured earlier than inferior children. Other studies have shown that the feeble-minded as a group mature later. However, the age of pubescence varies not only with sex, living conditions, and general intelligence, but also with race and climate. It has been shown that girls from central temperate areas mature earlier than those from colder northern or even warmer southern regions. (See Figure 2.) Also, children of Latin stock appear to mature earlier than those of Celtic stock; colored children in America mature earlier on the average than white children of the same age; and children from a favorable geographical environment mature earlier as a group than those from a less favorable one.[6] There is evidence from studies by Gould and Gould that puberty appears earlier today than a generation or more ago. This is indicated in Figure 3, which shows the distribution and average menarcheal ages of 357 mothers and their 680 daughters.

Also, earlier European data on the onset of puberty in girls shows a median between fourteen and fifteen years among Southern Europeans to sixteen years or later among the Northern Europeans. However, more recent data indicate an earlier maturity for Europeans as well as Americans. The question of what elements are causative in the rate of maturation is not considered here; however, the associations suggested are rather definitely shown in various studies relating to adolescence and maturity.

Although the adolescent is rather distinct as such, having qualities peculiar to this phase of life alone, the adolescent age is not considered

[4] Katherine Simmons, "The Brush Foundation Study of Child Growth and Development. II. Physical Growth and Development," *Monographs of the Society for Research in Child Development,* 1944, Vol. 9, No. 1.

[5] L. M. Terman, *Genetic Studies of Genius,* Vol. 1, 1925, p. 205; Bird T. Baldwin, "Mental Growth Curves of Normal and Superior Children," *University of Iowa Studies in Child Welfare,* 1922, Vol. 2, No. 1.

[6] C. A. Mills, "Geographic and Time Variations in the Body Growth and Age at Menarche," *Human Biology,* 1937, Vol. 9, pp. 43–56.

by the present writer as wholly isolated from and unrelated to other periods of life. Developmental periods are marked off somewhat arbitrarily, and an examination of the various classifications of these shows

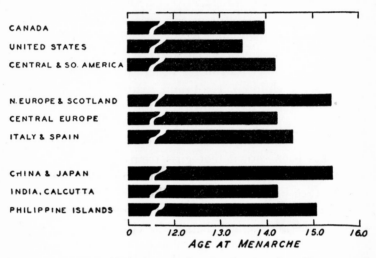

FIG. 2. *Average age at first menstruation in the Americas, Europe, and Asia.* Data are presented separately for northern, central, and tropical or semi-tropical areas. Note that the menarche is earlier in the central temperate areas and delayed in the colder northern and even warmer southern areas. Menstruation is earlier according to this chart in the United States than in any other area. (From F. K. Shuttleworth, "The Adolescent Period, a Graphical and Pictorial Analysis," *Monographs of the Society for Research in Child Development,* 3, No. 3. This is based upon data from C. A. Mills, "Geographic and Time Variations in Body Growth and Age at Menarche," *Human Biology,* 1937, 9, pp. 43–56.)

many inconsistencies and much overlapping; moreover, it is always to be remembered that adolescence covers a range of several years. This important truth, and the fact that there is a rather wide variation in the time of the onset of adolescence, have helped to determine the trend and organization of thought in the following chapters on the development of adolescents.

Pubic ceremonies. Before a more detailed study of the characteristics of the adolescent is undertaken, it should be of interest as well as value to note some social customs concerned with the passage of youth from childhood to maturity. The universality of pubic rites and the solemnity of their observation are evidence of the recognition, even in earliest times, of the importance of this stage of life. Consciously organized pubic ceremonies, sometimes very formal in execution, have been carried out by almost all primitive peoples, and, in many of these, tortures, humiliation, and various forms of instruction have had a place. Notably, the increase of formality in rites, and the lengthening of the

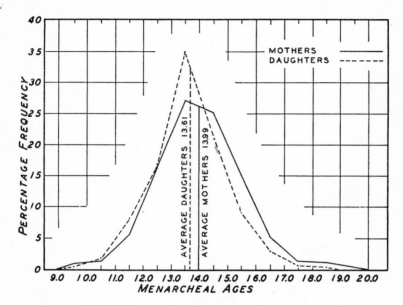

FIG. 3. *Percentage Distributions and Average Menarcheal Ages of 357 Mothers and Their Daughters.* (Reproduced from Figure 113 of F. K. Shuttleworth, "The Adolescent Period," *Monographs of the Society for Research in Child Development,* 1938, Vol. 3, No. 3.)

period of the adolescent's preparation for them, have constituted a fair index of the degree of development of various civilizations. The aims of education and the methods of teaching among the highly developed civilizations of the Greeks and Hebrews indicate that these peoples formally initiated individuals into manhood and womanhood; indeed, the simplest and probably the earliest type of systematized education among savages related to the preparation of children for such ceremonies. However, both secular and modern public education have broken away so completely from these practices of the past that it is difficult to recognize any vestiges in our educational processes.

Several authors, among whom G. Stanley Hall[7] is one of the most prominent, have given us full and vivid descriptions of these pubic ceremonies as carried out among the more primitive tribes. An inter-

[7] See his account in *Adolescence,* 1904, Vol. 2, Chap. XIII, pp. 233–249.

esting attempt to present a more or less subjective fourfold purposive classification of these ceremonies is presented by Boynton.[8] The first of these four types he listed as educational. "Ceremonies under this heading," he points out, "were designed specifically for the purpose of affording especial opportunities for the education of the youths, who were soon to become members of the adult milieu." A second type of ceremony described by Boynton was one involving physical ordeals which were based on the assumption that an individual who could not meet certain physical requirements demanded of him was not fit to become a member of the adult group. The third type of ceremony was designed to promote matrimony. Rites of this type concluded with the announcement to the members of the opposite sex that the individual was now ready for marriage, provided that another willing party could be found. The fourth and last of these types has been referred to as the vestigial group. "Here rituals were performed when so far as can be determined the one main idea was that of merely conforming to tribal customs. In this group are the ceremonies, frequently devious and often contradictory, whereby the tribe welcomed the children into

Courtesy, "National Geographic." Photograph by Lt. Col. L. E. Becher, D.S.O., Surry, England.

Primitive people believe that the strange ceremonies suddenly transform the young people into mature citizens. (Secured through the aid of the *Journal of the National Education Association*.)

[8] Paul Boynton, "Adolescent Initiations Among Primitive Peoples," *Peabody Reflector and Alumni News,* 1934, Vol. 7, pp. 89–90, 98–100.

adulthood without any apparent regard to why certain courses were followed or certain practices ever were initiated."

Some informal observances of modern civilization. Probably the most noticeable of the modern practices is the introduction of the young lady into "society." This usually takes place during the latter part of the adolescent period and signifies to the world—and particularly to the young men—that a daughter is about to enter woman's estate. Frequently the initiate-to-be is given a "house" or "coming-out" party, or makes her debut at a debutante ball or similar celebration. After this period in her life the girl is allowed certain privileges formerly denied her, such as having young men call and attending dances. Yet not only do we find her appearing in formal social activities; a changed attitude is assumed toward her as well. She is now addressed as "Miss" rather than by her maiden name, except among more intimate friends; and like the maturing boy she may often be admitted to a wide variety of adult life. But these, of course, are not the only tokens of maturity in modern society. The gift of a watch, commencement (a significant term) exercises at high school, the finishing school, the linking of the self with the church—these are all more or less socialized events related to the entrance into maturity.

It has been suggested that this recognition is informal, and that the youth has not wholly put away childish things. Yet the world at large, as well as the individual concerned, is advised of a person's becoming a matured social being. On the other hand, we may not ignore the fact that practices to which we have referred are mainly those of the more financially fortunate families. A survey of the life habits of people from a lower social stratum will show that many such observances are absent, and that today there is little here to indicate to the world and to the developing youth that the adult group, with its privileges and responsibilities, is receiving a new member. Frequently the "initiation" lessons are given by uncouth and unworthy elders. Often, indeed, the home is a very poor agent for developing and setting forth a responsible social being; and as a result of its neglect, inadequacy, or general unwholesomeness, undesirable psychological growths appear in the young.

Social demands upon adolescents. Lawrence K. Frank has emphasized the importance of social demands on the attitudes and behavior of adolescents. With the development into adolescence, the individual must face a *new self,* with added physical and mental abilities, and the emergence of the sex drive as a powerful force in his life. Also, he must learn to adapt to a society in which his role with the group has changed. New demands are made upon him. A few years ago he was excused from many acts because he was immature. Now, adults expect him to assume the role of an adult, even though he is still inexperienced in living and in participating as an adult in an adult society. Concerning the profound changes introduced at this age Frank has stated:

With puberty comes also a profound internal change, involving novel impulses and feelings and more sensitized social reactions. Almost suddenly the individual becomes aware of the peculiar characteristics of the members of the opposite sex and consciously regards them as the focus of his or her own interests and as the source of possible embarrassments or even dangers. These changes occur in the individual with greater or less rapidity and carry significance according to the rate and magnitude of the alterations and according to the individual's own past history.[9]

It is common observation that the adolescent is a source of perplexity and anxiety to adults, particularly to his parents and teachers. Thus, peer relations become very important at this stage, and will be given special study and consideration in Chapter X. Pressed both internally and externally to conform to specified modes of behavior, he develops one technique for adjusting to adult demands and another technique for adjusting to the demands of his peers. There is perhaps no period of life when individuals are so frequently misunderstood as the adolescent period—the transition stage of life.

The adolescent is not only misunderstood by adults, but is also oftentimes a real problem to himself. This will be shown in the case studies presented throughout the subsequent chapters. The brief case study of Thomas, referred to here as Tom, with whom the writer has had a number of close contacts, illustrates this.

Tom was slightly late in reaching physical maturity, although he was above average mentally and did better than average in his school work. At the age of fifteen, Tom's complexion was rather poor. This seemed to bother Tom when he was around adults, other than members of his immediate family. Thus, he was inclined to avoid visitors. When his mother asked him to come into the living room to speak to guests he was quite busy. He also showed very little interest in girls. He developed a keen desire for hunting and was most persistent in his desire for a special type of shot gun. Over the mother's protest, Tom was given a repeater shot gun for Christmas. During the next several months Tom seemed to get much pleasure in going

[9] L. K. Frank, "Introduction: Adolescence as a Period of Transition," *Forty-third Yearbook of the National Society for the Study of Education*, Part 1, 1944, Chap. 1.

hunting with a good friend of his, a young man of about the same age level.

Most of the usual habits and characteristics common to a large percentage of adolescents were to be found in Tom. He was rather careless, but with all good intentions. He would forget to clean his gun after hunting, despite his father's continuously calling this to his attention. Tom would wear a good shirt to school and forget about his clothing and enter into some tumbling or play activities, even though his mother had warned him on many occasions about this. He would leave the lights on in the basement, although he had been reminded of this a number of times. At times he would join with the family, consisting of his father, mother, and a sister, age twelve; at other times he would prefer to remain in his room listening to the radio or to be with his friends at the movies. Just two years prior to this Tom seemed to be very fond of his father in particular and they were together a great deal. The father was perplexed when Tom no longer wished to go with him to watch a practice ball game. Tom could not understand why he was always forgetting things, and why he changed in some of his interests and activities as rapidly as he did. Thus, Tom, a normal adolescent boy, became a problem understood neither by his family nor by himself.

Distribution of adolescents. There has been a continuous increase in the numbers of our population at all age levels throughout the major part of our history. However, beginning in 1921 and 1922 there has been a rapidly declining birth rate. This was reflected in an actual decrease in the number of adolescents in our population a decade or more later. One cannot, however, conclude that this will affect the long-time trend. Changed ways of living as well as changed goals and outlooks will affect population trends.

Owing to technological developments and the increased urbanization accompanying these developments, there has been a continuous migration from the farms to the cities and towns. This is reflected in a significant decrease in the number of adolescents among the farm populations, a decrease shown more clearly among the female group where the urban white female adolescents of 1940 outnumbered the males by over 70,000.

At all ages, and in all areas, there has been a considerable increase in the number attending school. As would be expected, there is a larger proportion of the urban than of the rural adolescents attending school. The distribution of the school age population for October, 1947 is presented in Table II.[10] The number of adolescents comprising the

[10] This table has been reproduced from D. I. Blose and E. M. Foster, "Children Not in School," *School Life,* 1949, Vol. 31, p. 3.

TABLE II

School Enrollment of the Civilian Noninstitutional Population, by Age, for the United States *

Approximate grade level	Age	Total non-instutional population	Enrolled in school		Not enrolled	
			Number	Per cent	Number	Per cent
1	2	3	4	5	6	7
Kindergarten	5 years	2,766,000	703,000	25.4	2,063,000	74.6
Elementary school—grades 1–8	6 years	2,522,000	2,366,000	93.8	156,000	6.2
	7–9 years	9,959,000	6,851,000	98.4	108,000	1.6
	10–13 years	8,570,000	8,451,000	98.6	119,000	1.4
	6–13 years	18,051,000	17,668,000	97.9	383,000	2.1
High school—grades 9–12	14 and 15 years	4,158,000	3,809,000	91.6	349,000	8.4
	16 and 17 years	4,334,000	2,928,000	67.6	1,406,000	32.4
	14–17 years	8,492,000	6,737,000	79.3	1,755,000	20.7
Elementary and high school—grades 1–12.	6–17 years	26,543,000	24,405,000	91.9	2,138,000	8.1
Kindergarten, elementary, and high school—grades K–12.	5–17 years	29,309,000	25,108,000	85.7	4,201,000	14.3
Junior college	18 and 19 years	4,137,000	1,007,000	24.3	3,130,000	75.7
Kindergarten, elementary, high school, and junior college.	5–19 years	33,446,000	26,115,000	78.1	7,331,000	21.9

* Data are taken from Current Population Reports, Population Characteristics Series P-20, No. 19, July 30, 1948. *School Enrollment of the Civilian Population; October 1947.* Department of Commerce, Bureau of the Census, Washington 25, D. C.

age group 14–17 attending school is expected to increase. The secondary school of the future will probably enroll all normal adolescents of this age group on a full-time or on a part-time basis.

The young workers 14 through 17 years of age numbered about two and one-half millions in September, 1948. This was about twice as many as were employed in March, 1940, although it was considerably below the peak number of almost three and one-half millions employed in April, 1945. With the pressure of the labor market eased there has been a growing decline in the number of persons under 18 years of age employed. These data are shown in Figure 4; however, the census total does not include workers under 14 years of age—a group made up primarily of pre-adolescents and adolescents who assist on the farms and around the home or at the neighbors' on such jobs as mowing the lawn, cleaning, and assisting at other tasks. Additional data and

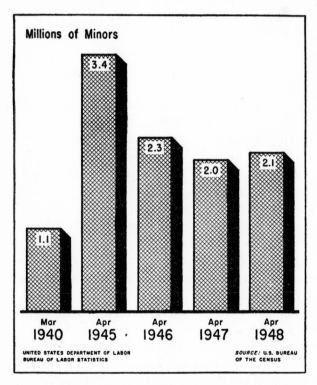

Fig. 4. *Children at Work, Fourteen to Seventeen Years of Age.* (E. A. Merritt and E. S. Gray, "Child Labor Trends in an Expanded Labor Market, 1946–1948, *Monthly Labor Review,* December, 1948.)

materials bearing on employment problems during the adolescent and post-adolescent years will be presented in Chapter XX.

METHODS OF STUDYING ADOLESCENTS

A number of different methods have been used in gathering information about adolescents. The methods used have depended in a large measure on the problem being studied, since certain types of approaches are more applicable to some problems than to others. Some of the most generally used methods are described here.

Adolescent diaries. It has already been suggested that G. Stanley Hall made use of adolescent diaries for gathering data about the nature, interests, and activities of adolescents. The adolescent period is characterized by a pronounced interest in diaries of various activities. This is perhaps an important reason why high school annuals have such an appeal to high school students, and why students persist in getting statements from each other written into the annuals. The diaries are usually written by more intelligent adolescents. Also, they are highly tinged with emotion and tend to express adolescent sentiments under emotional strain.

Retrospective reports from adults. During the past decade a number of students have made use of data gathered from adults relative to their adolescent years. The adult is asked to think back upon his adolescent years and to give some desired information. This method may be thought of as a questionnaire procedure, since it usually involves the use of questions on certain problems or areas. It seems rather safe to say that older subjects are more willing to give a true account of their adolescent years than most adolescents would be, since an appraisal of their present status is not involved. However, the problem of forgetting looms as an important source of error. Also, many of the early experiences have become highly colored during the course of years, and this causes the retrospective reports to be inaccurate. However, like diaries, these reports have some value when used with caution and understanding.

The observation technique. This method, in which one-way screens and motion pictures are brought into use, has been used most successfully with young children. The writer has made use of this method to introduce educational psychology students to the study of children and adolescents. A disadvantage found in the use of this method with adolescents is that it is not desirable to take notes at the time of the observation of the individual. Each adolescent is suspicious that he is the subject being studied. Some will, then, tend to put on an act, while others will withdraw completely into their own little world. High school teachers have observed this, and thus do not feel that it is advisable for observers to make notes on adolescents being observed. With lower elementary grade children this does seem to present a serious problem. Observations made on adolescents should not be recorded at the time of the observations, but at the convenient period following the observations. Comparisons between individuals making observations should then be made in order to insure as great an accuracy as is possible in recording the activities and characteristics observed.

The written questionnaire. This method enables the experimenter to secure much information about the adolescent in a short period of time. It has been frequently used in gathering data on high school students. This procedure is likely to be untrustworthy because of the motives prompting the adolescent to answer the questionnaire. If the subjects are properly conditioned for the questionnaire, the results become more reliable.

The personal interview. Sometimes the interview and the questionnaire are combined in the study of certain adolescent problems. Many students of adolescent psychology have used this method for gathering data. The extent to which such data are reliable will depend upon the training and upon other qualifications of the experimenter and the cooperation he is able to secure from the adolescents being studied. Many of the studies referred to throughout this volume have made use of this method combined with the questionnaire.

Tests and rating devices. Following the development of intelligence tests, a number of experimenters began studying other areas of growth and development. Tests and rating devices have furnished a good basis for comparing adolescent boys and girls at different age levels. Such comparisons have made use of tests of interests, personality, emotional maturity, social development, aptitudes, attitudes, and beliefs. Considerable data about the growth and development of adolescents have been gathered by means of these devices.

Projective techniques. The projective technique is described briefly in connection with the study of the adolescent's personality. This method makes use of the free association procedure, and of the involuntary responses of adolescents in a test situation. The use of this method requires an experimenter who understands the nature of adolescents and who has been trained in interpreting data gathered by this particular technique.

The use of anecdotes. Considerable data have been gathered about the nature and characteristics of adolescents by means of anecdotes. Anecdotes kept over a period of time have been found to be useful in showing behavior trends and growth trends in the life of a single individual. This method combines the observational procedure and the case study technique in the study of behavior activities of adolescents over a long period of time. Daniel Prescott and his students have made much use of this method in their work at the Collaboration Center for Human Growth and Development at Chicago University and at the University of Maryland. It offers much promise when used by individuals carefully trained in understanding children and in the use and interpretation of anecdotes.

Genetic case studies. The genetic case study procedure appears to be the most satisfactory method for studying the growth, development, and behavior activities of adolescents. This method studies a single individual over a long period of time, rather than a cross section of individuals over a very short interval. One of the best examples of use made of this method is the study conducted by Jones with "John Sanders" as the subject.[11] The adolescent growth studies conducted in the Institute of Child Welfare at the University of California are noteworthy in this connection. This method probably provides the best data available at the present time on the adolescent; however, it should be pointed out again that no one method can be applicable to all situations or used in the study of all aspects of adolescent growth and behavior activities.

SUMMARY

Since the first appearance of the momentous work of G. Stanley Hall, much has been written and said about the age of adolescence. There are many who would consider this period of life as separate and distinct from other periods, holding it up as a dramatic stage that justifies all the phrases and titles that have been built up around it. We hear the expressions, "Flaming Youth," "Coming-Out Parties," "The Age of Accountability," and the like. These are merely terms used to express ideas formerly conceived of in connection with various

[11] H. E. Jones, *Development in Adolescence.* New York: Appleton-Century-Crofts, Inc., 1943.

pubic ceremonies. The importance of this transition period was recognized by the early primitive tribes, but the conception of the nature of the transition has not always been in harmony with the notions that will be presented in the next several chapters dealing with growth. Students of adolescent psychology have made use of various methods for gathering data on the development, activities, and problems of teen-age boys and girls. These studies have provided valuable data for use in interpreting the growth, development, and special characteristics of adolescents.

With the advancements that have taken place in our social order, there has come about a greater necessity for continuing schooling over a longer period of time. This, combined with a number of other elements related to our industrial civilization, has effected an increased enrollment in our schools, so that today approximately seventy per cent of those of high school age are enrolled. The final chapter of this volume gives some notions of the role that adolescents of the present day will play in life's drama tomorrow. It should be the goal of those concerned with the direction and guidance of adolescents to direct them in such a way that they will be able to function as citizens of tomorrow with the greatest efficiency and satisfaction. In doing this we must never lose sight of the operation of issues, values, and dreams in these eager, energetic, growing adolescent boys and girls of today.

THOUGHT PROBLEMS

1. Look up several definitions of adolescence and note the points of similarity in each. (See Appendix A for a bibliography.)

2. What is meant by pubic ceremonies? Do you notice any points of similarity between the various ceremonies? Show how differences in the practices represent different folkways or general cultural patterns.

3. What factors are associated with the time of the beginning of pubescence? How would you account for the fact that pubescence is earlier today than it was a generation or more ago?

4. Look up further data on the distribution of adolescents. What factors affect this?

5. An annotated bibliography of popular literature involving adolescents is presented in Appendix B. Read one or more of these books along with your readings from this text. Note the characteristics of the individuals involved, and the problems encountered by the adolescent. Has the writer presented a description of the adolescent and his problems which is in harmony with the materials presented throughout this book?

6. Study the methods listed in this chapter for collecting data on adolescents. List several problems that would make use of each of these methods.

SELECTED REFERENCES

Briggs, Thomas H.; Leonard, J. Paul; and Justman, Joseph; *Secondary Education*. New York: The Macmillan Co., 1950, Chap. III.

Frank, Lawrence K., "Introduction: Adolescence as a Period of Transition," *Forty-third Yearbook of the National Society for the Study of Education*, Part I, 1944, Chap. 1.

Hall, G. Stanley, *Adolescence: Its Psychology and Its Relations to Physiology, Anthropology, Sociology, Sex, Crime, Religion, and Education*, Vol. 1. New York: Appleton-Century-Crofts, Inc., 1904, Chaps. I, II, XIII.

Hollingworth, Leta S., *The Psychology of the Adolescent*. New York: Appleton-Century-Crofts, Inc., 1928, Chaps. I and II.

Hurlock, E. B., *Adolescent Development*. New York: McGraw-Hill Book Co., 1949, Chap. I.

Mead, Margaret, *From the South Seas; Studies of Adolescence and Sex in Primitive Societies*. New York: William Morrow and Co., 1939, p. 1072. This is a one-volume edition of three anthropological works: *Coming of Age in Samoa, Growing up in New Guinea,* and *Sex and Temperament*.

Miller, Nathan, *The Child in Primitive Society*. New York: Brentano, 1928.

Sadler, W. S., *Adolescent Problems*. St. Louis: C. V. Mosby Co., 1949, Chap. I.

A selected annotated bibliography on the psychology of adolescence is presented in Appendix A.

II

Problems of Adolescents

The attainment of maturity brings with it many problems. Any period of change is likely to be a period fraught with them, and since the adolescent period is concerned with growth and development as well as changed interests and aspirations, it will be accompanied by many potential difficulties. A more complete discussion of these will be presented in the subsequent chapters bearing on different phases of the life and development of adolescents. The increased complexity of our social and economic order has introduced many problems that did not exist at an earlier period. Any training program that is going to be effective must take into consideration the problems and needs of those with whom the program is primarily concerned.

Classification of problems. Various techniques have been used in studying the problems of adolescents. It was pointed out in the previous chapter that the early studies of G. Stanley Hall made use of adolescent diaries as a means of gathering information about adolescent characteristics and problems. Later studies modified this method and introduced the questionnaire or check list procedure.

According to Laycock the problems of the adolescent throughout the ages may be grouped around the following major tasks: .

(1) Making adjustment to his changing physical growth and physiological development; (2) becoming emancipated from his family and free from too great emotional dependence on his parents; (3) accepting his own characteristic sex role and making adjustments to the opposite sex; (4) finding and entering on a suitable vocation; and (5) forging some sort of philosophy which will give meaning and purpose to life.[1]

---

[1] S. R. Laycock, "Helping Adolescents Solve Their Problems," *The Educational Digest,* November, 1942, p. 32.

21

The problems of high school pupils were studied by Pope through the use of an essay submitted by each of 1,904 pupils enrolled in Cleveland High School, St. Louis.[2] As a preparation for this work a statement was first read to the pupils indicating what was to be done, and this was followed by a discussion of what was meant by personal problems. They were then asked to consider their individual problems, carefully and seriously, until the next day, when they were requested to write essays on what they considered were their own personal problems. A total of 7,103 problems were listed in the 1,904 essays written. The problems were then grouped into six areas. The per cent of problems listed in each of the areas by the boys and girls of the different grades is presented in Table III. Study-learning relationships

TABLE III

PER CENT OF PROBLEMS LISTED BY HIGH SCHOOL BOYS AND
GIRLS COMPRISING EACH OF SIX AREAS

Areas		Nine	Ten	Eleven	Twelve	Total
				GRADES		
Study-Learning Relationships						
	Boys	49	44	47	45	46
	Girls	44	47	39	39	42
Occupational Adjustment						
	Boys	18	21	23	35	24
	Girls	16	15	23	34	22
Personal Adjustment						
	Boys	12	10	7	5	8
	Girls	15	13	15	11	13
Home-Life Relationships						
	Boys	9	14	12	7	11
	Girls	15	12	12	7	11
Social Adjustment						
	Boys	8	8	6	7	7
	Girls	9	12	8	7	9
Health Problems						
	Boys	3	3	3	2	3
	Girls	1	2	3	3	2

presented the greatest number of problems for both boys and girls at all grade levels.

The greatest concern of youth has to do with their relationship with their teachers. This was shown by the fact that nearly fifty per cent of the total number of problems mentioned by pupils were in this area. They were

[2] Charlotte Pope, "Personal Problems of High School Pupils," *School and Society,* 1943, Vol. 57, pp. 443–448.

concerned with the amount of home study, the teacher's unfairness, and his stern attitude.[3]

Other problems listed under study-learning relationships were in order of frequency: *study attitudes, educational guidance, physical conditions,* and *habits of study.* This and other studies have shown that girls are slightly more concerned about marks and grades than are boys. On the whole, pupils in the first year of high school were most concerned with school progress, whereas students of the upper high school years were especially concerned with *educational guidance.*

In this connection the pupils were especially concerned with their relationship with their teachers. Girls were more concerned than boys over school progress marks, tests, and criticisms. There was a special feeling among the twelfth grade pupils of insufficient time in school to prepare work and of a lack of ability in the tool subjects. The problems of occupational adjustment offered little concern for most of the high school pupils, but became more important as they advanced from the ninth to the twelfth grades. The financial problems, grouped with home life relationships, seemed more serious to the high school boys of this study than for the junior high school students reported by Olive Yoder Lewis. Feelings of inferiority and superiority were problems most frequently listed in the area of personal adjustments. The slight importance boys and girls attached to health is revealed by the small number of problems listed in this area; however, boys were more keenly aware of problems in this area than were the girls. "Pupils in the upper grades manifested much more concern than the younger pupils in problems referring to sex guidance."[4]

Olive Yoder Lewis suggests that the "average adolescent in junior high school has at least four fears: (1) Opinion of parents; (2) opinion of teachers; (3) opinion of his peers; and (4) fear of the unknown."[5] In order to secure further information about the personal problems of adolescents, a survey was conducted among the students of Franklin Junior High School, Vallejo, California. Each student was requested to list any personal problem he might have at this time. Of the 701 blanks received, 339 were submitted by boys and 362 by girls. These problems were classified into seven general groups; however, it is pointed out that many problems overlap into two or more fields.

[3] *Ibid.,* p. 445
[4] *Ibid.,* p. 448.
[5] Olive Yoder Lewis, "Problems of the Adolescent," *California Journal of Secondary Education,* 1949, Vol. 24, pp. 215–221.

A comparison of the number of problems reported by girls and boys is presented in Figure 5. This figure shows that home life and social problems were reported by more than twice as many girls as boys. Health and development and the future presented problems for many more boys than girls; twenty-eight boys listed as a problem, "School takes too much of our money." They displayed a resentment against school drives for Red Cross, Community Chest, and Cancer Research. The policy of the homeroom groups competing for 100 per cent participation in a drive oftentimes presents an embarrassing problem to the boy or girl who has little if any extra money to spend, and who may be having a difficult time providing for his minimum needs.

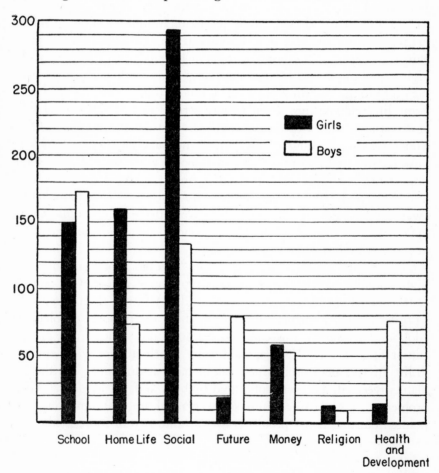

FIG. 5. *Comparison of Number of Problems Reported By Boys and Girls.*
(After Lewis)

The use of problem check lists. A number of problem check lists have been used in studying the problems of adolescents. The *SRA Youth Inventory* was constructed under the auspices of the Purdue University Opinion Poll for Young People with the cooperation of many high schools and over 15,000 teen-agers throughout the country.[6] The 298 questions making up the *Inventory* were developed from essays submitted by hundreds of students stating in their own words the problems that bothered them most. The needs and problems of these boys and girls were studied and classified into eight major areas. These areas are referred to in the *Inventory* as follows: (1) My School, (2) After High School, (3) About Myself, (4) Getting Along with Others, (5) My Home and Family, (6) Boy Meets Girl, (7) Health, and (8) Things in General.

Norms for boys and girls have been developed, based on a national sample of 2,500 cases. The results from this sample were analyzed for various subgroups. However, no marked differences in the area scores appeared; although significant differences were noted in the responses to the individual items of a particular area. The number of items in each category, the mean, and the standard deviation of the total scores for each area of the *Inventory* are presented in Table IV.

A problem check list devised by Ross L. Mooney[7] was administered

TABLE IV

NUMBER OF ITEMS IN EACH AREA, MEAN, AND STANDARD DEVIATION FOR THE NATIONAL SAMPLE OF 2,500 CASES OF THE *SRA Youth Inventory*

Area	No. of items	Mean	Standard deviation
My School	33	7.38	4.49
After High School	37	12.05	7.09
About Myself	44	9.42	6.10
Getting Along with Others	40	10.40	6.32
My Home and Family	53	5.76	6.59
Boy Meets Girl	32	6.64	4.98
Health	25	3.94	2.77
Things in General	34	6.36	5.06

[6] The *SRA Youth Inventory* and the *Examiner Manual* are published by Science Research Associates. These are copyrighted by Purdue Research Foundation.

[7] Ross L. Mooney, "Surveying High-School Students' Problems by Means of a Problem Check List," *Educational Research Bulletin,* March 18, 1942.

to 603 students of the Stephens-Lee High School, Asheville, North Carolina. The students responded favorably to the check list, and a large number asked for a special conference with someone to talk over certain problems suggested by it. The 330 items making up the check list were classified into 11 areas; each area contained 30 items. The 11 areas are:

> Health and Physical Development
> Finances, Living Conditions, and Employment
> Social and Recreational Activities
> Courtship, Sex, and Marriage
> Social-Psychological Relations
> Personal-Psychological Relations
> Morals and Religion
> Home and Family
> The Future: Vocational and Educational
> Adjustment to School Work
> Curriculum and Teaching Procedures

PROBLEMS FREQUENTLY REPORTED BY ADOLESCENTS

Educational adjustments. The various studies reveal that educational difficulties loom large among the problems of high school students. The study by Pope, referred to on page 22, indicated that fifty per cent of the problems reported by pupils were in the study-learning area. The fact that most adolescents are in school would provide an opportunity for this to be a major issue. Furthermore, insufficient and sometimes undesirable educational guidance results in many such problems; a fairly complete presentation of these is given in Chapter XII. That they are of frequent occurrence was revealed in a study of ninth and tenth grade boys and girls from several high schools of Connecticut.[8] Educational problems were checked by a larger percentage of the boys and girls than any other classification of problems on the check list, and among these the items most frequently checked were, in order of frequency: "don't like to study"; "being a grade behind in school"; "afraid of failing in school work"; "so often feel restless in class"; "getting low grades in school"; "afraid to speak up in class"; "not smart enough"; "teachers expect too much work"; and "don't like school."

√These results are somewhat in harmony with those obtained by

[8] Karl C. Garrison, Unpublished Study, 1945. In this study the problem check list devised by Ross L. Mooney was given to more than 400 boys and girls, representing a cross section of the boys and girls of Connecticut.

Mooney in his study of the problems of high school pupils of the Stephens-Lee High School of Asheville, North Carolina. Eighty-seven per cent of these high school pupils indicated a marked concern for the problems in the area *Adjustment to School Work*. Several items were marked by more than twenty per cent of the group. These were: "being a grade behind in school"; "fear of failing in school"; "worrying about grades"; "trouble in mathematics and physics"; and "not spending enough time in study." An analysis of individual cases from such a check list reveals different combinations of problems, although the percentages checking certain problems, such as "being a grade behind in school," show that there are some problems common to a large group of high school students. These problems are not valued by adults and adolescents on the same scale with respect to their seriousness; but a problem that gives adolescents much concern is a serious one and should be given definite consideration by those concerned with their guidance, even though it doesn't seem important to the teacher, parent, or counselor.

Home adjustments. Since most studies of adolescents' problems are made by people concerned with or interested in their educational program, problems related to the home are often not discovered or are neglected. These problems, however, are likely to be discovered in the psychological clinic. The characteristic listed as "parental troubles" ranks first among a list of symptoms manifested by boys and girls referred to the Educational Clinic of City College, New York.[9]

These may take the extreme form of a sharp emotional rejection of the child by a parent or both parents. Or it may be manifested in the uneven administration of discipline as between the two parents or by the same parent at different times. At times the trouble lies in an over-protective attitude of a parent. . . . The feature that ties all these forms together is the difficulty experienced in establishing a sound relationship between child and parent or the two parents.

The results of the *SRA Inventory* survey show that among teen-agers strong feelings against their parents are voiced by only a minority of the students; although ten to twenty per cent did indicate home and family problems. The three types of problems checked were (1) those indicating a lack of understanding between parents and adolescents, (2) those involving a limitation of their freedom, and (3) problems

---

[9] Harold H. Abelson, *Annual Report of the City College Educational Clinic Thirtieth Year, 1942–1943,* The School of Education, College of the City of New York, p. 13.

involving money or financial conditions. In harmony with other studies already referred to, problems connected with finance seemed more serious for the boys, whereas the other problems falling in this area were in general more serious for the girls.

All the home and family problems listed on the Mooney problem check list were checked by two per cent or more of the ninth and tenth grade pupils in the Connecticut study. However, only a few of these problems were checked by as many as ten per cent of the group. They were as follows: "being treated like a small child at home"; "never having any fun with father or mother"; "keeping secrets from my parents"; and "parents working too hard."

Financial problems. It has already been suggested that financial problems appear in connection with the home problems of adolescents. In the *SRA Inventory* survey, twelve per cent of the teen-agers complained that their allowance was too small; while eleven per cent said that they couldn't spend the money they earn without their parents interfering. It has also been pointed out that certain school policies and practices may aggravate the financial problems among certain students.

Throughout the Stephens-Lee High School, financial problems loomed very large. Over ninety per cent of the students marked one or more items from this area, and thirteen items were marked by ten per cent or more of the student group. The items frequently checked are as follows: "wanting to earn some money" (288), "learning how to spend money wisely" (168), "having to ask parents for money" (164), "learning how to save money" (157), "having no regular allowance or regular income" (108), "having no family car" (97), "needing a job during vacation" (94), "having less money than friends" (85), "too few nice clothes" (75), "living too far from school" (74), "needing to find a part-time job" (67), "too little money for school lunches" (66), "getting money for education beyond high school" (65). In our present social order it is impossible for young people to live a normal social life without some material expense, much greater than that of their parents in the days of their own adolescence. Parents often fail to realize this difference in economic standards, and admonish the adolescent boy or girl thus: "When I was your age, I didn't have any money to spend." This attitude may become an important source of home conflict. Means for providing for these material expenses may come from one or more of three sources: (1) parents, kinspeople, or friends, (2) employment, or (3) illegal or illegitimate sources.

Health adjustments. The lack of sensitiveness to the problems of health, referred to in the study by Pope, is not simply a characteristic of one group, but on the contrary is found among a majority of adolescents. This observation no doubt reflects in a large measure certain attitudes that have been fairly widespread in our social order. Boys seem to sense the necessity for practicing good health habits more than do girls, a fact that is, perhaps, closely related to their greater participation in physical activities and athletic contests. Girls manifest more concern over sex guidance than do boys, and pupils of the upper grades express this as a problem more frequently than do those of the first year of high school.

In the study by Mooney, fifty-five of the students showed concern over health and physical conditions. Other students, though not organizing so many responses around health and physical development, checked some items from that area. The frequently checked items from this area (marked by ten per cent or more of the students) were, in order: "weak eyes," "not as strong and healthy as I should be," "frequent headaches," "underweight," "poor teeth," "too short," "frequent sore throat," "tiring very easily," "poor complexion," "frequent colds," and "not getting enough exercise."

Vocational adjustments. Slightly less than one-fourth of the pupils in the study by Pope were concerned about their future vocations, although there was an increased concern about this among the students of the eleventh and twelfth grades. The problems listed comprising this area were, in order of frequency, as follows: *selection of a vocation, vocational preference, preparation for a vocation, necessity for part-time employment,* and *admittance to a vocation.*

The fact that 288 (forty-eight per cent) of the high school students of the Stephens-Lee High School want to earn some money of their own is proof that this is a pressing problem among adolescents. This view is further supported by the number checking the items, "having to ask parents for money," and "having no regular allowance." Because of changed social conditions, many problems involving money exist in the lives and activities of adolescents that were not present a generation or more ago, and this new factor must be given careful consideration in any program that is concerned with improving the adjustments of adolescents; a point that is very well illustrated in the case of a boy of the writer's acquaintance who was interviewed and asked to complete the check list. This boy, referred to as J. W., marked thirty-eight problems, a large number of which are in the area of *Finances, Living*

Conditions, and *Employment.*   These are the problems marked within this area:

> Wanting to earn some money of my own
> Having to ask parents for money
> Having no car in the family
> Needing a job in vacations
> Having less money than friends have
> Having to watch every penny I spend
> Too little money for school lunches
> Getting money for education beyond high school
> Too little money for recreation
> May have to quit school to go to work
> Needing to find a part-time job now
> Too few nice clothes

Personal-social problems.   Results from various inventories indicate that perhaps as many as ten per cent of the students in the typical high school may have fairly serious personal difficulties.   Although this represents a relatively small percentage, it is sufficiently serious to give those concerned with the guidance of adolescents a problem and a challenge.   Some of the personal problems checked by teen-agers in the *SRA Inventory* survey [10] are:

> 35% say they worry about "little things."
> 35% can't help daydreaming.
> 29% must always be "on the go."
> 27% report that they are nervous.
> 26% have guilt feelings about things they have done.
> 25% are ill at ease at social affairs.
> 24% of the students report "I want to discuss my
> personal problems with someone."

A problem frequently encountered by those dealing with adolescents who are socially maladjusted is that of lack of friends.   This area seems to present more problems for girls than for boys.   More students, in their daydreams, long for someone to be friendly to them than wish for riches or fame.   Some of the most urgent problems reported in the *SRA Inventory* survey in this area are suggested by the following:

> 54% say they want people to like them more.
> 50% want to make new friends.
> 42% wish they were more popular.

[10] The materials here quoted from the *SRA Inventory* survey are taken from the *Examiner Manual for the SRA Youth Inventory,* Form A, 1949, p. 3.

Additional materials bearing on personal and social problems will be presented in subsequent chapters dealing with various phases of the personal and social development of adolescents. These chapters will reveal that personal-social problems appear in connection with all phases of growth and development. Furthermore, these problems are directly related to personal and social adjustments, referred to in Chapters XV and XVI.

Table V gives problems found by Mooney among a significant number of high school pupils. It is apparent from these data that high

TABLE V

ITEMS FROM THE PERSONAL, SOCIAL, AND MISCELLANEOUS
AREAS CHECKED BY TEN PER CENT OR MORE OF THE
HIGH SCHOOL PUPILS (*After Mooney*)

Problems	Number Marking
Personal Psychological Relations:	
Forgetting things	176
Not taking some things seriously enough	174
Losing my temper	143
Afraid of making mistakes	132
Taking some things too seriously	131
Nervousness	104
Worrying	85
Sometimes wishing I had never been born	83
Cannot make up my mind about things	79
Daydreaming	73
Social and Recreational Activities:	
So often not allowed to go out at night	139
Taking care of clothes and other belongings	107
Wanting to learn how to dance	87
Wanting to learn how to entertain	80
Too little social life	68
Too little chance to go to shows	65
Too little chance to do what I want to do	65
In too few school activities	64
Courtship, Sex, and Marriage:	
Wondering if I'll ever find a suitable mate	104
Wondering if I'll ever get married	72
Not being allowed to have dates	70
Girl friend	69
Deciding whether I'm in love	68
Social-Psychological Relations:	
Wanting a more pleasing personality	113
Being disliked by certain persons	97

TABLE V (*Continued*)

Problems	Number Marking
Disliking certain persons	84
Feelings too easily hurt	80
Lacking leadership ability	61
Morals and Religion:	
Can't forget some mistakes I've made	148
Wondering what becomes of people when they die	103
Being punished for something I didn't do	96
Trying to break off a bad habit	84
The Future: Vocational and Educational:	
Wondering what I'll be like ten years from now	312
Wondering if I'll be a success in life	192
Deciding whether or not to go to college	140
Wanting advice on what to do after high school	138
Choosing best courses to prepare for college	114
Not knowing what I really want	108
Needing information about occupations	76
Needing to know my vocational abilities	66
Choosing best courses to prepare for a job	66
Needing to plan ahead for the future	62

school students are concerned over their future, and especially has this been found to be the case of senior high school boys and girls. Other characteristics noted by Abelson as occupying a prominent place in the list of manifestations noted for 1942–43 are: "social-emotional immaturity, 'nervousness,' motivational lack, aggressiveness, speech defect, and sibling rivalry. Characteristics associated with deliquency—truancy, lying, stealing, and sex—are noted relatively infrequent." [11] This is in harmony with findings obtained at the Educational Clinic of City College in the previous year.

Heterosexuality. Many problems connected with boy-girl relationships appear with the onset of adolescence. These are discussed quite fully in a subsequent chapter. During the early period of adolescence, problems pertaining to teasing, bashfulness, and wanting to ask a girl for a date, or to the unwillingness of parents to let the individual date, loom large. The problems of "how to go about dating," and "knowing when I'm in love" were checked by a fairly large percentage of the ninth and tenth grade pupils in the Connecticut study.[12] A fourteen-year-old boy in the ninth grade made the following notation:

[11] *Op. cit.,* p. 13.
[12] Unpublished materials on file with the writer.

Girls are a problem of mine. I like a girl and I don't know whether or not she likes me or another boy, a friend of mine.

I would like to have a date with this girl but my mother won't let me.

At a later stage, problems related to "going steady," "wondering if I'll ever get married," and "not enough dates" are very prevalent. Among college girls, the writer found "wondering if I'll ever get married" checked more than any other problem of the Mooney problem check list, a finding in harmony with the results of the study by Mooney of senior high school students in Asheville. (See Table V.) The *SRA Inventory* survey revealed some of the problems and confusions that beset teen-agers. A number of these problems involved boy-girl relationships. A few of the dating problems listed by these young people are as follows:

Boys

48% seldom have dates.
41% don't have a girl friend.
34% are bashful about asking girls for dates.
26% don't know how to ask for a date.
25% don't know how to keep girls interested in them.
23% wonder whether anything is wrong with going places "stag."

Girls

39% seldom have dates.
30% don't have a boy friend.
23% feel they are not popular with boys.
33% don't know how to keep boys interested in them.
36% would like to know how to refuse a date politely.
29% wonder whether it is all right to accept "blind dates."
22% don't know how to break up an affair without causing bad feelings.
20% wonder whether they should kiss their dates the first time they go out together.[13]

As youngsters grow older, these problems tend to diminish in importance, while other problems related to dating, courtship, and marriage take on an added significance. One-fourth of the students studied wanted to know what causes so much marital trouble. Looking ahead to the likelihood of marriage, about one-fourth wanted to know

[13] *Examiner Manual for the SRA Youth Inventory, Form A,* 1949, p. 4.

what things one should consider in selecting a mate, and how to prepare for marriage and family life.

A philosophy of life. Growth through adolescence is accompanied by wider social contacts, increased mental ability and understanding, and a changed physiological self. Thus, morals, religion, social problems, economic problems, and the future come to be thought of in a different light from that in which they were earlier viewed. The extent to which the individual's attitudes and behavior are controlled by ideals rather than impulses becomes a good basis for evaluating his maturity. The importance of attaining a satisfactory philosophy of life in relation to desirable personal and social adjustments will be emphasized in Chapters XV, XVI, and XVII. Many problems related to morals, conduct, the future, and one's role in the world appear at this age. These problems appear in connection with religion; they may appear in connection with any situation in which important decisions need to be made.

The attainment of a satisfactory philosophy of life will be discussed in later chapters in connection with attitudes, character, morals, religion, and mental hygiene. It should be pointed out here, however, that such an attainment follows the developmental growth process, as do other aspects of growth and development. A satisfactory solution to problems in this area that arise during adolescence should have its beginning during early childhood.

SUMMARY AND CONCLUSIONS

The increased complexity of our social order has brought about a greater demand for guidance and training, if growing boys and girls are to be able to meet satisfactorily the conditions that they will face tomorrow. However, *growing up* itself is accompanied by many problems. These relate to various aspects of the adolescent's life and are very real and significant to the individual boy or girl concerned, although they may appear trivial to the mature adult. A number of students of adolescent behavior have concerned themselves with the problems of adolescents. Studies show that home and school problems loom large in the lives of growing boys and girls. The consequences of these problems are important in connection with adequate personal and social adjustments of adolescents.

By means of check lists that have been developed a more accurate and complete inventory of adolescent problems has been made possible. These inventories show that there are distinct age differences.

Also, the nature of adolescent problems varies with social and living conditions. Problems among pre-adolescents will be related, in a large degree, to their personal needs, whereas those of older boys and girls are more often connected with social needs. In the subsequent chapters the needs of adolescents will be given special consideration in connection with the special phases of growth and personality characteristics of the group under discussion.

THOUGHT PROBLEMS

1. Report on the types of problems found in the lives of some adolescents in fiction or in the movies. Do you regard these as true to life? (See Appendix C for a bibliography.)

2. Show how the problems of adolescents have changed as a result of *industrialization* and *urbanization*.

3. Compare the problems of pre-adolescents with those of adolescents.

4. How do the problems among the high school boys and girls at Asheville compare with those you have noted in your observations and experiences?

5. If it is convenient for you to do so, administer some check list such as the *SRA Youth Inventory* to some high school student. What are some of the major problems revealed?

6. What uses can a teacher or counselor make of the results secured from such an inventory as the *SRA Youth Inventory?*

SELECTED REFERENCES

Blos, Peter, *The Adolescent Personality.* New York: Appleton-Century-Crofts, Inc., 1941.

Cole, Luella, *Psychology of Adolescence* (Third Edition). New York: Farrar and Rinehart, 1948, Chap. I.

Hurlock, E. B., *Adolescent Development.* New York: McGraw-Hill Book Co., 1949, Chap. I.

Newman, F. B., *The Adolescent in Social Groups.* Stanford University, California: The Stanford University Press, 1946.

Sadler, W. S., *Adolescence Problems.* St. Louis: C. V. Mosby Co., 1948.

Taylor, K. W., *Do Adolescents Need Parents?* New York: Appleton-Century-Crofts, Inc., 1938, Chap. VI.

PART II
GROWTH AND DEVELOPMENT DURING ADOLESCENCE

III

Physical and Motor Development

The growth of the individual from birth to maturity constitutes almost one-third of the normal life span. Genetic studies of the physical development of children from birth to maturity have furnished valuable information about the nature of growth and the factors that influence it. The materials of the subsequent chapters are designed to give the student a better understanding of how growth takes place. Also, materials will be presented throughout this and other chapters showing the effects of various forces and conditions on growth and personality development of adolescents.

Growth and development. All living things live in some sort of external environment. This, because of its nature, is always changing. Changes in the individual's environment influence him in many ways, and under these influences his behavior is modified. Growth refers to change of some organ or part of the body mechanism. It may also refer to change of behavior, such as the child's social growth, motor growth, and emotional growth. Development may well be differentiated from growth in that it is more inclusive and interrelated with the organism as a whole. Thus, the term physical development refers to the structural growth of the child in relation to the growth and development of the individual as a unified whole; while physical growth refers to growth in height, or weight, or some other structural characteristic.

Change goes on constantly in living cells—as someone has put it, "life is a process of changing." The growing child is constantly faced with new and different environmental forces of two special types. The one is organic, and is in essence the physiological process occurring in all living organisms; energy is being made available through the meta-

39

bolic processes related to food assimilation, and this is released through activity.   The other force is represented by man's external environment, which continuously stimulates him to reaction.   Concerning these as they relate to learning Boswell states:

Each living organism, in relation to internal as well as to external changes and conditions, tends to maintain itself as an integrated whole, as do also social organisms and a wide range of animate things.   Each is, then, not merely something happening, but is a complex, integrated, and unified system of activities.   Thus, definite internal changes are taking place within each living being in accordance with its character and mode of life; and all its vital mechanisms, however varied, combine to maintain a uniform dynamic state or "field" within each individual, in the face of fluctuating conditions of internal or external stimulation.[1]

How development occurs.   In an earlier paragraph it was stated that the individual is in close touch with the environment at all times. From the moment the egg is fertilized—the beginning of the individual —until death, we may say that the person is always responding to stimuli.   During his entire life he is continuously in a state of adjustment to environmental stimuli.   His behavior may very well be considered as composed of a series of responses to a continuous series of successive situations.

We should raise the question about what factor or factors in the relation of the individual and his environment make for development.   At present we are unable to give a complete answer to this question, and probably we will never be able to answer it as completely as we would desire.   From what we know of individuals and their development, it now seems certain that they are organized through their reactions to stimuli.   We frequently speak of the action of the environment upon the individual, yet we do not mean exactly what the statement implies. The individual always reacts to the stimulus.   He is active toward the environment and cannot be thought of as an inert, static thing merely being impressed by its surroundings.   Activity toward a situation tends to change the meaning of a situation so that it is never again the same for the individual.   Not only does activity toward a situation tend to give meaning to the situation, but it develops characteristic ways of behaving—habits.   It thus appears that the individual is organized or developed through his reactions to stimuli.   By control of the nature of

[1] F. P. Boswell, "Trial and Error Learning," *Psychological Review,* 1947, Vol. 54, p. 290.   Reproduced by permission of the journal and of the American Psychological Association.

the environment, the individual's reactions—and thus his development —may be controlled.

Concerning this problem as it relates to development during infancy, Gesell says:

The swiftness of development in infancy no one will be disposed to deny. The orderliness of this development is not so well recognized, but it is a fact of great significance. There are certain basic uniformities in the dynamics of development which apply to all infants, normal, abnormal, superior or inferior. There is a large system of uniformities which characterize all normal infants and keep them traveling on highly similar routes and on highly similar time tables. These uniformities are not stereotyped; they shade into small but important variations.[2]

The forces of heredity and environment are constantly operating in developing similarities and variations. It is commonly known that organisms arise through some process of reproduction, and it is with reference to this process that heredity is studied. Usually, there is a striking similarity between parent and offspring; invariably, heredity at least sets limits within which, whatever the environmental conditions may be, the individual's development will be confined.

Although it is often the custom to set the individual against environment, we find upon examination that he is in intimate relation with it at every turn; in fact, we find that only by definition can we separate the individual and his environment. There is a continuous interaction between them so long as the individual survives. Any statement regarding the individual which takes account of his biological nature must emphasize this mutual relationship. It may be emphasized thus: *the individual may be conceived of as protoplasm capable of maintaining itself by responding to a changing environment; during life, many of these responses become fixed or characteristic, so that we may consider an individual as a bit of protoplasm possessing more or less definite patterns of response.* Or, if we desire to think of him purely in terms of action, we may say: *the individual is a relational sum-total of behavior patterns developed in protoplasm in response to environment*— in which sense the individual is considered neither as protoplasm nor as environment but as the result of the reaction of the one to the other.

Importance of studying physical development. Physical development has been studied by various investigators by means of repeated measurements. Through the use of this method one is able to plot

[2] Arnold Gesell, *Infancy and Human Growth.* New York: The Macmillan Company, 1928, p. 124.

individual-growth curves as well as curves for different groups that would show race or sex differences. The greatest value of this method lies in the use made of measurements when kept over a period of time: they then furnish a permanent, objective picture, and the effects of various factors on development can be studied. Such measurements give a rather reliable index of the rate and periodicity of growth of boys and girls during adolescence.

In addition to the scientific value which data relative to growth may have, there are many applications that might be made of such data to a further understanding and guidance of growing boys and girls. Thus a study of the physiological differences in the rate of development of the sexes will give one a better perspective on the earlier changes of interest among girls during their passage from childhood to more adult activities. This variation within either sex for the same age is again important in its relation to the physical education program. Still further, it appears that variation should be considered in the general sectioning of pupils, since pupils who have the same degree of physiological maturity tend to play together, being more alike in their social interests and activities. And, of course, mental-hygiene problems, behavior disorders connected with problems of discipline, pathological disturbances, and other maladjustments also can be better understood by studying the pupil's physical development. On the whole, then, the knowledge of these facts may establish one of the bases for a program well-suited to individuals' needs, permit better sectioning for extracurricular activities, and foster more harmonious social relationships.

PHYSICAL DEVELOPMENT DURING ADOLESCENCE

Physical development—methods of study. Not only is it of interest to study the mutual relation of hereditary and environmental forces in development, but a knowledge of the general nature of growth itself is of value to those who wish to understand the changes appearing at the onset of adolescence.

In general, three methods have been used in the study of the physical development of children. The first in point of historical interest and frequency of use is the study of weight-height-age relationships. For the average parent and teacher, this is the most practical method to use. A study of either weight or height alone gives very little information, because children of the same sex, age, race, and environmental conditions vary greatly. Weight is probably a less reliable measure of physical development than height; however, when individual measurements only are considered, it is the one more generally used. A child may become heavy simply as a result of fat accumulation with no real growth in the number of tissue cells. Or just the opposite may happen: he may lose weight because he is using up adipose tissue, while at the same time the number of tissue cells is increasing.

Measurements of height furnish a much more accurate index of growth because they really indicate growth in terms of the length of the skeleton. This measure is fallible, however, because the bones may be growing in thickness, the cavities may be decreasing in size, and chemical constituents are, perhaps, being very greatly altered.

Because physical measurements are easy to make, they offer great possibilities if carefully interpreted. It is questionable, however, whether they are worth a great deal in the hands of a poorly trained school nurse or nutrition worker. Their chief value lies in the fact that they are easy to make and are objective, thus giving the teacher or school nurse a method of detecting extreme variations from the norm. Such cases should always be brought to the attention of competent medical authorities.

Baldwin [3] and others conducted some early studies of physical development by means of repeated measurements. By this method Baldwin was able to plot growth curves for individuals, as well as curves for the sexes and for different groups of individuals. While this method of studying physical development is not of so much immediate practical value, measurements kept over a period of time furnish a permanent, objective picture from which the effect of various factors on development can be studied. Such measures make possible both the scientific determination of how individuals grow and predictions about future growth. They also make an intelligent system of guidance feasible. This method has been used extensively in more recent studies conducted at Harvard University, the University of Minnesota, the University of California, and elsewhere. Some of the results from these studies are presented in this and subsequent chapters. These studies have furnished extensive and valuable information about the nature of growth and the special characteristics of individuals at the different age levels.

A third method of studying development relates to measurements that give results possible of interpretation in terms of ages. There are two of these ages, the anatomical and the physiological. Some workers

[3] Bird T. Baldwin, "The Physical Growth of Children from Birth to Maturity," *University of Iowa Studies in Child Welfare,* 1921, Vol. I, No. 1.

differentiate between these while others do not. *Anatomical age* has reference to the degree of physical development which a child has attained. It represents the point he has attained in his development toward physical maturity or adulthood. It does not have reference to size, weight, health, or strength. *Physiological age* is a term which has been largely used in connection with the development of the reproductive powers. In general, three physiological ages are spoken of: the prepubescent, the pubescent, and the postpubescent.

Growth in height and weight. Any table of averages is likely to be misleading, especially in respect to children's growth periods. Norms for height and weight should therefore be used with caution. Individuals differ widely in their physical make-up; some are referred to as tall, others as short, others as stocky, and so forth. Although the graphs obtained from statistics of large groups represent general tendencies, the attempt to apply the same formulae of growth to all individuals will meet with failure. Concerning this, Simmons has pointed out:

> In considering the dependency of weight upon height, it should be remembered that two children may be very dissimilar in the relative lengths of their trunk and legs and that trunk length is a greater factor in body weight than is leg length. When a child's ratings in transverse dimensions are also taken into consideration, his weight rating relative to the group norms become much more meaningful than when height alone is used for comparison.[4]

Children of the same age vary enormously when measured with respect to any developmental feature. They vary not only with respect to measurements made at any given time, but with respect to the rate and progress of development as well.

Just prior to the advent of puberty there is increased growth in height. Since pubescence appears earlier for girls than for boys, this increased rate of growth occurs earlier among girls. The amount by which the growth of girls between 10 and 13 years exceeds that in boys of the same age level is a good measure of trunk growth; and growth among girls after the thirteenth year is almost entirely trunk growth.

An examination of the materials in Table VI reveals the fact that there is a marked increase in the percentage of gain in weight at those ages where pubescence normally occurs. In addition to the increase in height and weight, there is a general change in the proportions of various bodily parts. The arms and legs change with the rate of

[4] K. Simmons, "The Brush Foundation Study of Child Growth and Development," *Monographs of the Society for Research in Child Development,* 1944, Vol. 9, No. 1, p. 57.

growth of different parts of the body. The arms and legs grow in length and become firmer; the hands and feet become larger. Other parts show equally important changes in the rate of growth. The most significant fact with respect to these growth changes is that there is wide variability among both boys and girls.

Courtesy, Los Angeles Public Schools.

Four twelve-year-olds; the same in age but different in physical development and other characteristics.

The only adequate way of finding out exactly when any acceleration in growth takes place is to secure individual growth curves for a few years before and after the advent of puberty. This was done and reported over two decades ago in a study that throws considerable light on this problem, although only 60 girls were included.[5] The greatest increase in height and weight occurred during the year before puberty. It is indeed noteworthy that the girls who matured at 12 years or younger, at 13, at 14, or at 15 had a greater increase in weight the year before puberty than either in the second year before puberty, in the year of, or during the year after puberty; essentially the same thing is true of

[5] G. E. Van Dyke, "The Effect of the Advent of Puberty on the Growth in Height and Weight of Girls," *School Review,* 1930, Vol. 38, pp. 211–221.

height. Anthropometrical data on 1,817 girls, ages 6 to 17 years, and on 1,884 boys, ages 6 to 18 years, who attended the Laboratory Schools of the University of Chicago were analyzed to compare rates of growth,

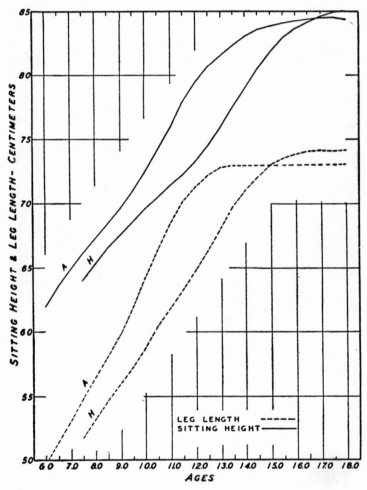

FIG. 6. *Contrasting Growth Trends. Showing differences in average sitting height and leg length of early (group A) and of late (group H) maturing groups of girls.* (After Shuttleworth.)

weights, heights, and height-weight relationships. The subjects were divided into three maturity groups on the basis of objective criteria. General conclusions were: (1) Differences in the height-weight relationship suggested differences in the bodily build of the three groups. (2) Statements concerning overweight or underweight should not neglect consideration of the maturity factor.   (3) "Growth as measured by height and weight is slightly accelerated before puberty."   (4) No significant differences were found in the heights of different female

TABLE VI

AVERAGE STANDING HEIGHT AND WEIGHT OF BOYS AND GIRLS
FROM AGE FIVE TO AGE SEVENTEEN (*After Simmons*)

| | BOYS | | GIRLS | |
C.A.	Height inches	Weight pounds	Height inches	Weight pounds
5	43.38	42.79	43.39	42.16
6	46.15	48.22	46.09	48.26
7	48.53	54.24	48.51	54.46
8	50.91	61.65	50.89	61.90
9	53.07	68.43	53.05	69.57
10	55.27	76.84	55.29	76.09
11	57.21	85.60	57.92	88.40
12	59.38	95.17	60.46	100.44
13	61.73	105.66	61.54	110.45
14	63.17	119.06	62.84	120.16
15	67.09	132.26	63.60	126.60
16	68.39	141.91	63.87	129.83
17	69.24	147.57	63.99	134.35

maturity groups after 15 years or in those of the different male maturity groups after 17 years.   (5) Girls maturing before 13 years of age were, as a group, heavier at each age from 6 to 17 years than those who matured later, and those who matured between their thirteenth and fourteenth birthdays were heavier at all ages than those who matured after their fourteenth birthdays.   (6) Boys who matured before their fourteenth birthdays were heavier than those who matured later, and those who matured between the fourteenth and fifteenth birthdays were heavier at all ages than those who matured after their fifteenth birthdays.[6]   These data are in harmony with results obtained from other studies.   There is evidence that boys who mature early are taller at all age levels than those who mature later.   Ellis[7] found from a study of 208 boys, ages 11 to 16 years, from two residential schools in

[6] "The Relation of Accelerated, Normal, and Retarded Puberty to the Height and Weight of School Children," *Monographs of the Society for Research in Child Development,* 1937, No. 8, pp. 1–67.

[7] Richard W. B. Ellis, "Height and Weight in Relation to Onset of Puberty in Boys," *Archives of Disease in Childhood,* 1946, Vol. 21, pp. 181–189.

England that differences between the growth curves for the early and late maturing boys could be demonstrated as far back as the sixth year.

Anatomical development. Anatomical development pertains primarily to the skeletal system, and especially to changes in the structure of the bones. With the advent of adolescence, as has been pointed out, there is an increase in height and weight. But there is a further change in the composition of the bones (of the osseous and cartilaginous materials, and so forth) as the individual matures. The ossification of the bones proceeds gradually but is rather far advanced at the beginning of adolescence. Early studies by Baldwin and others revealed that after five or six years of age, girls show more advanced ossification than boys.[8] The carpal bone of the wrist has been extensively used as a means of determining one's anatomical development. The *anatomic index* was used as the unit of measurement. An index of 10 indicates that 10 per cent of the area of the wrist shows ossification. At the age of 13, about 70 per cent of the area of the girl's wrist shows ossification, and there is a considerable slowing down in the rate of development at this period. A more recent method of assessing skeletal maturity utilizes the pattern of ossification at the epiphyses rather than a percentage of ossification of some specific area.

The child's age at the time of the eruption of the permanent teeth has been used as an indicator of anatomical development. Cattell[9] has presented data showing the number of permanent teeth erupted by boys and girls at different age levels. Dental age scales have been developed for boys and girls. The scales clearly show sex differences, and the pre-adolescence of both boys and girls is marked by a period of accelerated eruption of the permanent teeth.

In a study reported by Simmons and Greulich girls were divided into three menarcheal age groups. The three groups were found to be distinguished from each other by their skeletal age more consistently than by their height, weight, or height-weight index. This consistency is clearly shown in Figure 7 showing the mean skeletal age of the three groups of girls 7 to 17 years of age. Concerning terminal size, however, the investigators conclude:

[8] Bird T. Baldwin, Laura M. Bresby, and Helen V. Garside, "Anatomic Growth of Children, A Study of Some Bones of the Hand, Wrist, and Lower Forearm, By Means of Roentgenograms," *University of Iowa Studies in Child Welfare,* 1928, Vol. 4, No. 1.

[9] Psyche Cattell, "A Scale for Measuring Dental Age," *School and Society,* 1928, Vol. 27, pp. 52–56.

It is normal for some children to grow and develop toward maturity at a rapid rate with, consequently, an early attainment of terminal size; it is normal for some other children to grow and develop toward maturity at a slow rate with, consequently, a late attainment of terminal size. The rate of growth, in this series, has no reliable relationship to terminal size.[10]

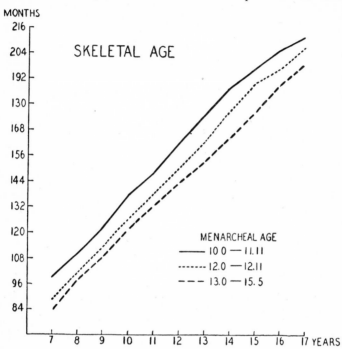

FIG. 7. *Mean Skeletal Age of Girls Seven to Seventeen Years of Age in Three Menarcheal Age Groups.* (After Simmons and Greulickh)

MOTOR DEVELOPMENT DURING ADOLESCENCE

Age and motor performance. It has been generally observed that older children are stronger and in general more proficient in motor activities than are younger children. The changes in motor performances that occur with age were studied by Atkinson.[11] He secured records from a series of tests given to 8,000 New York City high

[10] Katharine Simmons and W. W. Greulich, "Menarcheal Age and the Height, Weight and Skeletal Age of Girls 7 to 17 Years," *Journal of Pediatrics,* 1943, Vol. 22, p. 548.

[11] R. K. Atkinson, "A Motor Efficiency Study of Eight Thousand New York City High School Boys," *American Physical Education Review,* 1924, Vol. 29, pp. 58–59.

school boys. The improvement in test performance by chronological age was more consistent than that by height, weight, or physiological age. In a later study of the athletic performance of approximately 9,000 high school girls he found that performance tended to improve with height but to decline with weight.[12] In some events, namely hop step jump, basketball goal shooting, and distance events, the late maturing girls made the best performance.

Philip tested 165 girls and 146 boys and found a decrease in reaction time between the ages of 9 and 16.[13] There was a significant improvement with age in speed of reaction to light and sound. This improvement was apparent both for reactions with and without a warning signal. Boys were from 3 to 5 per cent faster than girls. This greater speed for boys had been noted in previous studies. Philip attributes this to previous experiences of boys calling for quickness of reaction time.

The development of physical fitness was studied by Jokl and Cluver[14] among a group of children from 5 to 20 years of age. No difference in performance was found among the different racial groups, although constitutional factors appeared to influence the development of physical efficiency more than environmental ones. In the case of endurance, measured by the 600-yard run, both boys and girls improved from 6 to 13 years. The improvement up to 13 years was about the same for both sexes; but after this stage of development boys continued to improve, whereas the girls lost in efficiency, so that in the range from 17 to 20 the girls' ability was about that of six- to eight-year-olds. This decline in efficiency was reflected not only in their running time, but was present also in their physical condition as revealed by their pulse rate, respiration, and fatigue. It seems likely that this early decline in motor ability among girls is a result of their habits and practices: that is, girls show an increased interest in social activities at a fairly early age, and a lack of interest in participating in athletics and other forms of muscular activities.

In the study by Espenschade,[15] measurements were made at intervals

[12] R. K. Atkinson, "A Study of Athletic Ability of High School Girls," *American Physical Education Review*, 1925, Vol. 30, pp. 389–399.

[13] B. R. Philip, "Reaction Time of Children," *American Journal of Psychology*, 1934, Vol. 46, pp. 379–396.

[14] E. Jokl and E. H. Cluver, "Physical Fitness," *Journal of the American Medical Association*, 1941, Vol. 116, pp. 2383–2389.

[15] Anna Espenschade, "Motor Performance in Adolescence," *Monographs of the Society for Research in Child Development*, 1940, Vol. 5, No. 1.

of six months, and averages were determined for each half year age level. These results for the fifty yard dash, the broad jump, and the distance throw are given in Figures 8, 9, and 10. On the fifty yard dash there is a continuous improvement for boys from age 13 to and beyond age 16. The best performance for the girls was reached at 13.25 and thereafter there was a gradual decline; however, the loss was slight during the next two years, but performance tended to increase after the latter part of the fifteenth year. In the broad jump there is a pronounced increase in ability for boys from age 13.00 to 16.5 years. This increase was most rapid during the ages of 14 and 15. In the case of girls, a gradual and continuous decrease in ability on the broad jump test was noted from age 13 to 16.5. These two activities are rather strenuous in nature, and no doubt offer real problems in motivation for girls. This lack of motivation on the part of adolescent girls combined with changes in body proportion seems to provide a logical explanation for the continuous decline in these abilities among girls from age 13 through age 16.5.

Scores in the throwing events, distance and target throws, showed a continuous increase for boys from 13 through 16.5 years. In the case of girls there was a steady increase to age 14.75 and 15.25, followed by a slow but gradual decline. On the jump and reach test Espenschade reports that the mean scores of girls were somewhat erratic but increased steadily, especially from 15.25 through 16.25 years.

The magnitude of changes in scores of boys from the fall of 1934 to the fall of 1937 (a three year growth interval from age 13.25 to age 16.25) was negligible in the target throw. On the different tests

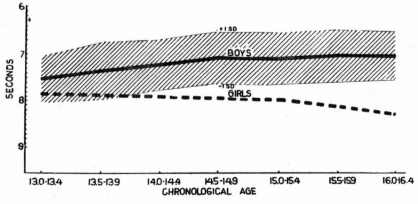

Fig. 8. *Comparison of Boys and Girls on the 50–Yard Dash.* (After Espenschade)

making up the Brace test the percentage passing all tests was greater
in the fall of 1937 than in the fall of 1934. For both boys and girls
there was little improvement in balance during this period, and in the
case of girls there was a decline in agility.

Sex differences. There were statistically significant differences found

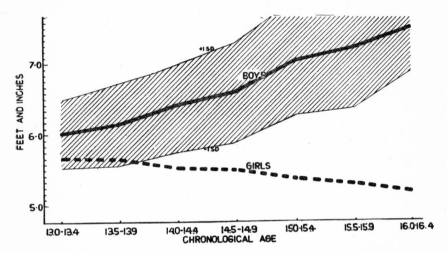

Fig. 9. *Comparison of Boys and Girls on the Broad Jump.* (After Espenschade)

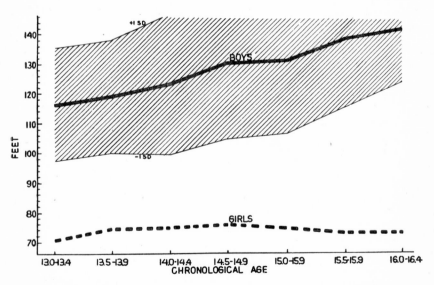

Fig. 10. *Comparison of Boys and Girls on the Distance Throw.* (After Espenschade)

in all the performances studied by Espenschade.[16]   These differences became greater at each age level.   The most extreme recorded was in the distance throw, although differences in the broad jump became as great among the older adolescents.   Sex differences in the Brace test were studied for the fall of 1934 and for the fall of 1937.   In the fall of 1934, 16 of the 20 tests comprising the *Brace Test* were passed or failed by more than 90 per cent of both boys and girls.   A larger percentage of girls passed six of these tests than boys, although the differences were not statistically significant, except for one event—the agility and control test.

In order to secure additional comparable data and also to extend the age range, 325 girls and 285 boys, ages 10.5 to 16 years, were tested during the years 1943–45.   The tests, comprising the *Brace Test,* were given at school during the school day.   The items making up the *Brace Test* are classified as: class I, agility; class II, control; class III, strength; and class IV, static balance.   The *Test* consists of twenty items selected to include a wide variety of coordinations not frequently practiced.   A comparison of the scores of boys and girls on the different tests showed that only slight differences appeared in total scores or in measures of the various classes before the age of 13.8 years.   After this age level boys excelled in all events and their superiority increases rapidly at each age level.   The most striking difference between the boys and girls was in the "push-up" test, which is classed as one of the tests of strength.   The greatest similarity noted was in the test, "sit, then stand again, with arms folded and feet crossed."   This is classed under control tests, and it is in this classification that girls make their best showing.   Their poorest showing was made on the tests of agility and strength.   The following items are given as a summary of the results:

1. The increment pattern for boys of total scores on the Brace Test battery is similar to that of adolescent growth in standing height. Scores for girls show little change after the thirteenth year.

2. The stunts of the Brace Test battery were placed in four general classes according to the predominant type of muscular action demanded. Additional elements necessary for performance were noted. "Dynamic balance," especially, is important in many events.

3. Boys show an increase in ability to perform events of all classes. The rate of growth is greater after 14 years of age than before and appears to be more rapid in "agility" than in "control."

4. All tests for boys in which "dynamic balance" is a factor show a marked "adolescent lag." . . .

[16] *Ibid.,* p. 59.

5. Girls improve in "agility" up to 14 years, then decline. In "control" and in "flexibility and balance," little change can be seen over the age range studied. . . .[17]

The physical characteristics of adolescent girls are different from those of boys in that the arms and legs are proportionally shorter, the trunk larger, the pelvis broader; the femur is attached to the pelvis at an oblique angle which is a mechanical disadvantage. Studies show that strength, speed, and coordination are essential for athletic performances. The shorter legs as well as other body features handicap the girl in competing with boys in motor activities of an athletic nature. Longitudinal studies of the growth in strength of boys and girls show that as early as age ten boys are stronger than girls at the same age level. This difference in strength increases markedly so that by age sixteen the average boy is considerably stronger than the average girl.

Relationship of strength to other traits. Significant relations have been reported by various investigators between dynamic and static strength and certain physical measurements. Dynamic strength refers to abilities involved in actions, such as track events, while static strength is usually measured by means of dynamometric tests. Some interesting similarities in the operation of these two aspects of motor performances are revealed in the results of a study by Bower and reported by Jones.[18] Correlations were obtained between these aspects of strength among a group of boys and chronological age; skeletal age, based on assessments of X-rays of the hand and knee; height; an evaluation of "good looks"; and intelligence, based on an average mental age obtained from the results of two forms of a group intelligence test. These correlations are presented in Table VII. An interesting feature of the results of Table VII is that while chronological age and physical measurements correlate highest with total strength, popularity and "good looks" are more closely related to the gross motor scores. This is to be expected, when one realizes the prestige value of motor performances among adolescent boys. The low correlation between motor performances and intelligence is also significant.

Evidences relative to the relationship between strength of ado-

[17] Anna Espenschade, "Development of Motor Coordination in Boys and Girls," *Research Quarterly, American Physical Education Association,* 1947, Vol. 18, pp. 30–43.

[18] See Harold E. Jones, *Motor Performance and Growth.* Berkeley and Los Angeles: University of California Press, 1949, Chap. II.

TABLE VII

MOTOR PERFORMANCE CORRELATIONS WITH OTHER DEVELOPMENTAL
TRAITS—BOYS (*After Jones*)

Variable	Total strength (grip, pull, thrust)	Gross motor scores (track events)
Chronological age	.39 ± .06	.18 ± .07
Skeletal age	.50 ± .055	.36 ± .06
Height	.65 ± .04	.40 ± .06
Popularity	.30 ± .07	.39 ± .06
"Good looks"	.21 ± .07	.38 ± .06
Intelligence	−.17 ± .07	.05 ± .08

lescent boys and physical growth have been summarized by Mac-Curdy.[19] There is a gradual increase in strength to the age of 12, followed by a very rapid rise. The maximum growth is reached for most boys around the eighteenth birthday. The study by McCloy confirmed the fact that the most rapid increase in strength for boys was between 13 and 16 years.[20] Only a slight increase was found after age 17. In girls the most rapid increase was between ages 12 and 14, there being no increase recorded for the average girl after age 15. For most girls, there was an actual decrease after this age.

Strength and physiological maturity. It has already been pointed out that there is a significant correlation between strength and skeletal age. Results from the Harvard growth study, reported by Shuttleworth,[21] show that the average size of early-maturing boys is superior to that of late-maturing boys as early as age six and is maintained until the age of 18, which was the terminal point of the measurements conducted. In comparison with the group means for strength of grip, each of the early-maturing boys studied was above the norm in strength at ages 13 to 16; while the late maturing boys tended to fall below the norm at these ages. The results for the boys and girls, classified as early-, average-, and late-maturing, are presented in Table VIII. The early- and late-maturing groups represent approximately the 20 per

[19] H. L. MacCurdy, *A Test for Measuring the Physical Capacity of Secondary School Boys.* Yonkers-on-Hudson: World Book Co., 1933.

[20] C. H. McCloy, "The Influence of Chronological Age on Motor Performance," *Research Quarterly,* Association of Physical Education, 1935, Vol. 6, pp. 61–64.

[21] See Frank K. Shuttleworth, "Physical and Mental Growth of Boys and Girls Ages Six through Nineteen in Relation to Maximum Growth," *Monographs of the Society for Research in Child Development,* 1939, No. 3.

cent at each extreme of a normal public school distribution at the different age levels; while the average group consists of those whose maturational level was approximately that of the norm for their age level. Jones states: "It is apparent that the three curves are more or less parallel, with some divergence of the early- and late-maturing groups between the ages of 13 and 15, and with a later convergence which, however, fails to bring them together at the end of the series of measures."[22]   A further study of the results of Table VIII shows

TABLE VIII

Mean Scores for Early-, Average-, and Late-Maturing Boys and Girls *
(Kg., right grip)

Age	Per Cent of Boys			Per Cent of Girls		
	Early $N=16$	Average $N=28$	Late $N=16$	Early $N=16$	Average $N=24$	Late $N=16$
11.0	27.1	24.0	22.7	21.1	20.9	20.6
11.5	29.3	25.9	25.2	24.4	23.2	21.2
12.0	29.3	26.9	26.0	26.1	25.8	22.5
12.5	31.3	28.4	27.0	29.1	26.8	23.7
13.0	33.3	30.4	28.1	30.3	28.8	25.7
13.5	37.6	32.5	30.0	29.3	30.3	26.8
14.0	44.2	34.3	30.2	29.7	30.7	26.4
14.5	47.1	38.6	33.3	31.0	32.2	28.4
15.0	50.0	43.0	36.3	32.5	33.3	31.4
15.5	52.2	47.6	41.1	33.4	35.2	32.7
16.0	54.3	49.0	43.9	33.4	35.8	32.4
16.5	55.9	50.9	48.4	34.7	36.1	34.4
17.0	57.2	53.5	51.3	34.3	36.5	34.8
17.5	......	55.8	54.3	33.9	37.8	35.3

* The boys are classified on the basis of skeletal maturing, the girls on the basis of age at menarche.

that the early-maturing girls, although superior at the age of 13, fail to maintain their superiority in the subsequent years, and actually drop below that of the average-maturing group.   This is in harmony with results obtained relative to height and weight.   Thus, precocious sexual development of girls appears to be associated with an early arrest in physical and motor development; this is not true for boys.

[22] Harold E. Jones, *Motor Performance and Growth.* Berkeley and Los Angeles: University of California Press, 1949, pp. 56-57. Table VII is also quoted from Jones, *op. cit.,* p. 57.

SOME GROWTH CHARACTERISTICS AND PROBLEMS

Changes in Voice. Closely connected with the question of muscular development are the obvious changes of voice in early adolescence. They are much more evident in the case of boys and constitute one of the external signs of the advent of puberty. They are the effect of the rapid growth of the larynx, or the "Adam's apple," and a corresponding lengthening of the vocal cords that stretch across it. These become approximately double their former length with a consequent drop of an octave pitch—an instance of the well-known law of physics that a taut string emits a lower tone on being lengthened. The voice of girls is not subject to such an outright transformation. In maturity, the female voice may be little lower, if at all, than it was in childhood, although it should be fuller and richer. In boys there is not only a change of pitch but there is also an increase in volume, and often the voice becomes more pleasant in quality.

It requires two or more years for the youth to achieve control of his voice in the lower register, and during that time he is often made self-conscious by the roughness of his own tones. He is mortified by the unexpected squeaks which punctuate his bass rumblings. Such whimsical "breakings" cause him to feel that he is making himself ridiculous —an opinion that is often confirmed, unfortunately, by the mirth with which others greet his vocal vagaries.

This difference in rate of physiological maturity may be a source of anxiety to the adolescent. When Bill's pal, Henry, suddenly surpasses him in physical development, develops a bass voice, and begins to shave, Bill may wonder whether he is normal in development. This difference between his appearance and that of his pal may become so pronounced that he finds himself seeking other pals, and may even resort to social behavior not wholly acceptable, in an effort to prove himself.

The voice change is not an accurate index for use in studies of the development of a boy, since there is no satisfactory way of evaluating it objectively. The voice change could be studied if a recording device was used for comparing the depth and other qualities of the voice at varying stages of development. In this connection, it should be pointed out that it is the progressive deepening of the voice, rather than the absolute pitch, which is significant as an indication of progress toward maturity.

It is generally recognized that the voices of young men at maturity will vary widely in pitch and other qualities. Also, pronounced differences will be found among adolescents of the same developmental level. In interpreting these differences, it should be pointed out further that there are no special relationships between the depth of voice and the degree of masculinity.

Unevenness of growth. The outstanding feature of growth, which sometimes continues up into the twenties, lies in the pronounced changes it effects in body proportion. The diagram on this page shows the nature of these changing proportions with growth toward maturity.

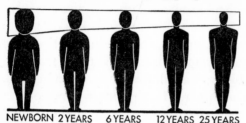

NEWBORN 2 YEARS   6 YEARS   12 YEARS 25 YEARS

PROPORTION CHANGING

In the end most of the early disproportions in growth are "smoothed out," and the individual reaches normal maturity.

Boynton [23] presents thirteen measurements of growth increments for anthropometric characteristics based on retests from ages 5.5 years to 16.5 years inclusive. His data show that although the 5½-year-old boy is 65 per cent as tall as he probably will be at 17.5 years, he weighs only 33 per cent as much and has only 18 per cent of the strength of grip he will possess twelve years later. The brain of the child at birth weighs a little less than one pound. The number of cell bodies is apparently complete, and the neurones increase in size and

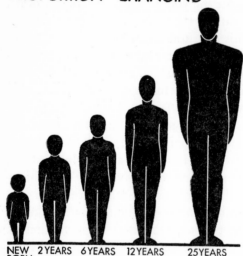

NEW   2 YEARS 6 YEARS   12 YEARS   25 YEARS
BORN

PROPORTION & SIZE CHANGING

richness of terminal ending up to the period of physical maturity. A small part of the growth of the nervous system consists in the medullation of fibers which, at birth, do not have the characteristic fatty white sheath. Medullation of cortical fibers continues through the periods of infancy and childhood, extending even into middle age. This may be

[23] Paul Boynton and Juanita Curry Boynton, *Psychology of Child Development.* Minneapolis: Educational Publishers, Inc., 1938, p. 114.

important in relation to the growth in ability to deal with more complex mental processes.

With respect to weight and strength of grip, the average girl 5½ years of age has approximately one-third of her 17-year development, but in the case of height, shoulder width, and ankle circumference she has approximately three-fourths of her ultimate development. Though the average girl is not completely developed in some respects at 17.5 years, there are certain elements of growth that are complete by the age of 15.5 years.

A further lack of uniformity in growth is in the development of lung capacity. Measurements made on groups of school children show that the increase of lung capacity is quite pronounced during the adolescent period. The greatest increase for girls occurs between the ages of 10 or 11 and 14, the greatest for boys from one to two years later. This increase, however, is considerably affected by the extent to which one engages in physical exercises. Baldwin's researches show that the physical curves for the development of lung capacity are quite similar to those for the development of weight and height in that they indicate an early pre-adolescent retardation followed by an increased adolescent growth. This same general curve of growth is to be found for the liver.

Tests for strength of arm, strength of back, tapping, and endurance all reveal that there is a great acceleration with the onset of adolescence. At the age of 8 the muscles constitute slightly over one-fourth the body weight, whereas at 16 they constitute approximately 45 per cent. At the age of 16 the strength of grip has become practically double what it was at the age of 11. However, following this period, growth continues for some time, but at a declining rate as the individual reaches complete maturity.

Body types. Among the more recent studies of this problem, the work of Sheldon [24] and his associates is outstanding. On the basis of their research, three polar types of human physiques are presented. These are (1) the *endomorph,* characterized by the tendency to store up excessive fat; (2) the *mesomorph,* showing extreme muscular development with broad shoulders and hips; and (3) the *ectomorph.* Individuals of the third type have light bones, long and slender muscles, and are characteristically thin. Sheldon's greatest contribution is that his method of classification is based upon the concept that variations in body build shows wide variations from individual to individual and

[24] W. H. Sheldon, S. S. Stevens, and W. B. Tucker, *The Varieties of Human Physique.* New York: Harper and Bros., 1942.

that these three types proposed should be regarded as dimensions on this large range of variability. Sheldon has proposed a seven-point rating scale for defining more exactly an individual's morphological characteristics. Thus, an individual may be rated 2 in endomorphy, 3 in mesomorphy, and 5 in ectomorphy. This provides for him a code designation of 235. From studies of college students he found seventy-six code somatotypes.

In addition to his studies and predictions of body types, he has developed a theory of temperament types and indicates that a close relation exists between morphological characteristics and temperament. Some of his concepts bearing on this will be presented in Chapter V.

The organismic age. Some students of child psychology have employed a combination of age units in an attempt to secure a more complete and accurate picture of a child's development through the years. The growth curves based upon single measurements show a great deal of fluctuation. In order to arrive at a clearer concept of the growth of the child as a whole, it becomes necessary to see the relation of separate phases of growth to each other as well as to the growth of the child as a whole. Olson and Hughes [25] arrived at an "organismic age" by taking a number of different growth values of given chronological ages. These values were averaged and given the name "organismic age."

The organismic age is based upon the average age unit obtained for a given child for various measures of growth. These measures would include such phases of growth as height, ossification, dentition, reading, mental ability, strength, coordination, and so forth. Such a measure is especially important in making a diagnosis of a child's growth level. A child may be found to be deficient in school achievement. By comparing his age in arithmetic or reading with his organismic age, one is able to determine whether or not his educational achievement level is within the matrix covered by most of the curves represented in the organismic age. In terms of educational diagnosis, there is no reason to label a child a reading disability case if his reading age is approximately the same as his organismic age.

The question of how many measures should be included in the organismic age cannot be answered with any degree of finality. In most cases, there are not enough measures available in age units to provide a stable organismic age. However, two measures are likely

[25] Willard C. Olson and Byron O. Hughes, "The Concept of Organismic Age," *Journal of Educational Research,* 1942, Vol. 35, p. 525.

to give a more accurate picture than one, while three or four measures would provide a still better picture. Olson and Hughes state, "Theoretically one would have determined a stable organismic age when no further additional values would cause it to fluctuate in a significant manner."[26]

Effects of nutrition and living conditions. A number of studies have been made of the relationship between the living conditions of children and their physical development. These studies indicate that children who live in undesirable slum areas average, during the elementary school years, from three to five inches shorter and from eight to twelve pounds lighter than children from good homes. No one can say just how much these differences are due to poor feeding, inadequate ventilation, unsanitary living conditions, ill-balanced rest and recreation, or deficient inheritance. Quantitative experiments were carried on as early as 1924 with large numbers of families of rats, continuing through successive generations and covering the entire life cycle of each individual.[27] These have shown that an improvement of an already adequate diet (1) expedited growth and development, (2) resulted in a higher level of adult vitality, and (3) extended the average length of life expectancy. There are evidences from many different kinds of experiments that this same sort of improvement of food supplies, that is, the taking of a larger proportion of the needed calories in the form of what has been called "protective foods," acts to support superior development in children and an increased number of years of good health and vigor among adults.

The elements that are most commonly deficient in the child and adolescent's diet are the minerals and vitamins. Evidence obtained by investigators shows a close relation between a deficiency of calcium and nervous characteristics manifested by children. Thus, dietary deficiencies affect the growing child's emotional life as well as his physical growth and health. The importance of diet in relation to physical status was shown in the study of Jeans and Zentmire.[28] When 404 malnourished elementary school children were supplied with the needed vitamins there was a significant improvement in their health and physical status. Further materials bearing on the development of

[26] *Ibid.,* p. 526

[27] H. C. Sherman and H. L. Campbell, *Journal of Biological Chemistry,* 1924, Vol. 50, pp. 5–15.

[28] P. C. Jeans and Z. Zentmire, "The Prevalence of Vitamin A Deficiency among Iowa Children," *Journal of the American Medical Association,* 1936, Vol. 66, pp. 996–997.

the digestive system and the breakfast habits of adolescents are presented in Chapter IV.

Weight modification during adolescence. The unevenness of the growth process during adolescence makes it easy to overemphasize the relationship between height and weight at different levels of development. It has already been suggested that one cannot single out individuals as too fat or too thin simply by relying on height-weight charts. However, those who are obviously above or below reasonable standards for their height and age level can be shown the difference between normal variations in weight and being abnormally fat or abnormally thin. In this connection, the influences of such factors as hereditary differences, body build, and temporary glandular imbalance should constantly be borne in mind.

Various attempts have been made for giving consideration to these dimensions of body build. Norman C. Wetzel has developed a grid-like chart which takes into account seven types of body build. Seven "physique channels" are set forth on this chart.[29] A child's growth in height and weight is considered normal so long as he advances steadily in his own channel. When periodic examinations show cross-channel progress, growth problems are indicated. If the child moves to a left channel his weight is perhaps too great for his height; whereas if he veers to the right of his channel his weight is likely to be too light for his height. The Wetzel grid chart is a useful instrument for indicating the presence of growth problems.

Any health program, concerned with the improvement of health, must take into consideration the weight of the individual children. A thorough health examination is essential to any efforts toward the modification of weight. Likewise, any health examination will take weight into consideration. It may be expected that ten to fifteen per cent of adolescents from an average group will require some attention to their weight. Furthermore, this problem will be found to be closely related to diet. Also, the problem of diet is a complicated one, since many factors enter into the eating habits of adolescents. The problem, therefore, is not always as simple as it might appear at first. It has been found that efforts directed toward the improvement of diet are most beneficial when the individual is able to record and observe his own progress. Thus, the adolescent should be given

[29] N. C. Wetzel, "Physical Fitness in Terms of Physique, Development and Basal Metabolism," *Journal of the American Medical Association*, 1941, Vol. 116, pp. 1187–1195.

systematic guidance and instruction in methods of controlling his diet in order to improve his weight standard and reach a higher level of health.

Good posture habits. The importance of good posture habits in maintaining the organs of the body in their correct position and in enabling them to function to the maximum of their efficiency has been stressed by physicians in recent years. Posture charts have been devised, posture exercises recommended, and posture clinics held, all in an effort to provide for the development of good posture among children, adolescents, and adults. Although posture training should be begun in early childhood years, adolescence is the period when so many bad posture habits are formed. There are several reasons why this age is a period of susceptibility to incorrect posture habits. In the first place, the early adolescent years are years of rapid growth. Much of the energy from food is used up in providing for growth. Thus the adolescent seems inclined to slump from a tired feeling. Secondly, the adolescent years are active years. Adolescent boys and girls use up a great deal of energy in their social, recreational, and play activities. This, again, brings on a feeling of tiredness and the tendency to slump while sitting. Thirdly, the individual may feel self-conscious over his long legs, or general height. This attitude tends to lead the individual to assume an unhealthy posture in an effort to cover or hide his height or gangliness. This is perhaps much more prevalent among girls than among boys. If the desired habits are not acquired during the earlier years of life, they must be learned during the adolescent stage.

Appraisal of posture. What constitutes good posture is a problem with which every school should be concerned. The examination for posture is not as simple as it was once thought to be. In the first place, most teachers are not aware of the factors involved in posture, and would consider them in too limited a manner. Secondly, it should be emphasized that individual differences in body build must always be taken into account. Where teachers are available—usually teachers in physical education and health—measurements and careful evaluations based upon observations may be made by them. Where such help is not available, other avenues and means should be resorted to in order to get a more accurate appraisal of the pupils' posture.[30] In general the appraisal should include standing posture, sitting posture, and posture in walking. The most important thing about standing

[30] See Charles H. McCloy, *Tests and Measurements in Health and Physical Education.* New York: F. S. Crofts and Company, 1939.

posture is that the individual should stand erect, with the head, trunk, hips, and legs well aligned. All children should be observed for sagging shoulders, proper standing, correct manner of walking, and lateral curvatures.

Problems related to physical development. The unevenness of growth of different parts of the body, coupled with the fact that the rate of growth and time of onset of puberty vary with different individuals, often presents difficulties and problems for boys and girls during adolescence. The degree of asynchrony of development as between leg length and body length, or hip width and shoulder width becomes more pronounced at this stage and is itself in many cases a source of disturbance. Also, there is a lag in the increase in the size of the muscles. This presents a significant problem to many boys.

It has already been suggested that there is considerable variation among individuals of the same sex as to the chronological age at which puberty may be said to begin, the rate at which it proceeds, and the age of physiological maturity. The importance of these variations as a factor which disturbs adolescents is worthy of attention. The adolescent boy is particularly conscious about lack of height; whereas tallness may be an extremely disturbing condition for the adolescent girl.

The systematic appraisal of the physical development of adolescents conducted by the Institute of Child Welfare at the University of California presented some of the conditions or factors relative to growth and body variations that were disturbing to a number of adolescents.[31] The physical manifestations that disturbed boys are listed as follows:

Physical Manifestations That Disturbed Boys [32]	*Number of Boys*
Lack of size—particularly height	7
Fatness	7
Poor physique	4
Lack of muscular strength	4
Unusual facial features	4
Unusual development in the nipple area	4
Acne	3
Skin blemishes, scars	2
Bowed legs	2
Obvious scoliosis	2

[31] See H. R. Stolz and L. M. Stolz, "Adolescent Problems Related to Somatic Variation," *Forty-third Yearbook of the National Society for the Study of Education,* Part 1, 1944, Chapter 5.

[32] *Ibid.,* p. 86.

Lack of shoulder 1
Unusually small genitalia 1
Unusually large genitalia 1

Among the eighty-three girls included in the study there were thirty-eight who gave evidence of being disturbed by the following physical characteristics:

Physical Manifestations That Disturbed Girls [33]	Number of Girls
Tallness	7
Fatness	7
Facial features	5
General physical appearance	5
Tallness and heaviness	3
Smallness and heaviness	3
Eyeglasses and strabismus	2
Thinness and small breasts	2
Late development	2
Acne	1
Hair	1
Tallness and thinness	1
Big legs	1
One short arm	1
Scar on face	1
Brace on back	1

The "sex appropriate physique" applies quite differently to girls from the way it applies to boys. Perhaps the phrase "sex appropriate face and figure" is more applicable to the girls. This is clearly differentiated from the health, strength, and muscular abilities so prominent in the appropriate physique concept among boys. For the girl too much muscular strength is regarded as undesirable, and the rugged appearance of a healthy individual is looked upon as masculine rather than feminine. Both boys and girls are in many cases frustrated because of their lack of a "sex appropriate physique" or "figure." In connection with any frustration existing, girls are more likely to do something about the condition than are the boys. This is reflected in their greater willingness to diet in order to give them a more appropriate figure. However, boys are often motivated to exert greater effort and endure continued exercise in order to develop a masculine physique. It seems likely, then, that even temporary deviations from the "sex appropriate" development may pro-

[33] *Ibid.*, p. 86

duce significant problems of adjustment for adolescent boys and girls.

SUMMARY

Growth begins with the fertilization of the egg cell, and birth merely extends the sphere of activity. There are many factors determining the nature, direction, and amount of development that will take place. There are certain biological laws that enable us to predict development when it occurs in average environmental conditions. Although varying circumstances may alter the direction and amount of development, it is still characterized by its unity. The growing child is a unified whole, and the nature of development of one part of the body must be considered in connection with its relation to other parts. Even though there is a lack of uniformity in the growth of different parts of the body, a continuous, interrelated form of growth is ever-present.

Height-weight charts, the anatomical index, dental age norms as well as other physical measurements have been used in the study of the growth process. Repeated measurements made on boys and girls from year to year show a distinct sex difference in the age for the onset of puberty and accelerated growth. The pre-adolescent decline in the general rate of growth followed by the adolescent spurt is about one and one-half years earlier on the average for girls than for boys. However, individual growth curves show that this will vary considerably with different individuals of the same sex. There is a wide variation in the age of the onset of puberty, as well as the variation referred to in the general physical development. Variations in growth may be affected by such extraneous factors as exercise, living conditions, and diet. Since growth is affected by so many factors, it becomes very difficult to set forth simple formulae or predictive procedures to estimate it for different stages of life.

The results from the developmental studies reported at the University of Iowa, at Harvard, and at the University of California may be summarized as follows:

1. There is a period of relatively slow growth prior to the prepubertal growth spurt.

2. A prepubertal spurt in growth is from 1½ to 2 years earlier for girls than for boys. In the case of girls the twelfth year frequently is the time at which they make their largest annual gain in height.

3. A decrease in rate of growth following puberty.

4. Sufficient consistency in stature rank in the group during elementary-school years for competent prediction in the groupings, "tall" and "medium," and to a less extent in the classification of "short."

5. A seasonal variation in weight gain. In general, weight increase of children is greatest in autumn, somewhat less in summer, and least in winter and spring. Seasonal variation in height tends to favor the spring months.

6. Individual differences are prevalent and important. An individual not only differs from other children; he also is different from himself from time to time. Although there are important general trends, there appears to be slight uniformity in the development of his various traits and abilities. The result of this variability in growth is that in the intermediate grades— Grades V, VI, and VII—there are, in general, a few pupils who are still in the stage of fairly uniform rate of growth, many who are at the beginning of the prepubertal growth spurt, and a small number who have passed through the accelerating phase and are beginning to slow down in their rate of growth.[34]

THOUGHT PROBLEMS

1. Consider your own adolescent years. What problems related to physical development appeared in your life at this stage?

2. Summarize briefly the sex differences presented in this chapter in motor ability at age 13.25. How would you account for the early cessation and actual decline in many cases of motor development of girls during the adolescent years?

3. What are the different methods used in the study of physical development? Give the advantages of each.

4. If they are available, study some data on physical development secured from a group of students, and note the variations existing. How do these variations relate to their interests? To their personalities?

5. In your observations, have you detected in yourself a spurt in growth with the onset of adolescence? What other pronounced changes occurred rather rapidly?

6. What is the significance of the lack of uniformity in growth discussed on pages 58 and 59?

7. What are the different methods of measuring anatomical development? Which of these do you regard as the most accurate? Which most useful in general? Give reasons for your answers.

SELECTED REFERENCES

Averill, L. A., *Adolescence*. Boston: Houghton Mifflin Co., 1936, Chap. II.

Boynton, Paul, and Boynton, Juanita Curry, *Psychology of Child Development*. Minneapolis: Educational Publishers, Inc., 1938, Chap. V.

[34] See "Pupil Personnel, Guidance, and Counseling," *Review of Educational Research*. Washington, D. C.: American Educational Research Association, 1939, Vol. 9, p. 148.

Cole, Luella, *Psychology of Adolescence.* New York: Farrar and Rinehart, 1948, Chap. II.

Espenschade, Anna, "Motor Performance in Adolescence, Including the Study of Relationships with Measures of Physical Growth and Maturity," *Monographs of the Society for Research in Child Development,* 1940, Vol. 5, No. 1.

Hurlock, E. B., *Adolescent Psychology.* New York: McGraw-Hill Book Co., 1949, Chap. II.

Jersild, A. T., "Education in Motor Activities," *Child Development and the Curriculum, Thirty-eighth Yearbook of the National Society for the Study of Education,* Part I, 1939.

Jones, Harold E., *Motor Performance and Growth.* Berkeley and Los Angeles: University of California Press, 1949.

Keliher, A. V., *Life and Growth.* New York: Appleton-Century-Crofts, Inc., 1941, Chaps. VI, VII, and VIII.

Stolz, Herbert R., and Stolz, Lois Meek, "Adolescent Problems Related to Somatic Variations," *Forty-third Yearbook of the National Society for the Study of Education,* Part I. Bloomington, Ill.: Public School Publishing Co., 1944.

Zachry, C. B., and Lighty, Margaret, *Emotions and Conduct in Adolescence.* New York: Appleton-Century-Crofts, Inc., 1940, Chaps. II–IV.

For a good summary of studies on motor and physical development see: American Educational Research Association, *Review of Educational Research,* 1944, Vol. XIV, No. 5; 1947, Vol. XVII, No. 5; 1950, Vol. XX, Vol. 5.

IV

Physiological Changes

The momentous work of G. Stanley Hall, referred to in Chapter I, gave very little attention to the physiological changes occurring during adolescence. This was a result of the lack of understanding, on the part of the physiologists of that time, of the differences between children, adolescents, and adults in the physiological functioning of the organism. It is generally recognized by child psychologists today that the young child is not a miniature adult; likewise, that the adolescent is neither a child nor an adult in his physiological reactions. Adolescence was referred to in Chapter I as a transition age. It may well be regarded as a period of physiological and behavioral changes. This concept will be clarified in the discussions throughout this chapter, which deals with various physiological changes occurring during adolescence and the effects of these changes on the activities of the adolescent.

Studying physiological changes. The problem of determining the degree and nature of physiological changes occurring during adolescence is one of the main concerns of this chapter. Again, children and adolescents frequently suffer from the "tyranny of the norm." The materials presented throughout this chapter are not given for the purpose of providing a standard by means of which to judge the physiological development of the individual. A recognition of the wide individual differences in the time of the onset of puberty and the rates of changes produced in the organs of the body should temper the acceptance of the averages given and emphasize the extent to which many boys and girls vary from the average and still stay within the realm of "normal" development.

Physiological measurements are not as easily determined as height, weight, eruption of the permanent teeth, and body build measurements. The pulse rate is one of the easiest of the physiological measurements to be made, while the basal metabolic rate requires laboratory procedures for its measurement. Another measure referred to in this chapter is that of blood pressure. A study of glandular secretions at the different age levels, although valuable as a basis for determining the status of sexual development, requires the services of a laboratory technician. Also, the results secured from physiological measurements are influenced by a great number of conditions both within and without the body. Thus, if results are to be obtained which will be comparable with results obtained at another period, careful controls must be established over the subjects for a period of twelve or more hours prior to the measurements. These difficulties in securing reliable and interpretable data account for the small number of longitudinal studies conducted on physiological changes at the different age levels.

Endocrine factors in relation to development. Recent studies reveal a definite relationship between developmental changes and hormones produced by the pituitary gland. Two hormones from this gland are especially important in this connection. One of these is the growth hormone, which enables the healthy, well-nourished child to attain his normal body size. If there is a deficiency of hormones from this gland, normal growth will be retarded, and a form of pituitary dwarfism will result. On the other hand, if an excess of the growth hormone is produced during the growing period, pituitary gigantism will follow.

Another pituitary hormone of special importance in maturation is the gonad-stimulating hormone. The action of these hormones in a normal, healthy child will cause the immature gonads to grow and eventually develop into mature ovaries or testes. This hormone, furthermore, helps to sustain the normal function of the ovaries or testes of the individual after maturity. An inadequate production of the gonad-stimulating hormone during pre-adolescence would interfere with the normal growth and development of the ovaries or testes; whereas too much of the gonad-stimulating hormone would tend to produce a type of precocious sexual development.

The gonad-stimulating hormones act upon the pituitary gland in such a way as to reduce the effects of the growth hormone, and thus retard growth. Greulich has pointed this out in the following statement:

If the testes or ovaries begin to function at the requisite level too early in life, growth is arrested prematurely and the child ends up abnormally short. If, on the other hand, the adequate production of the ovarian and testicular hormones is unduly delayed, growth, particularly that of the limbs, continues for too long a period and the characteristic bodily proportions of the eunuch are attained. It appears, therefore, that normal growth and development are contingent upon the reciprocal and properly timed action of pituitary and gonadal hormones.[1]

A number of studies have been conducted relating to gonadotrophic hormone secretion in children. In general these studies indicate that the excretion of gonadotrophic hormone in early childhood in both sexes is too low to be detected by the methods used; these studies indicate that measurable amounts first appeared in the urine during adolescence. Data are reported by Greulich and others on the results of 120 urinary gonadotrophin assays performed on 64 boys. Concerning the significance of gonadotrophin excretion in adolescence, they conclude:

The results show that with advancing age and with advancing developmental status there is a general tendency for gonadotrophin to increase in amount from the undetectable levels of early childhood to levels more characteristic of the adult. There is as yet no direct evidence as to the biological nature of this gonadotrophin; on the other hand, it does not seem likely that it differs from the hormone found in the urine of the adult male. The properties of hormones of this type have been described earlier, and it seems reasonable to suppose that the primary changes of puberty, namely an increase in size of the testes and the initiation of spermatogenesis are related to the action of this gonadotrophin upon the seminiferous tubules. Secondary sex changes related to the secretion of the steroid sex hormones may be ascribed to the action of the hormone upon the interstitial gland of the testes.[2]

Nathanson and others have reported somewhat similar results.[3] Average curves for boys and girls are presented in Figure 11. During early years the amount of androgens secreted into the urine is only slightly less for girls than for boys. This difference becomes more pronounced after age 11. Before the ages 10 or 11 both boys and girls excrete measurable amounts of male and female hormones. Slightly greater amounts of the male hormones are secured from the boys; while a greater amount of the female hormone are obtained from the girls,

[1] W. W. Greulich, "Physical Changes in Adolescence," *Forty-third Yearbook of the National Society for the Study of Education,* Part 1, 1944, Chap. II, p. 16. (Quoted by permission of the Society.)

[2] W. W. Greulich, *et al.,* "Somatic and Endocrine Studies of Puberal and Adolescent Boys," *Monographs of the Society for Research in Child Development,* 1942, Vol. 7, No. 3, p. 62.

[3] I. T. Nathanson, L. E. Towne, and J. C. Aub, "Normal Excretion of Sex Hormones in Childhood," *Endocrinology,* 1941, Vol. 28, pp. 851–865.

although the differences are slight. After the age of 10 and beginning with the age of 11, the excretion of female sex hormones is markedly increased in girls; while the excretion of the male sex hormones by boys is correspondingly increased, but usually at a later date.

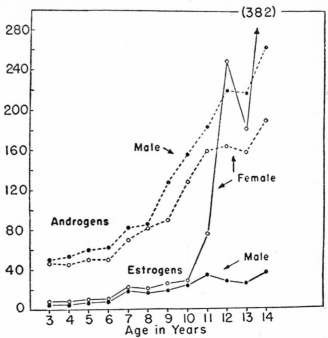

FIG. 11. *Age Changes in Excretion of Sex Hormones. The female sex hormone, produced by the ovary, is the chief estrogenic hormone.* (After Nathanson *et. al.*)

PHYSIOLOGICAL CHANGES RELATED
TO PUBESCENCE

Blood pressure, heart, and pulse rates. The blood pressure of both boys and girls increases with age. During early childhood, there is little difference between the sexes, but between the ages of 10 and 13 blood pressure is higher in girls than in boys; after the age of 13 the pressures of boys exceed those of girls, the differences increasing with age.[4] This is an example of the general trend toward an earlier incidence of maturity among girls, which has been pointed out in connection with other developmental characteristics. Blood pressure probably

[4] H. G. Richey, "The Blood Pressure in Boys and Girls Before and After Puberty," *American Journal of Diseases of Children,* 1931, Vol. 42, p. 1328.

decreases in girls after the age of 16—during the postpubescent period. There is some evidence that high blood pressure is related to the onset of puberty. However, blood pressure has little relationship to the height of the individual and only a moderate relationship to weight.

An exaggerated fear that the adolescent, through exertion, will overtax his heart still exists, and is usually explained on the basis of a lack of harmony between the development of the heart and that of the blood vessels. This idea has persisted for some time and was pointed out as early as 1879 by Beneke.[5] On the basis of data collected, he pointed out that the volume of the heart increases in proportion to the body weight, the circumference of the aorta and the pulmonary artery increases in proportion to the body length. These observations are essentially correct but the interpretations and generalizations based on them are misleading. As late as 1931 a text translated from German stated:

. . . the heart of an adult man is three times the size of the child's, while the proportionate circumference of the aorta (close to the heart) remains the same. . . . We can readily see that no system of exercise can meet the first principles of practical hygiene, unless it recognizes the physiological condition described.

The California studies revealed that with the beginning of menstruation a pronounced change occurs in the trend of average growth curves for a number of physiological variables.[6] The age at which menstruation first appeared was taken as a reference point, and values were computed for physiological measures at six-month intervals in each direction from the menarche. The results of this analysis for a total of 52 cases are presented in Figure 12. The continuous increase in systolic blood pressure ceases near the menarche and maintains a fairly uniform level after that stage. There is also a rise of pulse rate during the premenarcheal years, with the maximum reached during the year prior to the menarche. After the menarche there is a gradual decline in pulse rate.

Exercise has a pronounced effect on blood pressure as well as on pulse rate. Shock has also developed data showing the pulse rate for boys and girls 13 to 17.5 years of age one minute after exercise.[7] This is

[5] F. W. Beneke, *Uber das Volumen des Herzens und die Weite der Arteria pulmonalis und Aorta ascendens.* Marburg: V. Theodor Kay, 1879.

[6] N. W. Shock, "Physiological Changes in Adolescence," *Forty-third Yearbook of the National Society for the Study of Education,* Part 1, 1944, Chapter IV.

[7] *Ibid.,* p. 68.

shown in Figure 13. Pulse rate, as shown here, increased more in girls than in boys at all ages. For both sexes this increase becomes less as the individual becomes older. Since the increase in pulse rate is a

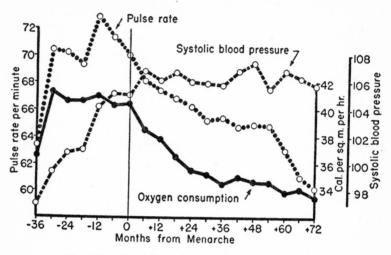

FIG. 12. *Basal Functions as Related to Maturity.*

method of bringing more oxygen to the working muscles, a diminished rate would result in a slower rate of recovery from exercise as the age of the individual increases beyond maturity.

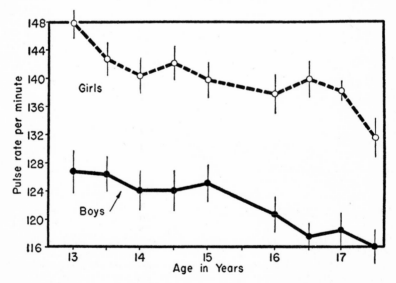

FIG. 13. *Mean Pulse Rate One Minute After Exercise.*

Age changes in basal metabolism. Probably the most striking non-sexual physiological change appearing at the time of the menarche is the rather sudden decline in basal metabolism.[8] There is furthermore a cessation of growth increase in respiratory volume at this stage of development. This presents quite a contrast with the change pointed out earlier in blood pressure with age. While there is little change registered in blood pressure after the menarche, there is a continuous decrease in basal metabolism throughout the teen years for both boys and girls. This is shown in Figure 14, which is based upon materials from the California study. Shock points out that the individual curves are less uniform than the average curves presented. Over one-half of the cases showed a pronounced decrease, as illustrated in Figure 14.

The digestive system. There are, also, pronounced changes in the organs of digestion during adolescence. The stomach increases in size and capacity and undergoes qualitative changes. Because of the rapid growth of the adolescent, he needs more food than formerly. The increased size of the stomach is perhaps closely related to his strong cravings for food. During the adolescent years this craving persists, and it appears that adolescent boys and girls are able to assimilate foods that they were unable to assimilate during the earlier years of life.

The extent to which this increased appetite is manifested by pre-adolescents, adolescents, and post-adolescents was brought forth in the longitudinal studies of adolescents at the University of California. The responses of these boys and girls to questions related to their health and appetite are presented in Table IX.[9] Over one-half of the pre-adoles-

---

[8] N. W. Shock, *op cit.,* Chapter 4. Shock states further:
"The basal metabolism or basal oxygen consumption indicates the amount of energy required to maintain the normal vital processes of the individual when in the 'basal' state. It has been found that this basal energy requirement is closely associated with the functional activity of the thyroid gland. When the thyroid gland is underactive the basal metabolism is reduced. In measuring basal metabolism, the subject breathes through a mask or mouthpiece so that the expired air can be collected in a large rubber bag or tank. The volume of air expired in an eight-minute period is measured and part of it analyzed for its oxygen and carbon dioxide content. Since the amount of oxygen present in the outdoor air which was breathed by the subject is known, the reduction in the oxygen content of the expired air represents the amount of oxygen consumed by the subject in the eight-minute period."

[9] The data from Table IX were taken from the *U. C. Inventory I: Social and Emotional Adjustment.* There are two forms of this: one for boys, and one for girls. These present a cumulative record of a group by items for a seven-year period. Complete records for the seven-year period were available for 71 boys and 72 girls.

cent and adolescent boys and girls indicated that they got sick at their stomach and vomited once in a while. Little sex difference was found in the response to this question; although girls appear to become sick

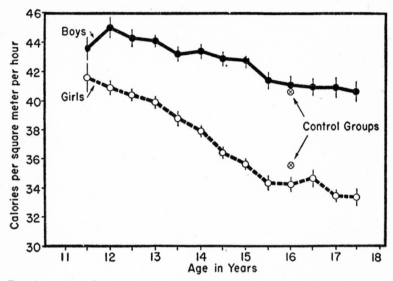

FIG. 14. *Age Changes in Basal Metabolism from Repeated Tests on Same Subjects.* (Smoothed data.)

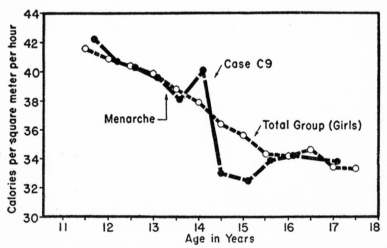

FIG. 15. *Individual Growth Curve of Basal Metabolism Girl.*

at the sight of foods to a slightly greater extent than boys. Both boys and girls at different age levels appear to enjoy their meals and to have a good appetite. No sex differences were discernible on their responses to questions about these items. Youngsters at this stage of life seem to always be ready to eat and are able to make entire meals from hot dogs, hamburgers, or pancakes. Digestive difficulties appearing in various forms at this age are doubtless due in part to the eating habits of these boys and girls. Likewise, many skin troubles may be traceable, at least in part, to an ill-balanced diet.

Good nutrition is closely related to the health and the digestion of

TABLE IX

RESPONSES OF ADOLESCENT BOYS AND GIRLS TO QUESTIONS DEALING WITH
THEIR HEALTH AND APPETITE

	H5L6		H8L9		H11L12	
QUESTION	Boys	Girls	Boys	Girls	Boys	Girls
Do you worry about your health?						
Never	52	62	52	50	59	54
Once in a while	38	31	46	47	38	42
Quite often	3	3	1	1	3	4
All the time	7	4	0	1	0	0
Do you get sick at your stomach and throw up?						
Never	21	33	39	37	55	50
Once in a while	76	65	61	57	45	46
Quite often	1	1	0	6	0	4
All the time	0	0	0	0	0	0
Do you get sick at the sight of some foods?						
Never	39	46	62	33	56	47
Once in a while	61	54	38	64	42	52
Quite often	0	0	0	3	0	0
All the time	0	0	0	0	0	0
Do you enjoy your meals?						
Never	3	1	3	0	1	1
Once in a while	3	1	3	3	0	1
Quite often	32	24	23	21	15	18
All the time	62	74	70	76	83	79
Do you have a good appetite?						
Never	1	7	4	3	0	0
Once in a while	7	8	7	11	6	7
Quite often	31	22	17	26	18	31
All the time	61	61	72	60	76	62

H5 refers to the high fifth grade, L6 to the low sixth, and so forth. The figures presented in the table refer to the percentage of boys and girls responding to the questions in the manner indicated.

foods by adolescents. The importance of adequate and nourishing food during the growing period cannot be overemphasized. Schools are becoming more keenly aware of this. The consumption of milk from the school lunchrooms over the country presents a good omen. However, an examination of the breakfast-eating habits of junior and senior high school boys and girls does not present an encouraging picture. A survey of these habits, conducted through the Institute of Student Opinion by *Scholastic Magazine,* revealed that 18 per cent of these boys and girls do not eat any breakfast on any given morning of the school week.[10] This survey reported the results from more than 150,-000 students in approximately 300 schools scattered over the United States. Of those surveyed, a large group who reported eating breakfast did not report a well-balanced diet. Marked regional differences, conforming perhaps to adult customs, were found. Breakfast appeared to be more leisurely in the South and most hurried in the Northeast. Hurry and lack of appetite were the main reasons given for not eating breakfast. The survey showed that 84 per cent of the boys spent 15 minutes or less at breakfast, and that 87 per cent of the girls spent 15 minutes or less at this meal.

SPECIAL GROWTH CHARACTERISTICS
AND PROBLEMS

The skin glands. During the adolescent years marked changes take place in the structure of the skin and in the activity of the skin glands. The soft, delicate skin of childhood gradually becomes thicker and coarser as the individual matures sexually. There is an enlargement of the pores of the skin, a characteristic that is closely related to some of the problems of adolescents associated with skin disturbances and blemishes. There are three different kinds of skin glands, each of which is distinctly separate from the other. These are (1) the *merocrine glands,* which are scattered over most of the skin surfaces of the body, (2) the *apocrine sweat glands,* which are limited primarily to the armpits, mammary, genital, and anal regions, and (3) the *sebaceous glands,* the oil-producing glands of the skin.

a. *The apocrine sweat glands.* There is a marked increase in perspiration of the armpits as the child reaches puberty. The sweat glands become enlarged. This occurs even before there is a growth of hair

[10] "Breakfast-eating Habits," *The Journal of School Health,* 1949, Vol. 19, pp. 138–139.

within the axillae (armpits). The characteristic odor of axillary perspiration is usually not detectable in boys prior to puberty and becomes more pronounced during the early adolescent years. Among girls the apocrine sweat glands appear to undergo a cycle of secretory activity during the menstrual cycle.[11] This is thought to be closely associated with increased perspiration in the armpits experienced by many girls and by young women at this time.

b. *Sebaceous glands.* The increased size and activity of the sebaceous glands during puberty is thought to be closely associated with skin disturbances during adolescence. There is a disproportion between the size and activity of these glands and the size of the gland ducts during puberty. When the secretion from the sebaceous glands is unable to drain properly it forms hard plugs in the pores at the openings of these glands. These are generally referred to as blackheads, and are most often found on the nose and chin. The glands continue to function, even though the opening has been blocked, and raised pimples then appear on the surface of the skin.

The sebaceous glands are also associated with hair follicles, and are absent from the skin in some regions where there is a lack of hair, such as on the palm of the hands. During puberty the sebaceous glands are associated with disproportionately small hairs. This causes a temporary maladjustment, and is regarded as the major reason for acne. There is some evidence that an excess of male hormones may be an important factor in the causation of acne.

Changes in hair. Changes occur in the hair as well as in the skin during early adolescence. Three kinds of hair succeed each other during one's life span: *lanugo,* unpigmented hair, which appears during the last three months of intra-uterine life; *vellus,* down hair, which replaces lanugo and persists during infancy and childhood; and the *terminal,* which replaces the childhood hair and becomes the dominant type in the adult. This replacement is greatly accelerated during puberty and continues at a less rapid rate throughout adult life.

There is a distinct change in the shape of the hair line on the forehead as the individual begins to mature. This has been referred to as a secondary sexual characteristic. The hair line of immature boys and girls follows an uninterrupted bow-like curve. This is illustrated in the upper row of Figure 16. In the case of mature males, this curved hair line is interrupted by a wedge-shaped recess on each side of the fore-

---

[11] Josef Klaar, "Zur Kenntnis des weiblichen Axillarorgans beim Menschen," *Wiener Klinische Wochenschrift,* 1926, Vol. 39, pp. 127–131.

head.   Greulich and others found this character to be a late rather than an early developmental feature.[12]

a. *Facial hair*.   There are no marked sexual differences during childhood in the vellus of the upper lips, cheeks, and chin.   Among boys, the downy hairs at the corners of the upper lip become noticeable about the time of puberty.   This development extends medially from each corner of the upper lip, and eventually forms a mustache of rather

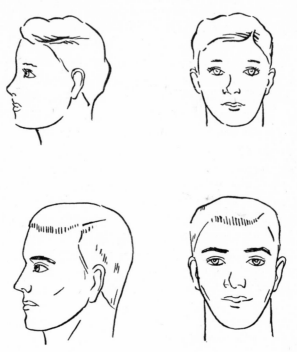

Fig. 16.   *Adolescent Changes in Hair Line and Facial Contour.*

fine hair which is perceptibly larger, coarser, and darker than the vellus hair it replaces.   This change begins with puberty.   The mustache becomes progressively coarser and more heavily pigmented as the individual passes through adolescence.   During the period when the mustache is developing, the vellus over the upper part of the cheeks increases in length and diameter.   It persists as long, coarse down until

[12] See W. W. Greulich, R. I. Dorfman, H. R. Catchpole, C. I. Solomon, and C. S. Culotta, "Somatic and Endocrine Studies of Puberal and Adolescent Boys," *Monographs of the Society for Research in Child Development*, 1942, Vol. 7, No. 3.

the juvenile mustache is fairly well developed. Somewhat later, a thin growth of long, rather coarse, pigmented hairs appears along the sides and lower parts of the chin and on the upper part of the face just in front of the ears. These, too, gradually become coarser and more heavily developed, eventually forming a beard.

b. *Axillary hair*. The axillary hair does not usually appear until the development of the pubic hair is nearly complete. The transition from vellus to terminal hair in the axilla is quite similar to changes in hair in other regions of the body, and the amount of axillary hair appearing is closely associated with the development of other body hair. Among boys, the development of terminal hair on the limbs and trunk begins to appear during the early stages of adolescence, with growth rather rapid at first. The development of terminal hair on the limbs begins on the upper part of the forearm, later on the sides of the lower arms, and still later on the back of the hands.

After the transition from long down to terminal hair has made considerable progress, a similar process begins on the distal half of the leg. It gradually extends upward toward the knee. The extension of the hair-covered areas from the centers on the trunk and limbs proceeds at different rates of speed in different boys. The amount of hair developed will vary considerably from individual to individual. By the age of 18 or 19 the growth of hair on the arms is fairly heavy for the majority of boys. Also, there is a moderate growth of terminal hair over the legs, thighs, and buttocks as well as a varying amount on the ventral surface of the trunk.

c. *Pubic hair*. Pubic hair is a secondary sex characteristic that appears during puberty. However, it is not until the growth of the genitals are well under way that the terminal hair appears to replace the vellus. It has been customary to associate the amount and extension of terminal hair over the body with the degree of masculinity; however, there is no indication that a close association exists, in the case of boys, between the degree of masculinity in terms of sexual potency and the amount of hair on the body.

Skeletal growth in relation to sexual development. Investigations, in which the same children were observed and measured repeatedly over a number of years, have provided much valuable information about the interrelations of growth and physical maturation. This information has furnished some basis for the prediction of the nature of growth changes that are likely to take place in an individual child. Observations of growing individuals, supported by experiments on animals, indicate that the development of the skeletal structure is closely related to that of the reproductive system. Clinical studies have revealed that normal skeletal development will not occur in the absence of adequately functioning gonads. In castrated or in hypogonadal individuals, for example, skeletal development is significantly retarded and epiphyseal fusion of the bones of the limbs is significantly delayed.

A study of some somatic and endocrine changes associated with puberty and adolescence among boys was carried on over a period of several years at Yale University.[13] The skeletal status of 476 private school boys was compared with the degree of development of their primary and secondary sexual characteristics. X-ray films were made of the hand and wrist and of the elbow of each of the boys. The "skeletal age" of each boy was determined by standards of skeletal development of the regions X-rayed. These boys were divided into five *maturity groups* representing successive stages in the transition from the degree of physical immaturity that exists just before puberty to the degree of maturity that is usually associated with late adolescence. A study of the skeletal development of each of the five maturity groups showed a gradual increase in the average skeletal age from group 1 (most immature group) to group 5 (most mature group). There was a tendency for skeletal age to increase with advancing maturity rating even among boys of the same chronological age.

It is evident from this and from materials presented in Chapter III relative to growth in height and weight that, by the time the maturation of the reproductive organs has attained a level producing changes in their functioning, the rate of growth in stature has already begun to decelerate. This deceleration is closely related to changes in the long bones of the legs. So intimate is the correspondence between sexual maturation and changes in the skeletal system that it is often possible to predict in the case of girls when the menarche will occur from an examination of an X-ray film of the hand and wrist during pre-adolescence.[14] It was pointed out in Chapter III that there is a distinct difference in the skeletal ages of three groups of girls of the same age levels but representing different levels of sexual maturity.

[13] W. W. Greulich, *et al.*, "Somatic and Endocrine Studies of Puberal and Adolescent Boys," *Monographs of the Society for Research in Child Development,* 1942, Vol. 7, No. 3.
[14] William W. Greulich, "The Rationale of Assessing the Developmental Status of Children from Roentgenograms of the Hand and Wrist," *Child Development,* 1950, Vol. 21, pp. 33–44.

Physiological maturity and strength. There is evidence from the California Adolescent Growth Study that bodily strength is associated with other growth phenomena.[15] In Figure 17 growth in right-hand strength is compared for two groups of girls representing contrasting extremes in age at menarche. Among these girls the earlier maturing group shows a rapid rise in strength of grip prior to age of twelve. The later maturing group is relatively retarded in strength, but the two groups eventually reach the same level. The greatest increment of growth for each group occurred near the time of menarche.

Comparisons of strength data for premenarcheal and postmenarcheal girls of the same chronolgical age have been reported by Jones.[16] The results for a "total strength" score, based on right grip, left grip, pulling

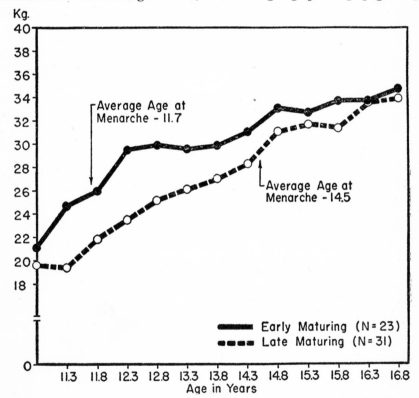

Fig. 17. *Manual Strength Development of Girls in Two Maturity Groups.*

[15] Harold E. Jones, "The Development of Physical Abilities," *Forty-third Yearbook of the National Society for the Study of Education,* Part 1, 1944, Chap. VI.

[16] H. E. Jones, "The Sexual Maturing of Girls as Related to Growth in Strength," *Research Quarterly, American Physical Education Association,* 1935, Vol. 6, pp. 61–64.

strength, and thrusting strength, are presented in Table X. At ages 12.25 to 13.25 more than 80 per cent of the postmenarcheal girls surpass the mean for the less mature girls of the same chronological age. After 13.25 the differences become smaller, since the less mature girls are continuing to grow while there is a reduction or cessation of growth among those who are well past the menarche. The anatomical basis for this greater strength among the more mature girls is to be found in the advanced skeletal development of the early maturing girls. The muscular and skeletal growth of the postmenarcheal girls have been shown to be significantly ahead of that of the premenarcheal girls. However, as Jones has pointed out, "The early- and late-maturing reach similar strength levels in later adolescence, but through different patterns of growth."

TABLE X

A COMPARISON OF PREMENARCHEAL AND POSTMENARCHEAL GIRLS
IN TOTAL STRENGTH (*After Jones*)

Age	Premenarcheal			Postmenarcheal		
	N	Mean (k.g.)	S. D.	N	Mean (k.g.)	S. D.
12.25	63	86.0	16.30	15	100.6	15.55
12.75	44	92.0	16.35	24	105.6	15.20
13.25	25	93.0	16.20	37	108.8	16.35
13.75	20	93.0	16.35	54	105.6	16.50
14.25	17	100.7	19.50	63	107.9	18.00

Special problems related to physiological growth. Any period of transition is likely to be fraught with problems. The physiological changes associated with adolescence present conditions the individual has not met with up to this time and, in many cases, is ill-prepared to meet when they appear. For the girl, the period of the first menarche can be a real problem if she has not been prepared for it, or, if she is prepared, if too much concern has been given to it. There has been a tendency among some groups to overemphasize the seriousness of the changes appearing at this time—thus causing the girl to limit her activities beyond the limitations called for by these physiological changes.

Problems of skin blemishes and acne disturb many boys and girls of this age. Also, closely related to this, is the problem of body odors.

Some adolescents and post-adolescents tend to go to an extreme in the use of perfumes, lotions, and other toilet articles in an effort to meet special problems appearing at this stage. The appearance of axillary hair is in some cases a source of disturbance for girls; while the lack of the appearance of hair on the arms and legs has been regarded by many boys as a weakness in the development of a masculine type. The appearance of hair on the chin and upper lip at this stage presents one more problem for the adolescent boy, which he must learn to meet by shaving. Needless to say, problems related to the physiological changes have been considerably aggravated by cultural forces. These changes are inextricably related to the sex roles to be played by adolescent boys and girls. Any change that interferes with the development and assertion of the masculine role on the part of boys is most likely to be a source of difficulty; while conversely, any change that interferes with the development and assertion of the feminine role on the part of the girl is likely to be a source of difficulty.

SUMMARY AND CONCLUSIONS

In recent years we have witnessed an increased attention to the physiological changes that occur during the adolescent years. Since growth in height and weight are quite observable and are easy to measure, it is but natural that such measurements would have first received the attention of students of adolescent psychology. Pronounced individual differences appear in the change of pulse rate, blood pressure, and glandular secretions. Thus, it has been pointed out that it is safer to evaluate the individual's development in terms of measurements made during the growing years than to rely upon the norm or average for a large group of individuals at the different age levels.

There is ample evidence that the endocrines play an important role in the physiological changes that take place during adolescence. It may be stated that they pave the way and initiate many changes connected with the sex drive as well as changes related to physical growth. The California longitudinal studies of physiological changes and other studies provide valuable data for arriving at a clearer understanding of the nature and characteristics of these changes in relation to various internal and external forces and conditions.

In our society, many problems, interwoven with customs and conventions, emerge as these physiological changes appear. A number of the most pronounced difficulties connected with these changes that adolescents face have been presented in this chapter. As a partial summary to this chapter, it is worth while to point out that these problems are *real*, and are not to be ignored or to be looked upon with ridicule or scorn by adults. Owing to the transition state of the adolescent, he is not always able or ready to accept these changes.

THOUGHT PROBLEMS

1. How is the pituitary gland related to gonadal stimulation?

2. What are some of the more pronounced physiological changes that take place with the onset of pubescence in girls? In boys?

3. Discuss the nature and amount of change in metabolism that takes place with age.

4. What are the different skin glands? Why are these especially important during the adolescent period?

5. Observe several children between the ages of eight and seventeen. What changes in the hair line appear with advancing age level?

6. Interpret the data presented in Table X showing a comparison of premenarcheal and postmenarcheal girls in total strength. What is the general significance of these data?

7. In the light of the discussions of this chapter, what would you suggest in the way of a physical education program for girls during adolescence?

SELECTED REFERENCES

Cole, Luella, *Psychology of Adolescence* (Third Edition). New York: Rinehart and Co., Inc., 1948, Chap. II.

Goodenough, Florence L., *Developmental Psychology* (Second Edition). New York: Appleton-Century-Crofts, Inc., 1945.

Greulich, W. W.; Dorfman, R. I.; Catchpole, H. R.; Solomon, C. I.; and Culotta, C. S., "Somatic and Endocrine Studies of Puberal and Adolescent Boys," *Monographs of the Society for Research in Child Development,* 1942, Vol. 7, No. 3.

Greulich, W. W., "Physical Changes in Adolescence," *Forty-third Yearbook of the National Society for the Study of Education,* Part 1, 1944, pp. 8–32.

Hurlock, E. B., *Adolescent Psychology.* New York: McGraw-Hill Book Co., 1949, Chap. II.

Jones, H. E., *Development in Adolescence.* New York: Appleton-Century-Crofts, Inc., 1943.

Shock, N. M., "Physiological Factors in Development," *Review of Educational Research,* 1947, Vol. 17, pp. 362–370. The author presents a brief summary of research on dietary and nutritional influences on development and the effects of disease on development.

Shuttleworth, F. K., "Sexual Maturity and the Physical Growth of Girls Age Six to Sixteen," *Monographs of the Society for Research in Child Development,* 1937, Vol. II, No. 5.

For a rather complete review of recent studies in this area, see "Growth and Development," *Review of Educational Research,* 1950, Vol. 20, No. 5.

V

Emotional Growth

Emotions and behavior. The changed concepts of the nature of the child and of the adolescent have brought with them changes in the methods and objectives of our schools and other educational and socializing agencies. There has been a shift from the consideration of the intellectual, moral, or social side of the child to that of the development of the total personality. An increased understanding of the development and importance of the emotions in the growing personality is of utmost importance to those concerned with the guidance of growing boys and girls. There is a tendency on the part of many people to regard the emotions as a stereotyped pattern of expression appearing with certain forms of stimulation. An adolescent's timidity is thus thought of as an expressive behavior pattern appearing on certain occasions. It has been suggested that in contrast with such a view, "the present tendency is to recognize that emotional components are in some form and to some degree present in all behavior."[1]

Emotional development, although treated in this chapter as a separate topic from that of physical development, must not be considered without reference to physical development. The case of Jo, reported by Zachry in a study of adolescents, illustrates how impossible it is to isolate certain elements as "physical" and others as "emotional." This case, herewith presented, shows the necessity of considering all the

---

[1] H. E. Jones, H. S. Conrad, and Lois B. Murphy, "Emotional and Social Development and the Educative Process," *Thirty-eighth Yearbook of the National Society for the Study of Education,* Part 1, 1939, Chap. XVIII, p. 363. (Quoted by permission of the Society.)

factors that enter into the nature and type of activities of an individual at any particular stage of life.[2]

Jo is a boy of twelve who has been feeling very much out of the family picture. He is the youngest child. His sister is soon to be married and his brother has just started to work, but Jo is at an age when he is not particularly interesting to any member of the family. He has been doing only fairly well in his school work and he has definitely neglected his arithmetic.

One morning he went down to breakfast and ate rather heartily: he had oatmeal with cream, eggs, bacon, jam, and milk; and while he was eating he recalled that he was going to have an arithmetic test that morning. He had a queer, twitchy feeling of excitement in his stomach at the thought of the arithmetic test. He started walking slowly to school, thinking more about the test, and his stomach felt queerer and the oatmeal weighed very heavily on it. He had a vague feeling, which was hardly a thought, that if his breakfast were to come up he wouldn't have to go to school, and the arithmetic test came to mind again. Suddenly he found it hard to keep the breakfast down.

Shortly after his arrival at school, it did come up. He was sent home by the principal with a clear conscience to have a day in bed. The principal telephoned his mother, who immediately became concerned. She put Jo to bed in the guest room and made a fuss over him such as he had not experienced since he was quite a small boy. His sister came in and showed him her wedding presents; his brother stopped and had a talk with him before going out in the evening, an event which had not occurred for months; and his father spent the evening reading to him.

This upset stomach had a high value: no arithmetic test, and solicitude from all the people from whom he had been wishing attention for some time. The next time Jo was faced with a difficult situation and there was a queer feeling in his stomach, it was no longer necessary to go through all the preliminary steps. Now meals just come up without further consideration on his part.

Conditioned versus unconditioned responses. Behavior activities have been given many classifications. One widely used classification grew out of the studies of Pavlov in the beginning of the present century.[3] While working on the physiology of digestion, using dogs as subjects, he introduced a minor operation, so that saliva could be conducted through a tube running from the dog's submaxillary gland through the cheek. He then noticed that the salivary response was promptly aroused—and by means of the apparatus devised could be

[2] Caroline B. Zachry and Margaret Lighty, *Emotions and Conduct in Adolescence.* New York: Appleton-Century-Crofts, Inc., 1940, pp. 69–70.

[3] R. M. Yerkes and S. Morgulis, "The Method of Pawlow in Animal Psychology," *Psychological Bulletin,* 1909, Vol. 6, pp. 257–273.

measured—when food was smelled. Furthermore, he noted that the salivary response was aroused by various elements in the environment, such as the sight of the food dish or the approach of one who usually brought the food. Again, in order to study more carefully the process by means of which the animal had come to respond to these elements in his environment, he arranged the now classic experiment in which an electric bell was sounded a little before the presentation of food. After this procedure had been repeated a number of times, the dog would exhibit the salivary reflex following the sounding of the bell as a stimulus, even though the food was not immediately supplied.

This salivary reaction to the sound of the bell as a stimulus was called by Pavlov a "conditioned reflex." The fundamental principle of the conditioned reflex is: *If an incidental stimulus is presented many times along with one that already arouses a specific reflex response, the incidental stimulus will cause the particular reflex response to appear.*

Conditioned behavior is distinguished from unconditioned behavior in that the response once made only to a specific stimulus is later made to a part of that stimulus or to another stimulus that was concurrent with the original.

It should be pointed out that conditioned behavior is not confined to reflex behavior. A review of the work of Watson shows the application of the conditioned response concept to emotional behavior. Further studies relative to this problem are discussed in relation to the various emotional stimuli as motivating forces tending to liberate certain emotional drives of the organism. Our concern is not to present the controversial issues of conditioned behavior, but to show the nature and function of these early conditioned and unconditioned processes. Conditioning is recognized as having a wide range of applications. The fundamental principle, however, is set forth in the statement presented as an outgrowth of Pavlov's classic experiment.

Biological and socialized responses. The effects of heredity and environment are so mutually influential that, even soon after the fertilization of the ovum from which the embryo and finally the young infant develops, one can hardly distinguish the contributions of either. At the moment of the fertilization not only does physical development begin, but various behavior patterns start taking form which are destined to function importantly throughout the individual's life. Any attempt to place these two factors (heredity and environment) in opposition to each other is unwarranted in psychology and biology. Concerning these, Conklin and Freeman state:

Biologically, what an individual inherits is not a structure or a function but, rather, the potentialities for developing certain structures and functions *under certain conditions during the period of development.* Psychologically, the individual does not inherit specific behavior: but he does inherit certain behavioral and functional potentialities, the details and extent of whose development will be dependent in part upon the organism's environment during its period of growth. It is for this reason that some biologists and psychologists subscribe to the doctrine that every structure, function, and behavior is *both* inherited and acquired.[4]

The outstanding characteristics of the infant years are marked by behavior activities directly resulting from biological drives. The control and direction of these drives are of tremendous importance, and are an outgrowth of training, experience, and maturity. The effect of socialization is to color and direct these primitive biological drives into socially acceptable and desirable channels. The period from ten to sixteen can be described as passing through the following categories: first, a period of continued socialization; secondly, a period of gang interest and loyalty; thirdly, the onset of puberty and an intensification of the sex drive; and, finally, the socialization of the sex drive. These are the stages present in the development and conditioning of biological drives, the broadening of the learning process, and the intensification of socialized responses.

THE DEVELOPMENT OF THE EMOTIONS

According to Bridges[5] the emotional reactions of the infant are not highly differentiated, but the most common response to emotional stimuli is that of general bodily agitation or excitement. Out of this general excitation develop during the first several months the differential responses of distress and delight. Here we note the negative and positive forms of emotional responses that have commonly been recognized and given varied classifications. Anger, disgust, fear, and jealousy emerge at an early age from distress; elation, affection for others, and joy grow out of delight. The different ages for the appearance of different forms of behavior during emotional episodes show that crying, screaming, restlessness, and struggling, as forms of behavior, appear during the first four months of life, and may be

[4] Edmund S. Conklin and Frank S. Freeman, *Introductory Psychology for Students of Education.* New York: Henry Holt and Co., 1939, p. 20.

[5] K. M. B. Bridges, "Emotional Development in Early Infancy," *Child Development,* 1932, Vol. 3, pp. 324–341; also see W. E. Blatz and Dorothy A. Millichamp, "The Development of Emotion in the Infant," *University of Toronto Studies in Child Development Series,* 1935, No. 4.

regarded as general bodily agitation appearing in the initial stages as a result of some sort of overstimulation. Following the period of infancy the child passes through a period of growth, coordinating and integrating each new stage with that which has gone on before. The emotional development is not so great, due to the slow rate of growth of the internal organs of the body that are controlled by the autonomic nervous system, and are thus closely identified with the emotional life.

Most of our fears or angers are acquired ways of responding to various stimuli and situations. Few of the stimuli that cause fear or anger among adults will frighten an infant. It is equally true that most of our other emotional patterns are the results of learning and maturation—particularly learning. Since emotions are learned, they may be unlearned, thus enabling one to avoid the handicaps of inefficiency, embarrassment, and annoyance that uncontrolled emotions produce. However, the term *emotion* should not be regarded as a name for a type of response that is entirely different from nonemotional behavior. Behavior is a continuous, complex process involving simultaneous activity in many parts of the body. Man does not respond now with an emotion, then with an instinct, and at some other time with a habit. These names do not designate distinctly different types of behavior; they are merely abstractions which are necessary for convenience of study. The behavior commonly called emotional is an *emotion* in pure form only within a textbook. The same is true of a conditioned reflex, or a *habit*. Some of the characteristics of emotional activity are present at all times in everyday life and comprise what is sometimes called one's *emotional tone*. These emotional elements intensify, inhibit, and otherwise modify the behavior in process at any given time and are integral parts of the whole pattern of behavior. The adolescent is likely to be in a hyperemotional state owing to the organization and repression of drives and as a result of the controls and educational forces operating at this stage of life.

The word *emotion* was derived from the Latin word *emovere,* "to move out." It is usually defined as a stirred-up state or condition. The child inherits as part of his constitution the disposition to basic emotions, and, although the exact nature of this disposition may not be known, genetic studies of child development show that differences in emotionality appear during the first months of life. A complete discussion of the nature and development of the different emotions would be a lengthy one. The earliest emotional response of the infant is a general agitation or excitement produced by a variety of stimulating conditions. Out of this diffuse behavior differentiated emotions emerge. That the viscera play a prominent role in man's emotional life not only has been recognized from the days of antiquity, but is surely substantiated by the great amount of research on the problem. The *autonomic nervous system* is the specialized portion of the nervous system that controls most of the visceral responses. It lies chiefly outside the central nervous system, but is adjacent to it and is connected with the spinal cord. The autonomic system has been divided into the cranial, sympathetic, and sacral divisions. Most internal organs are connected by nerve fibers with either the cranial or sacral divisions and with the sympathetic division.

One set of nerves augments, whereas the other inhibits, the activity of an organ. According to such an antagonistic structural arrangement the visceral organs are, then, both inhibited and driven through the functioning of the autonomic system; Figure 18 shows its structural arrangement. The cranial and sacral divisions function in the normal activity of the internal organs of the body, whereas the sympathetic division is responsible for the changed operation of the internal organs during an emotional state. Some bodily responses resulting from a stimulation from the sympathetic division may be listed as follows: retardation or bolting of the peristaltic movements of the stomach and the flow of gastric juices from the glands; an increased secretion of glycogen into the blood stream; an increased rate of breathing; increased pulse rate; expansion of the periphery of the body; and increased muscular tension. The stimulation of the adrenal glands is very important in this connection, since the presence of an increased amount of adrenalin in the blood stream characterizes much of the emergency emotional states.

The emotional changes just described are extremely important in relation to certain emotional situations. The organs involved are fundamental to the life of the individual, and their increased activity is especially helpful in meeting difficult physical conditions. This fact forms the basis of Cannon's emergency theory of emotions. However, the physical changes described are not valuable in meeting many present-day emotional situations. Increased blood sugar is not helpful to the timid adolescent boy asking for his first date, nor to the jealous adolescent girl trying to outwit her rival through a feminine appeal.

Description of three major emotions. A discussion of the nature and development of each possible emotion would be a very lengthy one, even supposing there were adequate data and materials available for such a treatment. Furthermore, there is no accepted list of emotions, inasmuch as various classifications have been made from time to time. Descartes refers to the "six passions," James to the "four emotions," and Watson to the "three original emotions"—fear, love, and anger. Some psychologists refer to a single "stirred-up state" that gradually

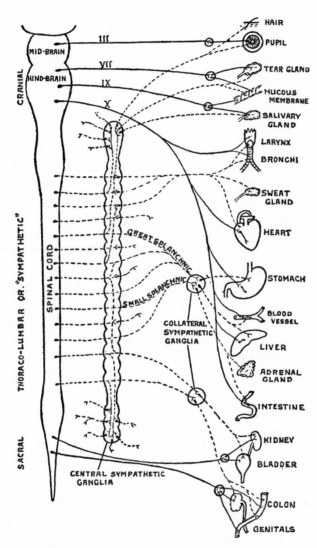

FIG. 18. *Diagram Showing the Structural Divisions of the Autonomic Nervous Systems and Their Relation to the Internal Organs.* (From F. H. Lund: *Psychology.* Ronald Press, 1933, p. 207.)

becomes differentiated in response to varied stimuli. A rather common classification is twofold, and is based upon the division of the autonomic nervous system in such a way that the visceral tissues are supplied with a double set of nerves, as was described on page 92. A great deal of research has centered around the development and nature of the three emotions presented by Watson. These emotions are sufficiently differentiated and have been subjected to sufficient study to provide material for special consideration here.

(a) *Love*. The emotion of love is directly related to the sexual impulse, and, like the emotions of fear and rage, is the consequence of physiological disturbances, especially in the visceral and glandular parts of the body. The ordinary response is a relaxed state of the body, and in the infant child especially is accompanied by gurgling and cooing. But as Gilliland has stated: "The earliest loves are not sexual in character, Freud to the contrary notwithstanding. However, sex stimulation gives pleasure and becomes a large factor in love responses."[6] The pattern, however, like that of fear, joy, and grief, seems to be inborn, noncoordinated, and unconditioned. It is somewhat similar in many different species, and is present at birth and through maturation. The importance of favorable affectional relationships during early childhood has been emphasized within recent years. Growth into adolescence brings forth to a much greater extent than formerly realized the relationship between the affectional state and the love emotions. And, although the exact nature of this relationship may not be clear, there is evidence for the general conclusion that there is a redirection of the love emotion at this stage. Because of social customs and group sanction, love responses are indirect and often rather subtle in nature, and this results in repressions and various forms of substitutions.

Since the various structures concerned with sex, and thus with love behavior, further develop and finally mature during adolescence, it is to be expected that there should be an excess of physiological manifestations of this emotion at this time. One must not infer either that love behavior appears suddenly in adolescence or that it is confined wholly to sex as sex is ordinarily thought of; for in the study of infants it has been found that the stimulation of the genital organs may cause smiling, which results from sexual experience but not from a sexual experience similar to that of the adolescent or adult. However, the

[6] A. R. Gilliland, *Genetic Psychology*. New York: The Ronald Press Co., 1933, p. 293.

fact that the emotions are so closely connected with visceral and glandular activities makes it apparent that, with the maturation of the sex and related glands, there will be a stronger tendency toward the love attachments and emotional manifestations associated with them. In this tendency probably the powerful social drives of adolescents, and their extreme loyalty and ideals, play an important part. This subject we shall consider further in connection with the various socializing influences to which growing boys and girls are constantly exposed.

(b) *Anger.* Anger and fear are so closely related that any attempt to develop a frame of reference concerning the nature and causes of one would in some manner or another comprise elements or causes to be found in the other. Concerning the nature and source of these emotions Hebb states:

> The fundamental source of either emotion is of the same kind, a disruption of coördinated cerebral activity. Flight and aggression are two different modes of reaction tending to restore the dynamic equilibrium, or stability, of cerebral processes. . . . Each of these modes of response tends to restore integrated cerebral action, one by modifying the disturbing source of stimulation, the other by removing the animal from it.[7]

Both anger and fear are somewhat opposed to love. Anger is characterized by stimulation of an inhibitory or negative type in which the subject's activities are interfered with. The physiological changes have already been mentioned as having certain protective values in that they tended to aid man in ridding himself of the inhibitory, obnoxious, or dangerous situation. It is well known that with the development of rational habits, conduct takes a higher form, being guided by reason and insight to a greater degree; but when an inhibitory situation is suddenly confronted, the individual's rational powers will often not operate to the maximum extent, and it is quite probable that his conduct will be prompted by emotion rather than reason.

The ready giving way to anger becomes a matter of habit in the life of the individual, but various social, recreational, and intellectual pursuits will cause the subject to develop habits of forethought and to inhibit certain emotional tendencies that developed prior to the age of adolescence. Thus, these pursuits will aid the individual in his social participation. However, in the development of emotional habits of a desirable nature, too much emphasis has been placed upon the inhibi-

---

[7] D. O. Hebb, "On the Nature of Fear," *Psychological Review,* 1946, Vol. 53, pp. 273–274. Reproduced by permission of the *Psychological Review* and the American Psychological Association.

tion of emotions after they have arisen rather than upon the avoidance of the arousal of the emotions. The best kind of self-control is the latter. The establishment of such habits is, of course, a slow process, and should begin during the days of early social experiences prior to the establishments of contrary habits.

The moral value of "righteous indignation," such as socialized anger, must not be overlooked. For here the anger tendency is on a rational basis—if, say, it results from a desire for the betterment of the group. But even these forms of anger should not be encouraged too far, since harmful behavior is likely to ensue. In his discussion of the emotion of anger in relation to moral and religious experience, Stratton[8] points out that the anger tendency may operate throughout an individual's life, and may develop certain forms of prejudice and lead to attempts to convert others to the "call." Anger, if developed as a protest against evil practices, is likely to carry the individual to an extreme in his attempt to right such practices. A rational form of control, only at times supplemented by anger, is more desirable. The idealistic nature of the adolescent, especially, is likely to carry his anger too far. It is partially for this reason that there are always large numbers of boys in the "teens" ready at the first call to battle the enemy in time of war.

(c) *Fear.* Fear is one of the most pronounced emotions experienced by man, and it has ever played an important role in his conduct. A study of primitive races will reveal the great influence it has had on the development of standards of conduct, beliefs, and man's innumerable institutions. Our ancestors constantly resorted to various rites and practices that were founded upon the appearance in one form or another of this primary emotion of fear. This emotion is oftentimes even today misunderstood and misinterpreted by the masses. The work of the psychoanalysts has done much to give us a fuller insight into the influence of this emotion upon various phenomena of conduct as well as upon physical well-being.

Fear has come more and more into disrepute as a method of social control. Especially in the adolescent are its evil effects found to exist, as fears begin to appear in certain social situations in a more permanent and stable form. To control by fear is, in the main, a negative method. Positive means of control, which tend more to bring into account such factors as suggestion, imitation, and guidance in the building up of acceptable modes of adjustment, are more desirable.

[8] G. M. Stratton, *Anger: Its Religious and Moral Significance.* New York: The Macmillan Company, 1923.

Moreover, as a child reaches adolescence it can be observed that fear as a means of control loses its influence; in fact, the exaggerated fear tendency established in early life will leave an imprint upon the growing child, but it does not insure that his conduct in certain situations will ever be really desirable.

During early childhood fears are mostly personal and pertain to things in the child's immediate environment. The child is afraid of things he can see, feel, or hear. The largest single category of fears at this age relates to animals; however, an early study of childhood fears by Jones and Jones[9] revealed that prior to the age of two years children displayed no fear of a live snake. The changing nature of fear manifestations with age is interwoven with other aspects of the child's development. Growth in knowledge and understanding brings forth new fears—first for real objects and conditions about them, later for imaginary dangers. Mental growth and development combined with a wider range of experiences enables the child to recognize possible dangers on the one hand, and also to imagine certain dangers. Thus, the mentally superior child of age ten may be disturbed about forthcoming possibilities of a war or catastrophe, while the average ten-year-old is unable to see the possibilities and consequences of such an event.

EMOTIONAL MANIFESTATIONS OF ADOLESCENTS

Adolescent fears. Fears of animals and other concrete things in the immediate environment appears to decrease as the child develops from age five to age twelve, while fear of the dark, of being left alone at night, and the like increase. With growth into adolescence, fears of a social nature come to be very important. However, a study by Hicks and Hays[10] shows that 50 per cent of a group of 250 junior high school students reported they were afraid of something. Some of their fears, in order of frequency reported, were for snakes, dogs, the dark, storms, accidents, high places, strange noises, and being alone at home. This indicates that many childhood fears persist into adolescence, and there is good evidence that they tend to persist throughout life. Older adolescents become much concerned over social approval, failure in school, fear of being disliked, and other fears pertaining to their relations with

[9] H. E. Jones and Mary C. Jones, "Fear," *Childhood Education,* 1928, Vol. 5, pp. 136–143.
[10] J. A. Hicks and M. Hayes, "Study of the Characteristics of 250 Junior-High-School Children," *Child Development,* 1938, Vol. 9, pp. 219–242.

their peers. The fears of adolescents may be roughly classified into three general groups:

1. *Fears of material things.* These include many of the early childhood fears of dogs, snakes, storms, and the like.
2. *Fears relating to the self.* These include death, failure in school, personal inadequacy, popularity, and the like.
3. *Fears involving social relations.* These include embarrassment, social events, meeting people, meeting with a more mature group, parties, dates, and the like.

Worries of adolescents. A number of studies have dealt with the worries of pre-adolescents, adolescents, and post-adolescents. In one of these studies a "worries" inventory was administered to 540 children in grades five and six in New York City.[11] The inventory consisted of 53 items. When these were grouped into eight categories, it was found that both sexes worry most about family and school situations and conditions. Next as sources of worry were those items grouped under personal adequacy, economic problems, and health problems. Little concern was noted for the imaginary and ornamental categories. The 10 items ranking highest for the boys and the 10 ranking highest for the girls are presented in order of frequency of worries reported in Table XI.

In the study of Hertzler,[12] returns from an expressionaire of 400

TABLE XI

Items from the "Worries" Inventory Most Frequently Reported by Boys and Girls, in Order of Frequency

BOYS	GIRLS
Failing a test	Failing a test
Mother working too hard	Mother working too hard
Being blamed for something you did not do	Mother getting sick
Father working too hard	Being late for school
Having a poor report card	Getting sick
Being scolded	Father working too hard
Spoiling your good clothes	Being scolded
People telling lies about you	Being blamed for something you did not do
Getting sick	Doing wrong
Doing wrong	Father getting sick

[11] R. Pintner and J. Levy, "Worries of School Children," *Journal of Genetic Psychology,* 1940, Vol. 56, pp. 67–76.
[12] A. E. Hertzler, "Problems of the Normal Adolescent Girl," *California Journal of Secondary Education,* 1940, Vol. 15, pp. 114–119.

problems were secured from 2,000 girls in secondary schools from five Southern California communities. Certain problems appeared to be common to these girls whether they lived in a large or small community, or whether they came from comfortable homes or poor financial circumstances. The most frequently reported problems in each area were found to be nearly the same in each school. The problem areas are here presented in order of frequency.

1. School life
2. Home life
3. Boy and girl relationship
4. Recreation
5. Friends
6. Vocational choice
7. Religion
8. Health
9. Clothes
10. Money

A careful study of these sources of worry reveals that fear is the foundation of the tendency to worry. As the child passes from the preschool age to the elementary school age considerable anxiety relating to failure appears. It has already been suggested that the earlier fears related to bodily injury; but as the child grows in understanding and reacts to different social situations, he develops anxieties and fears relative to his status in the group. This does not mean that the earlier fears about bodily injury are suddenly eliminated. Zeligs [13] found that sixth graders were most frequently worried about matters pertaining to bodily injury, health, grades, and promotion in school. Growth into adolescence is accompanied by anxieties on the part of many boys and girls connected with appearance, popularity, and inadequacies related to their sex role. A study by Lunger and Page [14] dealt with the worries of post-adolescents. A *Worry Inventory* of 78 items was constructed and administered to 100 college freshmen of each sex. The items were constructed so as to be answered in one of three ways: *very much, some,* or *not at all*. Sex differences were found to be negligible with respect to both the incidence and intensity of worries. They state: "Roughly about one-half expressed some concern over such items as: general religious problems, physical defects, being late for appointments, familial obligations, inability to make friends, and vocational success." We note here much more concern over responsibilities and vocational success than appeared among fifth and sixth grade children. This is distinct evidence of social maturity. The items most frequently

[13] Rose Zeligs, "Children's Worries," *Sociological and Social Research,* 1939, Vol. 24, pp. 22–32.

[14] Ruth Lunger and James D. Page, "Worries of College Freshmen," *Journal of Genetic Psychology,* 1939, Vol. 54, pp. 454–460.

worried about and those least frequently worried about are reported in
Table XII. This study furnishes a basis for comparing the nature of
the worries of fifth and sixth grade children with college freshmen; it
does not provide a safe basis for specific generalization about the worries of youth. A college group is a select group. Students from the
lower social-economic group are highly selected and do not represent
the typical youth from this group. One would expect to find financial
worries and worries related to home conditions, and parental hardships
appearing more frequently among youth from the lower economic
groups than among those from the middle and higher economic groups.

TABLE XII

Items Constituting the Highest and Smallest Percentage
of Worries (*After Lunger and Page*)

The largest proportion of freshmen were worried about:	Per cent reporting specific worries	
	Men	Women
1. Not being as successful in their work as they would like to be	94	89
2. Hurting other people's feelings	85	85
3. The impression they make on others	76	84
4. Not working hard enough	80	76
The smallest proportion of freshmen were worried about:		
1. The possibility that they are foster children	3	0
2. Going insane	2	4
3. Dying	9	9
4. Growing old	14	6
5. Their home being too shabby to invite and entertain friends	15	4

Changes in emotional behavior. It has already been suggested that
conditions or situations producing anger and fear in early childhood
may lose their potency as the child matures and encounters new experiences. Also, this maturity and educational growth tends to bring forth
new fears as well as enhance certain earlier behavior patterns. Observations show that young children rapidly outgrow temper tantrums—
oftentimes substituting more subtle and less violent ways of responding.
Some of these changes have been observed and recorded by investigators. Figure 19 presents changes in four types of behavior.[15] Accord-

[15] W. E. Blatz, S. N. F. Chant, and M. D. Salter, *Emotional Episodes in the
Child of School Age.* University of Toronto Studies, Child Development Series,
No. 9, 1937.

ing to the findings here presented, fighting reaches its peak at ages seven and eight and declines rapidly thereafter. Timidity likewise shows a definite decline after ages thirteen and fourteen. On the other hand, impertinence and sulkiness continue to increase throughout the period of childhood. The rise of timidity during pre-adolescence and early adolescence is no doubt closely related to the development of social consciousness and the desire of social approval that are associated with the dawn of adolescence.

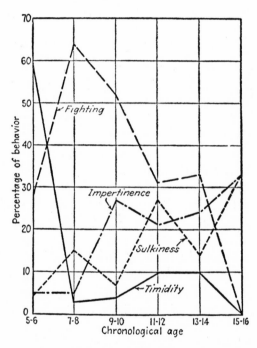

Fig. 19. *Changes of Behavior With Chronological Age for School Pupils.*

Fear of reciting in class is common among adolescents. This appears to be even more pronounced among boys than among girls, perhaps because of voice changes occurring at this stage in a large percentage of boys. The formality of the classwork operates to produce considerable fear among adolescents. When students are required to stand when they recite, or to follow in a parrot-like fashion some pattern of responding, shyness and timidity are very much in evidence. Another source of fears present in the schoolroom situation is that of the examination.[16] Pressure exerted from parents, teachers, or peers for making high marks is perhaps the major source of such fears. When the fear is sufficiently intense, learning and reciting are adversely affected. The stress given to the importance of a high score on the examination in order to secure good marks or grades produces considerable tension. Some degree of anxiety and tension may be unavoidable, and may actually be useful; however, much can be done by teachers and parents in the development of student attitudes toward the nature and function of examinations which would reduce this anxiety and tension and thus promote better personal and social adjustments.

PLEASANTNESS AND UNPLEASANTNESS

The nature of feelings. Taken together, feelings and emotions may be referred to as the affective experiences of the individual, as contrasted with his more purely unconscious physiological responses. But when we consider the *feeling state* of man let us not think of it as some qualitatively isolated form of activity. "The affective experiences may be wholly or partly visceral in origin. At moderate intensities they are vaguely felt and poorly localized, but in greater intensities they are less difficult to observe." [17] Again, any statement concerning feelings must be based largely upon introspection or subjective data; there is practically nothing of an experimental nature to offer as a proof for one theory or another. However, nearly all definitions describe feelings in terms of man's affective life and thus consider at least the two dimensions of pleasure and displeasure. Furthermore, it may be argued from the subjective standpoint, and with some justification, that feelings either are emotional accompaniments or are themselves experiences less vivid than an emotion. They are not states of knowing and reflecting. They are well-nigh impossible of analysis. The two complementary states or conditions result from one's becoming aware of some disturbance, and have been termed *pleasantness* and *unpleasantness*. It is a very difficult task to describe the feeling states here considered, but probably everyone has at some time been subject to them.

Feelings are quite often defined in terms of pleasure and pain, and

[16] See E. Liss, "Examination Anxiety," *American Journal of Orthopsychiatry,* 1944, Vol. 14, pp. 345–349.

[17] J. P. Nafe, "The Sense of Feeling," *The Foundations of Experimental Psychology* (C. Murchison, Ed.). Worcester, Mass.: Clark University Press, 1929, p. 411.

this definition confuses simple affective experiences with sensory processes or experiences. Actually, pleasure refers to man's affective life,
while pain refers to sensations. It is probably true that under ordinary
conditions pain sensations are accompanied by feeling states of unpleasantness, but this relationship does not imply either identity or a
necessary accompaniment. Most of us have experienced pain that
brought relief from an unpleasant stimulation; and in such an instance
the pain has been thought of as pleasant. If behavior is to be classified
under various captions, therefore, we must be consistent: pain and
unpleasantness are not synonymous terms and thus cannot be used
interchangeably. Of course, there are many other patterns of behavior
of an affective nature that result from rather localized physiological
conditions.

The feeling element during adolescence. Pleasantness and unpleasantness are not to be confused with certain sensations such as pain, a
pin-prick, a sweet or salty taste, and so forth, although one can well
recognize in retrospection that certain feeling states accompany these.
The feeling states are each correlated with a fundamental attitude existing in the organism independently of the particular stimulation, and
they must be recognized and dealt with accordingly. In interpreting
this fact, Henry T. Finck says: "Men will and must have their pleasures. Social reformers and temperance agitators could not make a
greater mistake than by following the example of the Puritans and
tabuing all pleasures."

The question may be raised: What are some of the most unpleasant
events experienced at different levels of maturity? This problem was
studied a number of years ago by Jersild, Markey, and Jersild.[18] They
found from their study of the "worst happenings" recalled by 399 children between 5 and 12 years of age that 74.5 per cent of all the responses
secured were concerned with bodily injury, possibility of bodily injury,
or illness. No significant age differences were discovered. An interesting discrepancy appeared between events that actually happened and
those most feared. The method used in this study was adapted by
Thompson and Witryol for studying unpleasant experiences of three
periods of childhood as recalled by adults.[19] The three periods of

[18] A. T. Jersild, F. V. Markey, and C. L. Jersild, "Dreams, Wishes, Daydreams,
Likes, Dislikes, Pleasant and Unpleasant Memories," *Child Development Monographs,* 1933, No. 12.
[19] G. G. Thompson and S. L. Witryol, "Adult Recall of Unpleasant Experiences during Three Periods of Childhood," *The Journal of Genetic Psychology,*
1948, Vol. 72, pp. 111–123.

childhood selected for studying were the first five years, the 6-12-year period, and the 12-18-year period. Such a division is in harmony with that commonly found in psychological and psychoanalytical literature. The subjects studied were female college students between 18 and 24 years of age. Fifty subjects were used in each of the three experimental groups. The results obtained were analyzed into 22 catgories plus an unclassified category. Available for analysis were 721 responses from the first five years of life, 803 from the age groups 6-12, and 735 from the age range 12-18. The results of the analysis into the various categories are presented in Table XIII.

The materials of Table XIII were further studied by grouping certain of the items into what appeared to be psychologically meaningful

TABLE XIII

PERCENTAGE OF RESPONSES IN THE 22 CATEGORIES OF UNPLEASANT EXPERIENCES

Types of unpleasant items recalled	During first five years of life	During 6-12- year period	During 12-18- year period
1. Painful Injuries	10.0%	7.6%	6.4%
2. Sensory Irritations	8.9	1.6	0.1
3. Illness	8.4	4.9	5.1
4. Loss of Personal Property	8.4	4.3	3.4
5. Corporal Punishment	7.3	7.1	1.0
6. Attacked by Animals	6.1	1.1	0.0
7. Forced to Do Unpleasant Things	4.9	15.2	3.4
8. Verbally Disciplined	2.8	9.9	4.1
9. Feelings of Guilt	4.9	7.6	2.1
10. Teased or Ridiculed	2.8	4.9	2.5
11. Persistent Fears	4.4	4.9	1.4
12. Fighting with Peers	2.2	4.3	1.4
13. Visit to Doctors	2.8	3.3	1.0
14. Deaths of Relatives and Friends	1.6	1.6	9.7
15. School Failure	1.7	4.9	8.3
16. Refused Desired Objects	4.4	4.9	8.3
17. Loss of Friends	2.8	3.3	8.0
18. Quarrels with Parents	0.0	0.5	5.4
19. Broke Up with Boy Friend	0.0	0.0	4.9
20. Feelings of Inferiority	0.0	0.0	4.9
21. Witnessed Accidents	2.8	0.5	3.4
22. Lack of Popularity	0.0	0.0	3.4
23. Unclassified (All Others)	12.8	7.6	11.8
Total	100.0	100.0	100.0

clusters. The first cluster comprised items relating to the following types of unpleasant experiences: painful injuries, sensory irritations, illness, corporal punishment, and attacked by animals. A comparison of the frequency of recall of unpleasant experiences constituting this cluster by the three age groups is presented in Figure 20. This cluster, referred to as the "unpleasant or painful sensory-and-emotional experiences," was recalled most frequently from the first five years of life. Thus, it appears that adults look back on these early childhood years as the ones fraught with painful sensory and emotional experiences.

The second cluster consisted of the following types of unpleasant experiences recalled: forced to do unpleasant things, verbally disciplined, feelings of guilt, teased or ridiculed, and fighting with peers. This cluster, referred to as "unpleasant experiences during process of learning to live in a social world," was recalled most frequently by the 6-12-year age group. Figure 21 shows a comparison of the extent to which the three age groups recalled unpleasant experiences included in this cluster. It is during this 6-12-year age period that the child is learning to live with others and to conform to established ways of living. Thus, he often finds himself at conflict with situations, conditions, and requirements set forth by the social group. This is a period of social learning sometimes referred to as "socialization."

The third of these clusters includes the following types of recalled experiences: deaths of relatives and friends, school failure, loss of

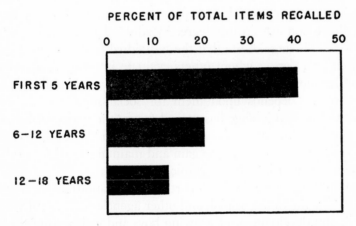

Fig. 20. *Unpleasant or Painful Sensory and Emotional Experiences*—includes: painful injuries, sensory irritations, illness, corporal punishment, and attacks by animals. (After Thompson and Witryol)

friends, quarrels with parents, broke up with boy friend, feelings of inferiority, and lack of popularity. This cluster has been referred to as "unpleasant experiences generating feelings of inadequacy and insecurity." A comparison of the frequency of recall of the items included

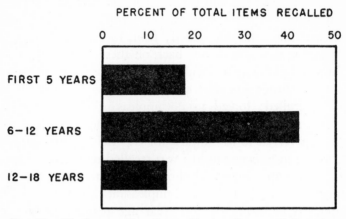

PERCENT OF TOTAL ITEMS RECALLED

FIG. 21. *Unpleasant Experiences During Process of Learning To Live in a Social World*—includes: being forced to do unpleasant things, being verbally disciplined, having feelings of guilt, being teased or ridiculed, and fighting with peers. (After Thompson and Witryol)

in this cluster by the three age groups is presented in Figure 22. The results here presented reveal that unpleasant experiences involving feelings of insecurity and social inadequacy during the 12-18-year age level are recalled to a much greater degree by adults than at other age levels. It is during this stage of life that the individual becomes keenly aware of himself as a person whom others evaluate in terms of accomplishment and achievement. He finds himself in freqent conflict with his parents, yet he depends upon them for social support and security. The individual at this stage has moved from "the period of aches and pains" and from the stage of conflicts involved in "learning to live in a social world" to that of trying to gain and maintain a satisfactory status in the world of which he is a dynamic part.

These findings have some important implications that should be considered by the home, school, and other agencies concerned with the guidance and development of growing boys and girls during these age periods. It seems likely that too much emphasis has been placed on social learning during the 6-12-year period and perhaps insufficient help has been given to the child in his social development. Problems appearing among boys and girls at this age often appear insignificant to the adult. He takes social learning at this stage for granted and looks upon any problems observed as a natural part of growing-up. Thompson and Witryol conclude:

Perhaps, the socialization process during this stage of development could be fostered just as effectively and with fewer psychological conflicts if approached from a more positive angle than from the restrictive and compulsory point of view mirrored by the present findings. It may be that in our culture the individual is being "forced" too rapidly to adapt himself to the current social rules and conventions—a parallel to this being the extreme emphasis on early toilet training for the young child that was accepted procedure some few years past.

It is also possible, as a variety of writers have pointed out, that many of the problems peculiar to the 12-18-year period of life may be partially a derivative of the educational, social, and economic philosophy prevalent in this country. "You can be at the top of your class," "it's personality that counts," and a host of similarly inane keys to successful living may be sufficient to make the half-boy-half-man during the 12-18-year period feel insecure and inadequate. Certainly modern advertising with its warning admonition against bad breath, offensive body odor, dull teeth, sallow complexion, etc., is enough to make even adults feel the threat of loss of social status. Perhaps parents, teachers, and other counselors of 12-18-year youth need to instigate counteractive measures to build up the individual's sense of personal worth and to help him set up levels of aspiration in work and social affairs that are commensurate with his abilities and needs.[20]

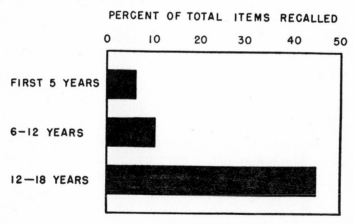

FIG. 22. *Unpleasant Experiences Generating Feelings of Inadequacy and Insecurity*—includes: death of relatives and friends, school failures, loss of friends, feelings of inferiority, and lack of popularity. (After Thompson and Witryol)

[20] *Ibid.,* p. 122.

HABITS AND CONTROL

The importance of emotional well-being in relation to personality adjustments and mental health will be given special consideration in later chapters. Since all aspects of growth are interrelated, the child's emotional growth affects and is affected by his physical, mental, and social growth. Thus, we have noted that the emotional life and behavior of the individual is affected by physiological changes as well as by social conditions and contacts.

Emotional and social development. There are many factors in the child's physical and social environment that affect his emotional and social development. The importance of class status as a factor affecting social development will be discussed in Chapter VIII. And, while class status does not leave the same emotional imprint on all individuals, there are certain experiences common enough to the different members of a class group to warrant consideration. From an analysis of over one hundred Negro adolescents from the deep South, Davis and Dollard[21] found that parents from the lower class group relied upon physical punishment to a much greater degree than did parents from the higher class groups. Furthermore, it has been observed by many investigators that children from slum areas engage in fights more than children from better living conditions. Thus, the ready giving way to emotions is part of the social environment of the child reared in an underprivileged home.

It should not be inferred here that all children from the lower class group lack emotional control. In this connection Davis and Dollard found that children within each of the social classes differed more widely from each other than one class differed from another. Thus, home conditions as well as biological factors operated to produce significant differences in emotional expressions within each class. There are unquestionably happy homes with little emotional tension in all class groups, just as there are unhappy, quarrelsome homes filled with tension in all class groups. However, Hyde and Kingsley[22] have presented data showing a significant increase in mental disorders and psychopathy from communities with inferior conditions to those

[21] A. Davis and J. Dollard, *Children of Bondage: The Personal Development of Negro Youth in the Urban South.* Washington: American Council on Education, 1940.

[22] R. W. Hyde and L. V. Kingsley, "Studies in Medical Sociology. I. The Relation of Mental Disorders to the Community Socio-economic Level," *New England Journal of Medicine,* 1944, Vol. 231, pp. 543–548.

where superior conditions exist.   The adolescent's emotional habits are, therefore, affected by many factors.   Adverse home and community conditions have a deleterious affect on healthy emotional growth and thus contribute to emotional instability and personality disturbances.

Habits as drives to action.   The importance of this topic in the study of the adolescent cannot be overemphasized.   Habits of a social nature are in their formative stage during later adolescence.   In considering the individual's emotionality as a drive to activity, one must not over-simplify the general development of emotional habits and their rela-tion to mental life.   Emotional habits should, furthermore, be viewed from the developmental point of view.   During the adolescent period they are still in an unstable state, and are found to be very transitory in their general manifestations.   Many mannerisms appear, being manifested in isolation from the individual's general habit patterns—which, in fact, are often inconsistent and changeable.   The extent to which a habit pattern once built up becomes a drive to action will depend mainly on the extent to which it becomes integrated in the individual's general habit patterns and finally becomes automatic.

It has been found that attentive repetition of an act tends to make for automaticity of the act.   Habits are continuous rather than periodic. A habit once formed is never completely eradicated from man's neural structure, for all changes which are effected must be built upon the structural patterns existing at the time in the individual.   James recog-nized this in his well-nigh classical statement:

Every smallest stroke of virtue or of vice leaves its never so little scar. The drunken Rip Van Winkle, in Jefferson's play, excuses himself for every fresh dereliction by saying, "I won't count this time!"   Well! he may not count it, and a kind Heaven may not count it; but it is being counted none the less.   Down among his nerve-cells and fibers the molecules are counting it, registering and storing it up to be used against him when the next temptation comes.   Nothing we ever do is, in strict scientific literalness, wiped out.   Of course this has its good side as well as its bad one.   As we become permanent drunkards by so many separate drinks, so we become saints in the moral, and authorities and experts in the practical and scientific spheres, by so many separate acts and hours of work.[23]

Such changes as are made may become automatic in nature, but the old habit system operates under special emotional conditions when rational behavior is not so much in evidence.   Even volition must be

[23] William James, *Psychology* (Briefer Course).   New York: Henry Holt and Co., 1892, p. 150.

studied in terms of learning and can best be thought of in terms of man's habit system. All these habit patterns which tend to contribute to the efficiency of the human mechanism become potent drives for the initiation and direction of action. (This subject we shall consider further in connection with the development of attitudes and social behavior.) .

Emotional control. If emotional activity results in prepotent action tendencies, it is certainly necessary that a control be exercised over both the emotions and their expressions. But here it should be noted that there is a definite difference between emotional control and emotional repression or elimination; for whenever a man reaches a point such that he experiences no emotions, he is no longer referred to as an ordinary man but rather as a case or subject for psychological or psychiatric treatment. Life would be a deadly monotony but for some emotional experiences—if, in fact, it continued at all, which would be highly doubtful in the event that all emotions, including those of sex, were eliminated from existence. Without emotions all family ties would vanish—love for wife, love for husband, love for children, love for parents: all would cease. Religion would disappear, for there would be no fear of God, no awe of God, no love of God. Governments would crumble without patriotism, feelings of security, and protection. To be sure, if emotions give us the bitters of life, they give us the sweets, also. The words of Tennyson imply the same thought: "The happiness of a man in this life does not consist in the absence but in the mastery of his passions."

Now we have noted that the emotions are closely related to bodily changes and are thus fatiguing, and that unless there is ample opportunity afforded for recovery following periods of emotional upsets, individual injury is the unavoidable result. Furthermore, every individual develops emotional habits to such an extent that an emotional response to a situation today will be repeated if he meets the same or some similar situation tomorrow. A well-unified habit system with the proper volitional, attentive, persistent, and imitative types of habits developed in harmony with the individual's innate physical and mental ability is impossible unless the boys and girls are given the opportunity to accept responsibility and are held accountable for consequences. This means self-control. The stage of maturity has been reached in which they can see, understand, and generalize from their home and school experiences; and in order that emotional control may be developed, they should be given the opportunity to participate in activities leading toward the acceptance of responsibility. This participation will foster a spirit of fair play and cooperation, habits of confidence, and a larger consideration of the rights of others. These are essential prerequisites for emotional control.

SUMMARY

The growth and development of the child into adolescence are accompanied by glandular changes closely related to the emotional life. The heightened emotional states during this period of life have constantly been recognized as a part of the nature of adolescents; however, in addition to a heightened emotional state at this time, there is also an expansion of the emotions into the social realm. Fears and angers related to social situations become very important; self-conscious feelings about one's own adequacy appear; and the adolescent becomes especially concerned over the approval of his peers. This increased fear is also observed in connection with classroom situations. Fear of reciting in class, fear of failure, and fear of ridicule become more pronounced at this age level.

The child is conditioned to many emotional stimuli that often interfere with a healthy development of the mental and emotional life. This conditioning process may have a very early beginning and thus affect the development of the emotions and attitudes during the adolescent years. The child's early habits have been described as mainly self-centered; later, through social contacts, he gradually arrives at a fuller realization of his relationship with others. If these early habits have not developed so that the individual may grow into independence and responsibility, we have a dependent creature maturing into the social group. He lacks character traits essential to a happy and successful adjustment to the new and sometimes strange situations he is constantly meeting. Emotional development, no less than the development of motor abilities, is dependent upon maturation and learning. Increased attention given to the development of desirable emotional habits during childhood and early adolescence will be helpful in the development of a well adjusted adolescent.

THOUGHT PROBLEMS

1. Distinguish by definition between pleasantness and unpleasantness. What difficulties are encountered in making such a distinction?

2. What conditions are necessary for the development of desirable emotional habits during childhood?

3. How would you account for the great amount of feelings of insecurity and inadequacy during the years 12 to 18 as shown in Figure 22?

4. Based upon your own experiences, what were the characteristics of the things or conditions that brought you the greatest amount of unhappiness during each of the age periods referred to in connection with the materials of Table XIII?

5. What is the significance of the changes in emotional manifestations presented in this chapter?

6. Look up the James-Lange theory of emotions. What evidence have you observed that would support this theory?

7. What values do the emotions have for us? Elaborate.

8. List some principles of emotional control. Which of these have you found most useful in your life?

SELECTED REFERENCES

Blos, Peter, *The Adolescent Personality*. New York: Appleton-Century-Crofts, Inc., 1941, Chaps. III, IV, and V.

Brooks, Fowler D., *Child Psychology*. Boston: Houghton Mifflin Co., 1937, Chap. XII.

Cole, Luella, *Psychology of Adolescence* (Third Edition). New York: Farrar and Rinehart, 1948, Chap. IV.

Conklin, E. S., *Principles of Adolescent Psychology*. New York: Henry Holt and Company, 1935, Chap. XIII.

Hurlock, E. B., *Adolescent Psychology*. New York: McGraw-Hill Book Co., 1949, Chap. IV.

Olson, W. C., *Child Development*. Boston: D. C. Heath and Co., 1949, Chap. X.

Sadler, W. S., *Adolescence Problems*. St. Louis: C. V. Mosby Co., 1948, Chap. V.

Schaffer, L. F., *The Psychology of Adjustment*. Boston: Houghton Mifflin Co., 1936, Chap. IV.

Thorpe, L. P., *Foundations of Personality*. New York: McGraw-Hill Book Co., 1936, Chap. VIII.

Zachry, Caroline B., and Lighty, Margaret, *Emotions and Conduct in Adolescence*. New York: Appleton-Century-Crofts, Inc., 1940, Chap. XI.

VI

Mental Development

INTELLIGENCE: ITS MEANING AND MEASUREMENT

During the past several decades there has been a continuously mounting tide of research on the nature of mental development. These studies, while providing valuable information about mental growth, have opened up new areas for further study. No clear-cut set of principles has been presented relative to the nature of mental growth during the adolescent years, although the results of research indicate certain characteristics of mental development. In any consideration of mental growth during adolescence, it is worth while for the student to recognize (1) that the principles of growth applicable to physical and emotional development are also applicable to mental development; (2) that mental growth during the adolescent years cannot be divorced from growth during the periods of infancy and childhood; and (3) that mental growth is an integral part of the total development of the individual.

The discussion in this chapter will be confined mainly to the adolescent stage of life, although it is widely understood that any interpretation of the mental characteristics of this period is directly related to and dependent upon the mental characteristics of childhood. Moreover, since many problems relating to mental development are strongly controversial, the materials presented here are given in the spirit of what scientific studies tend to point out. Some of the major problems to be studied are the meaning of intelligence, or mental ability; the physiological basis of mental ability; the problems encountered in the attempt to measure mental ability; and the general nature of mental growth. The latter will include further subsidiary problems relative to the uniformity and constancy of mental growth, the maturity age of subjects of varying abilities as well as those reaching physiological maturity at different periods, and the relation between the rate of mental growth and that of physical growth.

Concept of intelligence. The concepts of the nature of intelligence held by early students of psychology were quite simple. Intelligence was conceived by many students of educational psychology as a general mental power or a multiplicity of mental powers that could be measured on a vertical scale by a single score. There scores were further transmuted into mental ages.[1] The ratio between the mental age and the chronological age, then, provided the basis for the IQ (intelligence quotient). The IQ was, therefore, a measure of the rate of growth of mental ability, from infancy. Any significant changes in an individual's IQ from year to year were regarded as exceptions. Thus, the theory of "the constancy of the IQ" was developed and generally accepted.

Thurstone has set forth the hypothesis that intelligence consists of a number of primary mental abilities. Through intricate mathematical procedures his students have obtained evidence of certain independent primary mental abilities. *The Chicago Test of Primary Mental Abilities* is an outgrowth of some of these studies. This test measures the following abilities: Verbal Meaning, Word Fluency, Reasoning, Memory, Number, and Space. Correlations obtained between the results of this test and high school achievement show that Verbal Meaning is closely associated with achievement.[2] Reasoning ability is moderately related to high school achievement. Varying combinations of these mental factors will be associated with success in different activities.

The organismic concept of the individual would substitute for *intelligence* the term *intelligent behavior.* This term refers to the total personality, with motives, emotions, and intelligence signifying closely related phases of behavioral patterns that are never separable, except in the abstract. Martin has presented a good description of this concept.

Intelligent behavior is therefore more inclusive in its significance than the behavior measured by intelligence tests currently in use; it is more inclusive also than the comprehending of relationships in their semantic unities; it

[1] A child's mental age, according to the early Binet tests and revisions of his test was expressed in terms of the average age of children making that test score.

[2] See Duane C. Shaw, "A Study of the Relationships Between Thurstone Primary Mental Abilities and High School Achievement," *Journal of Educational Psychology,* 1949, Vol. 40, pp. 239–249.

implies attributes of personality described in the more or less vague, subjective terms of persistence, courage and determination in overcoming difficulties, pride of achievement, level of aspiration, and other designations of like import.[3]

Goddard, after reviewing many definitions of intelligence and analyzing what happens as a result of the operation of intelligence, arrived at what appears to be a sound and functional definition of it, which reads as follows: *"Intelligence is the degree of availability of one's experiences for the solution of immediate problems and the anticipation of future ones."*[4] This is the concept most widely used in the general construction and use of intelligence tests and should provide a working basis for studying mental growth during adolescence.

The measurement of mental development. Studies conducted during the past two decades have introduced important changes in the concepts and nature of intelligence and intelligence tests. In the first place there is ample evidence that intelligence test scores are affected by favorable environmental influences as well as by schooling and educative experiences. This has disrupted the generally held viewpoint that intelligence tests measure the level of mental ability with which a child is endowed by his inheritance. On the contrary it is generally held that a child's mental ability or abilities are a result of native endowment, growth, opportunities for educative experiences, and perhaps the nature and direction of his motives and drives.

Secondly, case studies of individuals with a similar test score revealed that their abilities to succeed at various mental tasks were by no means equal. Various combinations of abilities on the intelligence test could yield the same score. These results showed that equivalent mental ages of IQ's did not yield equivalent abilities at different tasks.

Two individuals with the same total score on an intelligence test may differ considerably from each other in terms of the factors making up the score. The one may have a high score on verbal tests with low scores on some of the other test items; while the other individual makes a low score on verbal items, but high scores on the items of a nonverbal nature. The *California Test of Mental Maturity* provides for a total numerical and a total verbal score, which should be useful in educational and vocational guidance.

[3] W. W. Martin, "Some Basic Implications of a Concept of Organism for Psychology," *Psychological Review,* 1945, Vol. 52, p. 338. Reproduced by permission of the *Psychological Review* and the American Psychological Association.

[4] H. H. Goddard, "What Is Intelligence?" *Journal of Social Psychology,* 1946, Vol. 24, p. 69.

CHARACTERISTICS OF MENTAL GROWTH

Mental growth. All psychological studies show that there is an increase in mental ability with age, up to a certain period of life— approximately the period of adolescence. However, there are many problems relating to the nature, amount, and causes of such an increase which inspire differences of opinion. Some of these problems will be brought out in the course of the following discussions.

Freeman and Flory [5] reported results from the Chicago growth study, in which tests were administered to several hundred children over a period of years. Many individuals were retested at the age of 17 or 18 years at the time of their graduation from the University of Chicago Laboratory Schools. Some of these were later retested in college. A composite of four standardized tests consisting of (a) vocabulary, (b) analogies, (c) completion, and (d) opposites was used. The growth curves drawn from the raw scores showed mental development continuing well beyond the age of 17 or 18 years. There was furthermore some evidence from these studies that the children of average ability might continue intellectual growth to a somewhat later age than the brighter pupil. This, however, is in all likelihood a result of the failure of the average environment to present opportunities for stimulation in such a way as to continue an accelerated rate of growth on the mental tests, and is in harmony with the Minnesota studies of the mental growth of children from 2 to 14 years of age.[6] According to these studies, the nonverbal tests surpass the

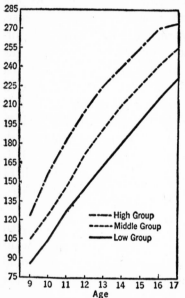

Fig. 23. *Mean Intelligence Test Scores of Three Groups of Pupils with Consecutive VACO Tests from 11 to 16 Years.*

[5] F. N. Freeman and C. D. Flory, "Growth in Intellectual Ability as Measured by Repeated Tests," *Monographs of the Society for Research in Child Development,* 1937, Vol. 2, p. 116.

[6] F. L. Goodenough and K. M. Maurer, "The Mental Growth of Children from Two to Fourteen Years: A Study of the Predictive Value of the Minnesota Preschool Scales," *University of Minnesota Child Welfare Monograph Series,* 1942, No. 19.

verbal in predicting scores that will be made at a later year. There is also evidence from these studies that the stability of the intelligence quotient is reached earlier for girls than for boys.

An interesting phase of this general problem of mental growth concerns the variation in individual-growth curves. According to early studies there appears to be a great deal of uniformity and a general continuity of growth, as is further shown by the correlation method.[7] Recent studies offer evidence that mental growth is affected by a number of factors. Terman retested gifted children after a six-year interval, and concluded as follows:

Making due allowances for complicating factors in measuring IQ constancy, one can hardly avoid the conclusion that there are individual children in our gifted group who have shown very marked changes in IQ. Some of these changes have been in the direction of IQ increase, others of them in the direction of decrease. The important fact which seems to have been definitely established is that there sometimes occur genuine changes in the rate of intellectual growth which cannot be accounted for on the basis of general health, educational opportunity, or other environmental influences.[8]

Constancy of mental growth. There has been much controversy over the general nature of mental growth curves. This has centered around the constancy of the ratio of mental age to the chronological age. The IQ is determined by dividing the individual's mental age by his chronological age. The mental age thus becomes the unit for measuring mental development, and mental growth curves are usually plotted in terms of the mental age.

Mental growth curves are influenced not only by the type of intelligence test used but by the units in terms of which the curves are plotted as well. If growth in mental age is plotted against chronological age, a straight line will be the result, provided that (1) the test scores are derived from an unselected group of persons, (2) the test results are not influenced by training more at certain ages than at others, and (3) the degree of brightness of the persons tested on the average remains constant. It must be recognized that months and years represent units of time, and not necessarily units of growth, for in a given individual the rate of growth will be different at different periods in

[7] See Baldwin and Stecher, "Additional Data from Consecutive Stanford-Binet Tests," *Journal of Educational Psychology,* 1922, Vol. 13, pp. 556–560; S. C. Garrison, "Additional Retests by Means of the Stanford Revision of the Binet-Simons Tests," *Journal of Educational Psychology,* 1922, Vol. 13, pp. 307–312.

[8] L. M. Terman, *Genetic Studies of Genius.* Palo Alto: Stanford University Press, 1930, Vol. 3, p. 30.

his life. Certain aspects of some of these problems are considered in the discussions that follow dealing with physical development and pubescence in relation to mental growth.

Correlations between test scores obtained one or more years apart are far from perfect.[9] Significant fluctuations are likely to occur in a large percentage of cases. The extent of such changes will depend upon such factors as the circumstances that have appeared in the child's life during the interim (such as emotional problems, illness, change of environment, opportunity for schooling, dynamics). In the guidance study at the University of California mental test scores were obtained upon 252 children at specified ages between 21 months and 18 years.[10] These scores were analyzed to show the extent of stability of mental test performance during this age period. The constancy of mental test performances of these children depended in part upon the age at testing and the length of interval between tests. That is, prediction was good over short age periods with the predictive value of scores increasing after the preschool years. However, the correlation between tests given at 2 and at 5 years of age ($r = .32$) is not indicative of much stability in test performance at this age level. When test scores earned at 21 months and 24 months are correlated a significant amount of stability is observed ($r = .71$).

The study by Houzig, Macfarlane and Allen indicates a considerable stability of mental test performance between 6 and 18 years; however, a study of individual cases showed that the IQ's of about 60 per cent of the group changed 15 or more points. The IQ's of 9 per cent changed 30 or more points. Thus, predictions of IQ's at age 18, based on 6-year mental test scores should be made with extreme caution.

Mental growth during childhood. Studies of adopted children indicate that the size of the IQ may be affected by the type of environment represented in the foster home. Skodak[11] has sought to extend the facts relating to the growth of intelligence by selecting two groups of preschool children from the lower socio-economic levels. The first

---

[9] A perfect correlation would be a correlation coefficient of 1.00. For a more complete understanding of the meaning of correlation and the statistical procedures involved in computing it, the reader is referred to any recent textbook on statistical methods applied to education or psychology.

[10] M. P. Houzig, J. W. Macfarlane and L. Allen, "The Stability of Mental Test Performance Between Two and Eighteen Years," *The Journal of Experimental Education,* 1948, Vol. 17, pp. 309–324.

[11] Marie Skodak, "Children in Foster Homes," *University of Iowa Studies in Child Welfare,* 1939, Vol. 16, Series No. 364.

group was placed in superior foster homes in early infancy; the children of the second group remained, for different periods of time, in their own homes or in environments known to be inferior, and were then placed in homes superior to those they had previously known. The children were given intelligence tests at different periods and these results were studied in relation to the environmental background. During the first year, no noticeable differences were observed, indicating that the causal relation between intelligence and environmental background at this period is negligible. Following the first year there was a continued increase in the average IQ's for those children in the superior environments with a continued decrease for those in the inferior environments. It is pointed out from this study that the mental level of the children is more closely related to environmental background than to that of the true mothers. The conclusions presented by Skodak emphasize the importance of a stimulating environment in raising the IQ level of children during the growing years.[12]

Certain broad generalizations emerge from a careful analysis of the results of these studies. The first is that intelligence is much more responsive to environmental changes than had previously been conceived. For practical purposes of education, it is the environmental stimulation or, in many cases, the lack of early environmental stimulation, that sets the limits to a child's mentality.

The hereditary constitution probably sets rather broad limitations, and when all the children have been placed in an environment sufficiently stimulating to develop them nearer to their hereditary limitations, its effects become very pronounced. As long as we are dealing with children who have not had an environment sufficiently stimulating to develop their abilities and talents, we shall find that responses to ordinary intelligence test items improve as a result of better environmental conditions.[13]

[12] These University of Iowa studies have created considerable controversy. Simpson has presented some interesting notions dealing with factors that may have affected the results ("The Wandering IQ: Is it Time for It to Settle Down?" *Jour. Psychol.*, 1939, Vol. 7, pp. 351–369). McNemar has critically evaluated the studies and points out certain methodological and statistical inadequacies; while Wellman, Skeels, and Skodak have defended their results against such criticisms (see the *Psychological Bulletin* for March, 1940, for the critical examination of the studies and the critical review of the examination).

[13] Hollingworth has pointed out that practically all long-time studies of predictability of intelligence support the conclusion that the results of tests administered during the first 6 or 8 years of life do not have a very great predictive value.

Mental growth during adolescence. We note that over long periods of time the prediction of the IQ from early test scores becomes more precarious. In this connection Anderson has pointed out that suitable tests for the measurement of intelligence are not available for the preschool child.[14] He reports correlations from the Harvard growth data slightly in excess of .50 between mental test scores at the seven- and sixteen-year levels. Some of the discrepancies found in growth curves of this study may be observed from the comparison of the individual mental growth curves shown in Figure 24.[15] Mental age curves of this figure are based on Harvard growth data for five boys whose scores were the same at the age of seven years—the IQ of each boy being 92. The progressive divergence of these curves with the IQ's at the seventeen-year age level ranging from below 80 to 110, suggested that repeated measurements must be made if accurate classifications are to be maintained. The fact that different tests were used and that these were group tests may account in part for these variations.

There are certain factors, in common, for all the curves of mental growth during the adolescent years. In the first place, there is greater constancy during these years than there was for earlier years. A number of factors that may account for this are: (1) the presence of a more nearly constant environment for each individual during this period than during the earlier years; (2) the nature of the tests used at the adolescent age levels; (3) the growth in complexity of the tests at the advanced age level; and (4) an increased constancy in the administration of the tests to the more advanced subjects. Another factor found in different mental growth curves is the decrease in the rate of growth during the teen years. This was given special emphasis earlier in this chapter.

There is evidence that continued exposure to different environmental or different social class groups favors the development of certain abilities and discourages the development of others. Consequently, abilities appear to become more highly differentiated as the individual moves

(L. S. Hollingworth: "Personal Reactions of the Yearbook Committee," *Thirty-ninth Yearbook of the National Society for the Study of Education*, Part 1, 1940, pp. 451–454.)

[14] J. E. Anderson, "The Prediction of Terminal Intelligence from Infant and Preschool Tests," *Intelligence: Its Nature and Nurture. Thirty-ninth Yearbook of the National Society for the Study of Education*, Part I, 1940, pp. 385–403.

[15] Harold E. Jones and Herbert S. Conrad, "Mental Development in Adolescence," *Forty-third Yearbook of the National Society for the Study of Education*, Part 1, p. 159.

in and through the adolescent years. This was shown in the study by Janke and Havighurst. All available sixteen-year-old boys and girls in a midwestern community were given the *Revised Stanford-Binet Intelligence Test,* the *Performance Scale* of the Wechsler-Bellevue

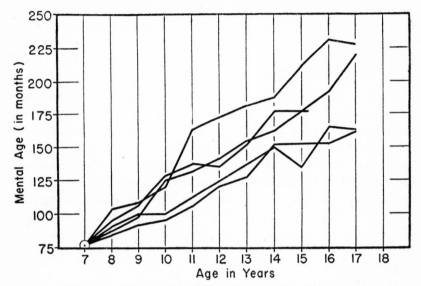

FIG. 24. *Diverging Individual Mental Growth Curves.*

Adult and Adolescent Scales, the *Nova Silent Reading Test,* the *Minnesota Form Board,* the *Minnesota Mechanical Assembly Test,* and the *Chicago Assembly Test for Girls.* They conclude:

(2) Boys and girls from families of higher social status tended to do better in all the tests than boys and girls of lower social position, with the exception of the Mechanical Assembly Test, where there was no reliable social class difference between the boys.

(3) Urban boys and girls tended to do better than rural boys and girls, but not significantly so.

(4) No significant sex differences were obtained.[16]

Growth differences among mental functions. There are some who have described childhood as the "golden age for memory," while adulthood is looked upon as the age of reasoning. Studies of the growth

[16] L. L. Janke and R. J. Havighurst, "Relations Between Ability and Social Status in a Midwestern Community: II. Sixteen-Year-Old Boys and Girls," *Journal of Educational Psychology,* 1945, Vol. 36, pp. 499–509. There is, however, a difference of opinion as to whether or not abilities become more highly differentiated with increased maturity.

of different mental functions show that memory, reasoning ability, interpreting ability, and other mental functions grow in an orderly sequence, and that growth among these functions is continuous in nature, beginning at a very early age and continuing to maturity. There are several reasons why misconceptions have developed about the memory ability of children. In the first place, a great deal of the child's earlier mental activities consists of mechanically remembering meaningless materials; in these, the child does compare favorably with adults. Again, children have not developed a wide range of logical associations; therefore, they are forced to rely largely upon mechanical memories, and in turn are not distracted by meaningful elements that a situation may suggest. Further, children often spend a great deal of time and go through a great deal of drill in memorizing certain materials. The developmental curve for memory ability is somewhat similar in nature to that for other phases of mental growth. Such a growth curve reveals that memory ability increases with age and experience. Maturation of those structures concerned with learning and mental life is important in affecting the ability to memorize.

Pyle[17] gave the "Marble Statue Test" to 2,730 persons ranging in age from eight years to maturity. An analysis of these data shows a rather steady improvement in memory for items until the thirteenth year. Shaffer's[18] study of the growth in ability to interpret cartoons shows a fairly early development with a continuous increase until maturity. There is evidence that mental development continues through the late teens, although this will vary with individuals and with the complexity of the function. In interpretative activities of a fairly complex nature, maturity comes considerably later than for less complex materials. Immature perception characterizes infancy and early childhood. The child notes things in large units and careful classifications are wholly lacking. The horse may be called a big dog. The courthouse is a big house, and the house is best represented in drawing as a square with perhaps a chimney on top. Studies in genetics show the gradual growth of discriminating ability resulting from maturation and experience. Guided experiences in harmony with the maturity level are highly important for the growth of perceptions.

[17] W. H. Pyle, *Nature and Development of Learning Capacity.* Baltimore: Warwick & York, Inc., 1925.
[18] L. F. Shaffer, "Children's Interpretations of Cartoons," Teachers College, Columbia University, *Contributions to Education,* 1930, No. 429.

Pubescence and mental growth. Some evidence has been advanced indicating that pubescence is preceded or accompanied by a fairly rapid rise of both the mental and the physical growth curves. Abernethy's study[19] indicates that high school girls who have matured between 10½ and 11½ years were superior in their schoolwork to girls who matured four or five years later. The median IQ's of the two groups studied were approximately the same, 114 for those maturing early and 112 for those maturing late.

In a study by Stone and Barker,[20] 175 postmenarcheal and 175 premenarcheal girls paired for chronological age were compared with respect to Otis intelligence test scores, personality, and socio-economic status of their parents. The postmenarcheal girls made a mean score on the intelligence test which was 2.25 points higher than that made by the premenarcheal girls. This difference is not statistically reliable. The *Pressey Interest-Attitude Test* scores showed the postmenarcheal girls to be more mature than the premenarcheal of the same chronological age. Postmenarcheal girls were also more mature when measured by the results obtained from administering the Sullivan Test for developmental age. Both of these differences are statistically reliable. The two groups were from families of about the same socio-economic status, and did not show a difference in their general personality traits as measured by the *Bernreuter Personality Inventory*. Apparently, then, pubescence has vastly more significance as a physiological change affecting various glandular secretions—especially those relating to sexual characteristics—and the rate of growth in height, weight, and other physical measurements, than it has as a criterion for mental growth. Physical and emotional changes are much more closely related to the onset of puberty than are the more specific mental abilities.

Gesell concludes:

The nervous system, among all the organs of the body, manifests a high degree of autonomy, in spite of its great impressionability. . . . *It tends to grow in obedience to the inborn determiners, whether saddled with handicap or favored with opportunity.* For some such biological reason, the gen-

---

[19] Ethel M. Abernethy, "Correlations in Physical and Mental Growth," *Journal of Education Psychology,* 1925, Vol. 16, pp. 438–466 and 539–546.

[20] C. P. Stone and R. G. Barker, "Aspects of Personality and Intelligence in Postmenarcheal and Premenarcheal Girls of the Same Chronological Age," *Journal of Comparative Psychology,* 1937, Vol. 23, pp. 439–445.

eral course of mental maturation is only slightly perturbed by the precocious onset of pubescence.[21]

Most of the discussions of intelligence deal with abilities to manipulate abstract symbols and deal with ideas. It is recognized that the ability to deal with words, ideas, and various symbolic elements and processes relates to mental abilities of a more or less academic type. The usual test of intelligence is, therefore, a measure of an abstract ability and deals with the question of how well an individual is able to do in performances involving symbols and different types of abstractions. The more extended mental development found in many studies among children of superior intelligence is probably a result of the nature of the instruments used for measuring intelligence. Children who are alert and continue to make progress in the intellectual activities of the school quite possibly do develop superior intellectual habits. They are, therefore, at a decided advantage in performances of an abstract intellectual nature over the child who is handicapped by deficient environmental stimulation.

MENTAL GROWTH CORRELATES

Relationship between mental and physical development. From the results of studies on the interrelation of abilities, it has been found that individuals who are superior in one aspect of physical growth are more than likely to be superior in others, although, to be sure, the correlations between some traits are not very high, indicating that there are many exceptions to the association of physical growth in one direction with physical growth in some other direction. Some questions of especial interest in the study of mental development are: What is the relation between mental and physical development? And, more specifically, is the child superior in physical development likely to be superior mentally? Also, is the child who is slow in general physical development more likely to be slow in mental development? Are some physical traits associated with mental growth, while others are not?

So far as the various features of physical and mental development are concerned, growth in one does not retard growth in another. This is contrary to general opinion, but is substantiated by scientific evi-

[21] Arnold Gesell, "Precocious Puberty and Mental Maturation," in "Nature and Nurture: Their Influence upon Intelligence," *Twenty-Seventh Yearbook of the National Society for the Study of Education,* 1928, Part 1, pp. 408–409. (By permission of the Society.)

dence.[22] Positive, though sometimes small, correlations are usually found between measurements of physical and mental traits.

However, the correlation between different measures of physical development and mental development among children, although significant, are not sufficiently high to make safe predictions from one to the other. Furthermore, the matter of socio-economic status affects both the physical and mental development scores. The child from an inferior home and neighborhood has often been undernourished or malnourished, and has often not had the medical care, recreational opportunities, cultural advantages, and even school advantages of the child from the average and superior home and community. Wellman's[23] summary of the various studies related to this problem indicated no close relation between mental development and such measures of physical development as width of chest, lung capacity, ossification ratio, weight, height, and grip.

Physical appearance and mental development. The belief that there is an intimate relationship between mental development and general appearance, although gradually disappearing, is still quite widespread. To review the history of this belief along with the investigations that it has inspired would carry us too far from our present general interest. (Donald G. Paterson[24] gives a very good summary of studies dealing with various phases of the problem.) However, in passing we may note that Binet and Simon, in their search for some reliable method of measuring intelligence, devoted much effort to cephalometry and described their findings between 1900 and 1910 in *l'Année Psychologique.* Dr. C. Rose[25] in Germany and Karl Pearson[26] in England also

[22] Bird T. Baldwin, "The Physical Growth of Children from Birth to Maturity," *University of Iowa Studies in Child Welfare,* 1920, Vol. 1, No. 1; James C. DeVoss, "Specialization of the Abilities of Gifted Children," *Genetic Studies of Genius,* Palo Alto: Stanford University Press, 1925, I, Chap. XII; E. A. Doll, "Anthropometry as an Aid to Mental Diagnosis," *The Training School,* Research Department, Vineland, N. J., 1916, No. 8; C. D. Mead, "The Relation of General Intelligence to Certain Mental and Physical Traits," Teachers College, Columbia University, *Contributions to Education,* 1916, No. 76.

[23] Beth L. Wellman, "Physical Growth and Motor Development and Their Relation to Mental Development in Children," *A Handbook of Child Psychology* (C. Murchison, Ed.). Worcester, Mass.: Clark University Press, 1931, p. 265.

[24] Donald G. Paterson, *Physique and Intellect.* New York: The Century Co., 1930.

[25] C. Rose, "Beiträge zur Europäischen Rassenkunde," *Archive für Rassen- und Gesellschafts-Biologie,* 1905, Vol. 2, pp. 689–798; 1906, Vol. 5, pp. 42–134.

[26] Karl Pearson, "Relationship of Intelligence to Size and Shape of the Head and Other Mental and Physical Characters," *Biometrika,* 1906, Vol. 5, pp. 105–146.

conducted early scientific studies, observing large groups of boys and girls; and they came to the conclusion—which has been borne out by many and more recent investigations—that in the judgment of intelligence no importance is to be attached to head measurements.

During the early part of the present century there was a general notion that diseased tonsils, adenoids, lack of energy, malnutrition, and other such *single* factors were direct causes of mental deficiency. The different studies conducted related to these problems have shown that no such direct relationship exists. A summary of studies concerned with the relationship between mental development and certain physical impairments indicates that these factors bear little relationship to each other.[27] In spite of these findings, there is recent evidence from different sources showing that a large number of subnormal children suffer from physical defects and afflictions. One of the most complete studies bearing on this shows that subnormal boys grow more slowly than normal children.[28] Furthermore, it was noted that the degree of physical abnormality or deficiency was related to the degree of mental deficiency. The results of this study were further corroborated by a study conducted in Los Angeles, comparing the physical characteristics of 900 educationally backward children with those of 2,700 mentally normal children.[29] In spite of the medical care and services supplied these subnormal children there was a higher incidence of physical defects of every kind among them than among the children of normal intelligence.

School achievement in relation to intelligence. Many studies show the overlapping achievement and aptitude of individuals with varying degrees of intelligence. Correlations obtained between achievement and intelligence test scores usually range between .45 and .60 Although these studies indicate that a significant relation exists between achievement and intelligence, there is evidence that school achievement is dependent upon a number of factors, some of which are quite intangible and unpredictable. Bradley[30] studied the relationship be-

[27] N. W. Shock and Harold E. Jones. "Mental Development and Performance as Related to Physical and Physiological Factors," *Review of Educational Research,* 1941, Vol. 11, No. 5, Chap. 3.

[28] C. D. Flory, "The Physical Growth of Mentally Deficient Boys, "*Monographs of the Society for Research in Child Development,* 1936, Vol. 1, No. 6.

[29] M. Goldwasser, "Physical Defects in Mentally Retarded School Children," *California and Western Medicine,* 1937, Vol. 47, pp. 310–315.

[30] William A. Bradley, "Correlates of Vocational Preferences," *Genetic Psychology Monographs,* 1943, Vol. 28, pp. 99–169.

tween intelligence and teachers' marks of 1,500 junior and senior high school students from the Philadelphia area. Correlations ranging from .33 ± .06 to .64 ± .04 were obtained. The relationship of marks and intelligence is more clearly revealed by a comparison of the percentage of pupils from the different intelligence levels who were assigned marks of A with those assigned marks of D. This comparison is shown in Table XIV.

It has already been suggested that many factors are usually involved in the assignment of a grade to a student. Furthermore, it is well known that teachers' marks are not highly reliable. The results of Table XIV do show, however, that high school students with an IQ in excess of 110 are not likely to be assigned marks of D, while students with an IQ below 110 are not likely to be assigned marks of A. The implications of these relations are of utmost importance in educational and vocational guidance, and will be given special consideration in Chapters XIX and XX.

TABLE XIV

RELATIONSHIP OF HIGH AND LOW MARKS TO THE MENTAL ABILITY OF HIGH SCHOOL STUDENTS (*After Bradley*)

I.Q.	Per cent of pupils	Per cent of A's	Ratio of per cent of A's to per cent of pupils	Per cent of D's	Ratio of per cent of D's to per cent of pupils
140–149	.2	3.2	16.0	0	0
130–139	1.3	10.6	8.1	0	0
120–129	7.5	25.5	3.4	0	0
110–119	22.5	35.1	1.6	0	0
100–109	32.5	18.1	.6	15.7	.5
90–99	25.3	4.3	.2	36.8	1.5
80–89	9.7	3.2	.3	42.3	4.3
70–79	1.0	0	0	5.3	5.3

Success in any of the school tasks is dependent in a large measure upon mental maturity (enabling the pupil to respond intelligently to ideas and problems), a background sufficiently rich in experiences, habits of concentration and attention, habits of persistence and self-reliance, and the tools for thinking and understanding sufficiently developed to interpret and use the ideas presented through the written symbols. The many studies that have been made on this problem reveal correlations[31] ranging from .20 to above .75 between grades obtained and intelligence test scores. The sizes of such correlations will depend upon a number of variables, among which may be stated

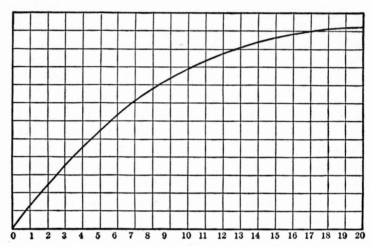

FIG. 25. *The General Nature of the Relation of Mental Development to Age in Years 0 to 20.*

motivation, nature of subject matter, study periods, teaching procedures, and criteria for grades.

Curves for growth in mechanical ability and information reveal that boys are superior to girls from the beginning of their school work to maturity. This difference can probably be accounted for on the basis of cultural influences. There is a significant relationship between intelligence and general knowledge, although sex differences will also be found. Girls are, in general, superior to boys in various linguistic activities, while boys are superior in general out-of-school knowledge.

An extensive study of the relationship between socio-economic status and scholastic performance was conducted by Coleman,[32] in which approximately 2,800 boys and girls from the seventh, eighth, and ninth

[31] The correlation ratio is a statistical term used to denote the degree of relationship existing between certain variables. For example, if we find that those who are advanced in mental ability are likewise advanced in vocabulary development, and vice versa, we would note a high correlation. There is a statistical procedure useful for ascertaining more accurately the extent of such a relation. A correlation of 0 is considered no relationship while one of 1.00 is considered a perfect one.

[32] H. A. Coleman, "The Relationship of Socio-Economic Status to the Performance of Junior High School Students," *Peabody College Contributions to Education,* No. 264, 1940.

grades were used as subjects. These adolescents were chosen from the highest 5 per cent, the middle 5 per cent, and the lowest 5 per cent in socio-economic status from a population of over 12,000 junior high school pupils. Those from the highest five per cent were found to be on the average superior in scholastic performance to those from the middle 5 per cent; while those from the middle 5 per cent were superior to those from the lowest 5 per cent. The same general trend was also found to be true in the case of intelligence test scores. In the case of achievement in relation to their intelligence, it was observed that those from the highest 5 per cent were on the average lowest. This is quite significant in its relation to motivation and the inculation of desirable work habits during this stage of life.

Some problems related to mental growth. Mental growth and development is usually accompanied by increased social and intellectual demands and the need for assuming more responsibility and for making decisions. This presents some difficult and baffling problems for adolescents who are ill-prepared through lack of guidance and experiences to make choices and assume responsibility. Some of these problems are discussed in later chapters dealing with guidance.

A common source of worry and anxiety to the adolescent in our contemporary society is the question of whether he will or will not be successful in meeting intellectual requirements. The competitive nature of our society is reflected in the procedures in our schools and the attitudes present in the homes. Passing examinations, making satisfactory marks in school, getting on the honor roll, and being admitted to college are among the common preoccupations and aspirations of pupils in the upper grades of our schools. The urgency of these demands is often unrelated to the values and aspirations the individual has set for himself; however, the adolescent does not want to be regarded as a failure by his parents, teachers, and peers. School success (particularly intellectual attainment) is held up as the important criterion of real success. Thus, many types of problem behavior and personal and social maladjustments are traceable to failure in school. Studies of adolescent groups reveal that intelligence offers little predictive value as to one's personal adjustment. There are a number of factors responsible for this. Some of these are suggested in the following statement:

These are the child's absolute level of intelligence; the level of intelligence required in the activities toward which he is being pointed, through the ambitions of his family and friends; the social pressures which arise from such ambitions; his own "felt needs" and level of aspiration; and his actual achievement. These factors are interconnected in a variety of ways and a great variety of complex patterns may result.[33]

SUMMARY

This survey of some of the more important findings concerning mental growth demonstrates that many prevalent notions of the nature of mental growth and its special characteristics at different stages are very much exaggerated. Since pubescence occurs in different individuals at different times, predictions and generalizations in individual cases are likely to be very inaccurate. A mental-growth characteristic of special interest relates to the age of maturity, or limit of mental growth. Problems of maturation, learning, and motivation affect the results. Also, the nature of the test materials may be the important factor in accounting for differences found. Also, individual growth curves presented in this chapter show that the special characteristics of mental growth are not uniform from individual to individual. The close relation between intelligence and size of vocabulary during adolescence indicates that mental ability is closely related to the ability to learn materials of an abstract and symbolic nature. Although mental growth, like other aspects of growth, is gradual and continuous in nature, research studies indicate that it tends to proceed rather rapidly during the early years of life, and even up to the ages of 14, 15, or 16; and then, according to Thorndike's analysis of H. E. Jones' data,[34] there is a slow increase up to 20. Whether the latter slow increase is an increase in actual mental ability or one of mental content is a matter of controversy. However, it is quite likely that many extraneous factors reduce the accuracy of the rates and limits derived for mental growth at any given period.

THOUGHT PROBLEMS

1. What experiences have you had with intelligence tests? On the basis of your experience, what do you consider they are actually measuring?

2. To what extent is mental development related to the age of pubescence? To physical maturity?

3. What do the various experiments appear to indicate relative to intelligence? What are the various correlations that have been obtained?

---

[33] H. S. Conrad, F. N. Freeman, and H. E. Jones, "Differential Mental Growth," *Forty-Third Yearbook of the National Society for the Study of Education,* Part 1, 1944, p. 180.

[34] E. L. Thorndike and others, *Adult Learning.* New York: The Macmillan Company, 1928.

4. How is mental ability related to learning? How should this relation affect the curriculum prior to adolescence?

5. How would you account for the relation between mental and physical development? What is the educational significance of this relation?

6. What mental expansion in yourself took place as you reached adolescence?

7. What are some of the different methods of measuring mental development? Why might an application of one of the methods to a group of adolescents be unfair to some?

8. Compare curves of growth in memory and reasoning. What would you conclude from such a comparison?

SELECTED REFERENCES

Arlitt, Ada H., *The Adolescent*. New York: McGraw-Hill Book Co., 1938, Chap. IX.

Boynton, Paul, and Boynton, Juanita Curry, *Psychology of Child Development*. Minneapolis: Educational Publishers, Inc., 1938, Chap. VI.

Cole, Luella, *Psychology of Adolescence* (Third Edition). New York: Farrar and Rinehart, Inc., 1948, Chap. XIII.

Cruze, W. W., *Educational Psychology*. New York: The Ronald Press Co., 1942, Chap. VI.

Garrison, Karl C., *The Psychology of the Exceptional Pupil* (Revised Edition). New York: The Ronald Press Co., 1950, Chap. IV.

Jones, H. E., and others, "Adolescence," *Forty-Third Yearbook of the National Society for the Study of Education*, Part I. Chicago: Department of Education, University of Chicago, 1944, Chaps. VIII and IX.

Segel, David, "Intellectual Abilities in the Adolescent Period," *Bulletin No. 6*, 1948, U. S. Office of Education.

Stoddard, G. D. (Chairman), "Intelligence: Its Nature and Nuture," *Thirty-Ninth Yearbook of the National Society for the Study of Education*, Part I. Bloomingdale, Ill.: Public School Publishing Co., 1940.

Stoddard, G. D., *The Meaning of Intelligence*. New York: The Macmillan Co., 1943.

Terman, L. M., and Merrill, M., *Measuring Intelligence*. Boston: Houghton Mifflin Co., 1937, Chaps. II–IV.

Wechsler, D., *The Measurement of Adult Intelligence*. Baltimore: Williams and Wilkins, 1939, Chaps. I, III, and IV.

For a rather complete review of the recent research on language development and mental growth; factors affecting mental development; and the relation of mental development to learning, sex differences, and cultural forces during pre-adolescence and adolescence see Worbois, G. M., "Mental Development During Preadolescence and Adolescent Periods," *Review of Educational Research*, 1947, Vol. 27, No. 5; *Review of Educational Research* 1950, Vol. 20, No. 5.

VII

Adolescent Interests

INTERESTS: THEIR NATURE AND DEVELOPMENT

The meaning of interests.   It has already been pointed out that the adolescent is in no sense a passive agent in a constant environment. The mode of reaction on the part of the adolescent is determined not only by the environment but by the specific direction, in accordance with changes that have been wrought in the neuromuscular system during the earlier years of experience, of the energies of the organism. Interest, then, is purposive insofar as a situation produces a response in the individual such that certain desires and strivings are channeled toward realization.

The word *interest* is derived from the Latin word, *interesse,* which means "to be between," "to make a difference," "to concern," "to be of value."   Interest has been described as that "something between" which secures some desired goal, or is a means to an end *which is of value to the individual* because of its driving force, usefulness, pleasure, or general social and vocational significance.   Interest is a form of emotional state in the individual's life which is interrelated with the general habit system of activity.   Moreover, during a state of interest, certain parts of the environment are singled out, not merely because of such objective conditions of attention as *intensity, extensity, duration, movement,* but because changes have been established in the neuromuscular system which cause the organism to favor some reactions to the exclusion of others.   The term *interest* has ordinarily been referred to in describing or explaining why the organism tends to favor some situations and thus comes to react to them in a very selective manner. Interest is directly related to voluntary attention, and when interest is not present, attention tends to fluctuate readily.

The organism must be considered in terms of the biological and social drives that have been referred to.   Hence, with growing knowledge, and experiences developing and integrating special habit patterns, the individual reaching adolescence has both *intrinsic* and *extrinsic* interests.   It is of course well that there be a balance existing between these interests.

The age of adolescence has been referred to by psychologists as the period of varied and peculiar interests.   It should be recognized, first, that all interests grow out of experiences, and the life experiences of the organism tend to guide and direct the development of further interests. In attempting to build some interest in the life of the child, it should therefore be recognized that any such interest must be established according to the laws of learning, just as other habit patterns are formed.   Over a long period of careful observations it becomes evident that different individuals have preferred ways of reacting to a specific phase of their environment, and these are somewhat characteristic of the organism concerned.   When the adolescent chooses some special book to read instead of pursuing an athletic game, we recognize that a special type of interest is present.   This interest is in itself a drive to a special type of action.   When a boy pursues a game for its own sake or for the amusement and fun that he gets from the exercise, then his interest is referred to as intrinsic or as "an end unto itself."   On the other hand, when a boy goes into athletics in order to keep himself fit or to develop certain desirable character traits, we have an example of extrinsic interests, or a means to arrive at some desirable element. Athletics, reading a book, driving an automobile, and practically any activity we might consider may be of either an intrinsic or an extrinsic type of interest.   Intrinsic interest is usually more spontaneous than extrinsic interest.

This differentiation in the nature of interests is a matter of importance to parents, teachers, or boys' workers who wish to regulate the overflow of restlessness in boyhood and youth.   An individual responds to an intrinsic interest, to the pleasure which his palate will take, for instance, in a fine dinner, more readily than to a plain meal which is good for his health.   At the same time, adolescence may also be rightly thought of as the period when individuals begin to look with a longer horizon upon the experiences of daily living as a means to an end.   Wise adults are accustomed to look beyond the immediate gratification yielded by an activity to discover its values.[1]

The growth of interests.   The early interests of the child are centered on purely personal relations.   When he sees an animal which he has

[1] W. R. Boorman, *Developing Personality in Boys*.   New York: The Macmillan Company, 1929, p. 41.   (Quoted by permission of the publishers.)

not seen before, he will ask, "What is it? Will it bite?" and these questions are not scientific in nature; neither are they prompted by the ideal of scientific inquiry. Nevertheless, even at this stage in the intellectual development of the child one sees evidence of individual interest in the structure, behavior, and life history of the animal. This represents the beginning of a scientific interest in life, and especially the life of animals. Interest is dependent on experience, but this does not mean that the native ability does not play a part in the development of interest. The physical growth of the organism, itself, is an important factor in the development of interests. Even visceral and glandular activities may affect the direction of one's interests.

The University of California Interest Record was used in a study reported by Jones.[2] A group of activities in which there is a decided decline in play interests for boys is shown in Figure 26. These curves show that beginning with the seventh grade boys, there is a rapid and continuous decline in early childhood interests in marble playing, collecting stamps, the magicians outfit, and reading *Child Life*. The extent to which a child replaces these types of interests with more mature interests such as dating, dancing, and parties on the part of girls and team activities, dating, dancing, and parties on the part of boys is a good measure of the social development of maturation of the adolescent boy and girl.

By the time boys and girls reach the age of adolescence and are beginning high school work, a very great range of interests and also a pronounced sex difference will be noticed. Careful, controlled observations have led many psychologists to emphasize the importance of the role played by experience in differentiating the interests of both races and sexes. Symonds[3] found that boys conventionally expressed greater interest in health, safety, money, and sex than did girls. Prominent among interests which girls expressed on questionnaires are: personal attractiveness, personal philosophy, daily schedule, mental health, and home relations. There were indications, as might be expected, that city pupils were more conscious of social skills than rural boys and girls. Environmental factors, intelligence, sex differences, maturity, and training combine as complex and integrated factors in

[2] H. E. Jones, *Development in Adolescence*, New York: Appleton-Century-Crofts, Inc., 1948, p. 104.

[3] P. M. Symonds, "Comparison of the Problems and Interests of Young Adolescents Living in City and Country," *Journal of Educational Sociology*, 1936, Vol. 10, pp. 231–236; also, "Life Problems and Interests of Adolescents," *The School Review*, 1936, Vol. 44, pp. 506–518.

affecting the growth of the interests of boys and girls from childhood
through adolescence to adulthood.   Strong's[4] studies of the maturity
of interests at different age levels reveal among other things that the
period of adolescence and youth is characterized by pronounced changes
in interests.

Interest in personal appearance.   When a boy begins to spend more

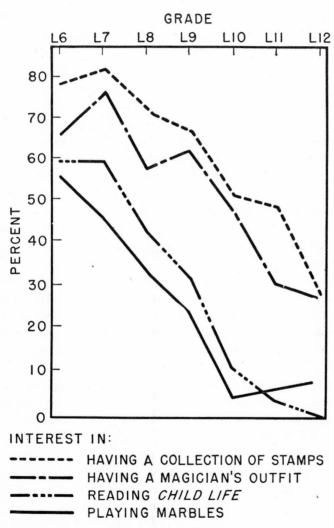

GRADE

INTEREST IN:

- - - - - - -   HAVING A COLLECTION OF STAMPS

———-———   HAVING A MAGICIAN'S OUTFIT

——-··-——   READING *CHILD LIFE*

————————   PLAYING MARBLES

Fig. 26.   *Changes of Four Childhood Interests with Change
of Grade Status.*   (After Jones)

[4] E. K. Strong, "Interest Maturity," *Personnel Journal,* 1933, Vol. 12, pp. 79–90.

time combing his hair and washing his teeth, when he calls for a clean shirt with a necktie to match, we can be sure that there is a dawning sex and social consciousness. As long as he is a boy he cares little about his personal appearance, except when his parents or friends have, through constant effort, made it otherwise This is very noticeable in the fact that his hair so often remains uncombed and his hands, face, and neck dirty. Life among other boys is too fascinating and adventurous to bother about trying to keep clean, so that dirty and wrinkled wearing apparel is often found preferable to clean and well-kept clothes.

In the previous chapter it was pointed out that when pubescence arrives, a pronounced change is manifested in the interest in the opposite sex, and this is all closely related to a keener consciousness about personal appearance. Before this time the girl, though somewhat less indifferent than her brother in the matter of personal appearance, has not shown much interest in styles or appearance except in a sort of imitative manner. With the onset of pubescence, she becomes more interested in the show window and the fashion sheet and visualizes herself dressed in a tailored suit according to the pattern of youth. The demand for sport clothes, beach pajamas, winter sport outfits, and sport jackets is characteristic of this period of life.

Although the boy is not provided with as many decorations nor as wide a variety of wearing apparel as the girl, he is very interested in making the most of the things he uses. The well-pressed suit and clean shirt become the order of this time; he turns his attention to cleanliness and to well-groomed hair and nails without being constantly reminded of these things by his mother. The taste concerning some of these things will vary with different localities, but the one that will be followed by most of the boys is the one that meets with approval, and especially approval from the opposite sex.

This increased interest in both boys and girls reaches a very great height toward the postpubescent period, and at an early stage is likely to bring adolescents into conflict with their parents. Some of the problems related to this will be discussed more fully in Chapter XI. The ten-cent stores have made it possible for boys and girls to find cheap imitations that aid them considerably in adorning themselves and in copying styles of others who are in better financial circumstances. By examining and recognizing the nature of these interests one can obtain a more accurate portrait of the teen years and the dawning social consciousness than through perhaps any other means available.

SPECIAL INTERESTS CHARACTERISTIC OF
ADOLESCENTS

Interest in play. A differentiation has already been made between extrinsic and intrinsic interests. Needless to say, both should have a place in the development of a well-balanced personality. Educators are recognizing more and more keenly the necessity for educating people in better means of using their leisure. With the increase of complexity in civilization and the decrease in hours of labor, much unoccupied time is left to the average citizen; but education has not yet prepared the citizens to use it wholesomely and worthily.

Play has an intrinsic value for the adolescent, but with further growth and development extrinsic values become more and more sought. Play activities tend to supply the adolescent with physique, health, neuro-muscular skills, and the desire for recreation. Pupil interests in play are conditioned largely during the adolescent age; such forces as environment, age, sex, race, custom, and intelligence operate to effect various differences. Some of these forces we shall review.

Some early studies by Lehman and Witty indicate that interest in play cannot be confined to early childhood. They gathered data from 6,881 children concerning activities in which the children had engaged during the preceding week and the number of activities in which they had participated alone. The data thus gathered led the investigators to conclude:

1. Attempts to differentiate certain C. A. (chronological age) periods in terms of differences displayed by children in diversity of play activities seem unjustifiable.
2. The play trends which characterize a given age group seem to be the result of gradual changes occurring during the growth period. These changes are not sudden and characterized by periodicity but are gradual and contingent.
3. Nor can any age or group of ages, between 8 and 19 inclusive, be characterized as disclosing play behavior primarily social or primarily individualistic. . . . Such a practice is unwarranted.[5]

Today play activity of some kind is recognized as of value in all stages of life. The time is past when, like our Puritan fathers, we turned from the play activities because they were a total "waste of time." Only the idle, daydreaming child who indulges in fantasy

[5] H. C. Lehman and P. A. Witty, "Periodicity and Growth," *Journal of Applied Psychology,* 1927, Vol. 11, pp. 106–116.

instead of wholesome play activities wastes his time. Two significant facts or conclusions are revealed from the various studies of play activities. These are: (1) play is a continuous process rather than an activity confined to the period of childhood; (2) there is an enormous overlapping in play interests for individuals of the same age but of different sex, of different racial groups, or different intelligence levels.

4-H Club Boys and Girls at Play.

a. *Strength and play participation.* Van Dalen[6] reports a study of the participation of adolescent boys in play activities. Strength tests were administered to 348 boys in the seventh, eighth, and ninth grades. The median age of the boys was 13.4 years with a range from 12 to 16 years. Strength index for the boys was determined from the results on the strength test, while the physical-fitness index was derived from comparing achieved-strength index with the norm based upon the individual's age and weight. This, then, is a measure of the immediate capacity of an individual for physical activity.

A comparison of the frequency and amount of participation of the

[6] D. B. Van Dalen, "A Differential Analysis of the Play of Adolescent Boys," *Journal of Educational Psychology,* 1947, Vol. 41, pp. 204–213.

high and low physical-fitness index groups showed that the boys in the high groups engaged in more play activities and devoted more time to play than did the low groups. This was true for all types of activities except for the reading and constructive categories. The great difference in participation in these was between the high and low physical-fitness index groups. In this comparison the low physical-fitness index group exceeded by a minimum of ten times the boys of the high group. The differences between the high and low strength index groups was not as great. The high strength index group engaged more frequently as spectators at physical activity events than did the low groups. Van Dalen draws these conclusions concerning the qualitative differences:

> Boys in the low strength groups participated in games which were somewhat individualistic in nature, involving some element of competition, but were distinctly of a lower degree of organization than activities in which boys of the high strength groups participated. Boys of the high strength groups participated more frequently in games and activities which required marked muscular strength and large muscle coordinations in comparison with the boys of the low strength groups. Play activities of a rather restricted range of action, and activities requiring little muscular strength and large muscle coordination were common to boys of the high strength groups but decidedly more common to boys of the low strength groups.[7]

b. *Sex differences.* Simpson reported several representative play activities that are best liked by boys and girls as they advance in age.[8] These are shown in Table XV. In the case of sex differences, boys' games are found to be a bit more vigorous and better organized. The differences appearing in different communities reveal that sex alone is not responsible for the nature and interests in play; but that customs, environmental conditions, size of the group, and educational level of the boys and girls affect the nature and extent of play activities.

Intelligence and play. There is no evidence that children of superior mental ability are lacking in play interests. The study of Lehman and Wilkerson is of special interest in connection with this problem.[9]

Through the use of the *Lehman Quiz Blank,* data were obtained relative to the play behavior of 6,000 children. The problem of the

[7] *Ibid.,* p. 212.

[8] R. G. Simpson, *Fundamentals of Educational Psychology.* Chicago: J. P. Lippincott Co., 1949, Chap. III.

[9] H. C. Lehman and Doxey A. Wilkerson, "The Influence of Chronological Age Versus Mental Age on Play Behavior," *Pedagogical Seminar and Journal of Genetic Psychology,* 1928, Vol. 35, pp. 312–321.

investigation was to compare the relative influence of mental age with that of chronological age as far as these affect the play behavior of children. From an analysis of the data gathered it appears that a vari-

TABLE XV

Play Preferences by Age Groups (*After Simpson*)

Boys		Girls
	6 to 8 year olds	
Playing marbles		Playing house
Riding wagons *		Playing with dolls *
Playing cowboys		Playing school
	9 to 10 year olds	
Roller skating		Roller skating
Playing ball *		Dressing up as adults *
Riding bicycle		Playing jacks
	11 to 12 year olds	
Basketball		Roller skating
Riding bicycle		Hiking
Scout (Hiking) *		Swimming *
Baseball		Reading
	13 to 15 year olds	
Baseball		Reading books
Basketball *		Social dancing and parties
Going to movies		Going to movies
Hunting		Having dates
Watching athletic contests		Watching athletic contests

* Denotes outstanding performance.

tion of one year in the chronological age exerted a greater influence on the subject's play behavior than did a variation of the mental age by one year. The various studies of play activities among the gifted reveal a greater tendency toward solitary types of play; they prefer games involving rules and systems, and engage less frequently in activities demanding muscular strength and endurance. Gifted children, like all other human beings, find success pleasing and prefer activities in which they can succeed. Since this is the case, they prefer games requiring mental ingenuity to those of pure chance.

An analysis of the data gathered by Lehman and Witty [10] regarding sex differences of bright boys and girls shows a great similarity in activities. A closer analysis reveals that the dull boys have a higher index of social participation and prefer activities of a motor type, although there is a great deal of overlapping. Successful competition

[10] H. C. Lehman and P. A. Witty, "The Play Behavior of Fifty Gifted Children," *Journal of Educational Psychology*, 1927, Vol. 18, pp. 529–565.

which in the end brings vicarious satisfaction is the most probable explanation for these differences in interest. The organism tries out various modes of behavior until success is attained to some degree. The bright pupil gains vicarious satisfaction in reading and is able to compete most successfully in activities requiring problem solving and thinking. The formula for both the bright child and the dull child is the same; but it seems reasonable to assume that the type of activity that satisfies the felt need of the adolescent is the one that is chosen.

Team activities. The apparent sudden change in the play activities of adolescence is not to be accounted for on the basis of the sudden ripening or maturing of some instinct or impulse. The growing child has matured in strength and prowess, and surplus energy acts as a biological drive. New social realms are ever broadening, and constant contacts with fellow members of the group contribute to the development of team play. The individual soon learns that through cooperative endeavors he may satisfy certain needs that cannot be satisfied in solitary play; therefore team activities develop in harmony with the satisfaction of certain *felt* needs. These needs have a biological basis but are socialized in accordance with the expanding social life of the individual. Interest in team games develops along with earlier individualistic play interests and tends to supplement rather than supplant them. The maturation of the sex glands and a consequent interest in the opposite sex are partially responsible for the change toward group activity in adolescence. At this period, many games have the social element involved to a greater degree than before. The sexes are beginning to mingle and to develop interests of a sexual-social nature; girls now become loyal to boys' teams, and boys to those of the girls. Also at this period of life, games for both boys and girls become more formal in nature, and definite rules are laid down in order better to standardize the playing. The play of adolescent girls is often similar to that of the boys, usually having some modification in order that it will not be too strenuous.

In both large and small high schools certain types of activity predominate. Athletics seem to be the most popular in the average high school. Baseball, basketball, and other games that do not require expensive equipment, and prolonged periods of training are usually found in the smaller high schools. An increased number of the small high schools are providing indoor space for team activities. Group cooperation and group competition through team activities tend to take the place of individual competition. Furthermore, there is a tendency on the part of the small high school toward interscholastic rather than intramural participation.

Much controversial discussion has been carried on concerning the effect of athletics on the scholarship and health of growing youths. Exactly how much of the discussion has included points worthy of consideration is hard to determine, since the results of any athletic program will depend largely upon the nature of the program and the manner in which it is directed. There is much evidence that participation in athletics has had a beneficial effect, by keeping a great number of boys in school who did not find sufficient inducement to remain in the rest of the school program.

Interest in competitive team activities among adolescents and postadolescents is encouraged not only through high school contests, but also through programs sponsored by the American Legion and other organizations, which have aided in creating a national interest in baseball, as well as in other types of athletics. The values to be derived from participating in athletics are many and diverse, and not the least of these is the development of interests likely to become permanent and to have recreational and mental hygiene value throughout the later years of life.

Since participation in athletics tends to make for muscular development, and since muscular development is not looked upon by our present social groups as feminine in nature, girls are less interested in participating in athletics, especially as they reach adolescence and are motivated to play the woman's role by becoming and remaining feminine in nature. This is a problem which must be reckoned with by those concerned with athletic programs for girls. There is need for a redefinition of feminine qualities and of human values and needs in connection with this problem.

Interest in movies. Studies reveal that boys and girls of high school age attend movies considerably less frequently than do the children of the fifth, sixth, and seventh grades. This is due to a great amount of social activity in high schools, in clubs, and committees, which allows them less leisure. The attendance of both grade school and high school children is very largely confined to the week-end—Saturdays and Sundays. During adolescent years, romantic attraction, "dating," develops and movie attendance increases on Friday evenings. Children at all ages attend more often in the evening. Boys place athletics above movies. Girls do not. Later, girls show a preference for dancing and "dates" over movies. According to the materials presented in

Figure 27, about 45 per cent of boys and approximately 60 per cent of girls 8 years of age attend the movies with the father or mother.[11] There is a constant growth with advancing age in attendance of the movies with their own friends and others. Thus, the socialization process is definitely operating in movie attendance. The kinds of movies preferred by high school boys and girls show sex differences.

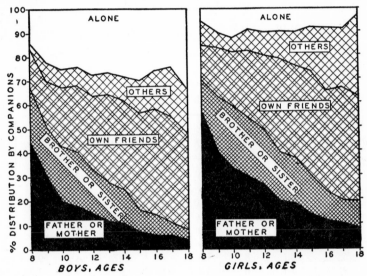

Fig. 27. *Percentage Distribution of Movie-goers According to the Companions Accompanying Them. Ages are 8 to 18, by Sex.*

Adventure pictures come first for high school boys, eighth for high school girls (see Table XVI).[12] Romantic pictures come seventh for high school boys, first for high school girls. Tragic pictures come eighth for high school boys, fourth for girls.

As growth progresses into adolescence, there is a change in their preferences. Western pictures come first for grade school boys, sixth for high school boys. Historical pictures come ninth for grade school boys, third for high school boys. Western pictures come first for grade school girls, seventh for high school girls. Historical pictures come seventh for grade school girls, second for high school girls. A study of the movie interests of 2,000 Catholic boys enrolled in parochial

[11] From Edgar Dale, *Children's Attendance at Motion Pictures*. New York: The Macmillan Company, 1935. (Reproduced by permission of the publishers.)
[12] A. M. Mitchell, *Children and Movies*. Chicago: University of Chicago Press, p. 167.

schools showed that the mystery type movie ranked first among high school freshmen but was surpassed by the musical comedy in the case of high school seniors.[13] The rank in order of interest for the four high school classes combined was found to be as follows: mystery, musical comedy, comedy of manners, historical, gangster-G-men, Western, news, love story, educational, and travel.

TABLE XVI

Frequency of Selection of Specified Types of Movies as First Choice (*After Fleege*)

	Boys				Girls			
	Grade School		High School		Grade School		High School	
	Per Cent	*Rank Order*	*Per Cent*	*Rank Order*	*Per Cent*	*Rank Order*	*Per Cent*	*Rank Order*
Adventure	13.7	2	13.7	1	7.2	5	6.1	8
Comedy	11.4	3	13.0	2	16.5	2	10.3	3
Educational	1.3	10	2.8	10	1.4	10	2.1	10
Historical	2.0	9	12.9	3	4.6	7	10.6	2
Mystery	6.3	4	8.9	5	6.8	6	8.8	5
Romance	4.7	7	7.0	7	13.3	3	22.8	1
Sport	5.3	5	11.9	4	3.7	8	8.0	6
Tragedy	2.1	8	5.6	8	7.7	4	9.9	4
War	5.1	6	3.8	9	2.2	9	2.2	9
Western	34.0	1	7.5	6	20.2	1	6.5	7

Order of rank correlations figured from this table give the following results:
Grade school and high school boys .... $r = .49$
Grade school and high school girls .... $r = .52$
Grade school boys and girls .... $r = .58$
High school boys and girls .... $r = .35$

Movies and the radio have a marked influence upon adolescence. In their earlier years, adolescents like pictures about love, war, and mystery, as well as adventure films and comedies. Generally, boys attend the movies more than girls, except when play interests keep them from attending. Interest in love stories on the stage and in the movies increases in the grades and is more characteristic of girls than boys. Ninth and tenth grade children like mystery plays. Often boys are interested in science pictures—such as those dealing with the atomic bomb. It is characteristic for girls to have a movie "hero" or "ideal."

[13] U. H. Fleege, *Self-Revelation of the Adolescent Boy*. Milwaukee: The Bruce Publishing Co., 1945, p. 251.

Interest in the radio. The studies made of the types of radio programs preferred by boys and girls at different age levels show these preferences to be in harmony with those obtained for reading activities and the movies. Brown reports a study made by graduate students of New York University during the spring of 1936.[14] A total of 2,500 boys and girls from grades 5, 8, 10, and 12 were asked to "check each of the types of programs to which you enjoy listening." The results of that study are presented in Table XVII. Mystery plays were liked by

TABLE XVII

Types of Radio Programs Liked Best by Boys and Girls of Different Grade Levels by Per Cent Choosing Each Type (*After Brown*)

Type of Program	Fifth Grade		Eighth Grade		Tenth Grade		Twelfth Grade	
	Boys	Girls	Boys	Girls	Boys	Girls	Boys	Girls
	%	%	%	%	%	%	%	%
Mystery plays	94.0	97.1	94.3	95.1	84.9	76.0	66.7	39.4
Comic dialogues and skits	86.4	96.2	92.0	98.5	87.9	93.0	76.8	78.3
Dramatic plays	88.0	90.0	82.8	84.7	67.1	86.2	55.0	89.8
Popular dance music	47.3	70.2	63.4	81.3	83.5	95.2	94.2	98.7
Popular song hits	45.1	53.6	52.4	60.2	78.1	81.3	70.5	82.4
Semiclassical music: orchestra and band	44.4	60.0	78.3	81.4	60.8	57.2	17.2	32.4
News, including sports	12.3	46.4	47.5	31.3	55.3	52.6	54.8	26.1
Political speeches	6.6	2.0	46.4	19.6	51.2	33.0	45.2	18.0
Classical music, including opera	5.7	10.2	16.4	25.0	12.3	23.1	10.8	20.2
Educational talks	1.4	3.6	11.5	8.1	22.0	12.2	13.4	13.4

a very large per cent of students at all grade levels, with the possible exception of twelfth grade girls. Comic dialogues and skits and dramatic plays were also liked at all grade levels. The increased interest in social activities is reflected in the growth of liking for popular dance music and for song hits. With an increase in age there is also an increased interest in political events and educational topics; however, the small percentage of those liking educational programs, like the small percentage interested in educational movies, presents a challenge to educators. Perhaps a few lessons from professional entertainers or playwrights could be used to an advantage in this connection.

[14] F. J. Brown, *The Sociology of Childhood.* New York: Prentice-Hall, Inc., 1939, pp. 327–328.

Responses to a comprehensive questionnaire dealing with their reactions to different types of radio programs were secured by Clark[15] from 505 children of the ages nine to eighteen. These children were representative of the white public-school population in Washington, D. C., the rural children in Fairfax County, Virginia, and the boys of the National Training School for Boys. A close agreement was found between the program listened to and the broadcast from the Washington radio stations. However, some interesting differences were noted for the three age groups studied. These are shown in Table XVIII. The decreased attention given by the older groups to drama and to children's programs and the increased attention given by them to the dance, comedy, and variety programs are significant and in harmony with the change in interests found during the adolescent period.

TABLE XVIII

RELATIVE LISTENING TO RADIO PROGRAMS BY WASHINGTON, D. C. CHILDREN
AND ADOLESCENTS (*After Clark*)

TYPE OF PROGRAM	PERCENTAGE FREQUENCY WITH WHICH PROGRAMS OF EACH TYPE WERE MENTIONED		
	Age Group		
	9–12	12–15	15–18
1. Classical and semi-classical music	1.8	2.8	3.4
2. Religious	.4	.5	.7
3. Dance, popular, and novelty type	12.9	14.9	23.5
4. Comedy and variety	25.8	36.4	39.5
5. Detective, crime, and mystery programs	3.7	6.5	2.6
6. Drama: General, historic, romantic	24.5	20.2	15.6
7. Travel and adventure	1.2	.7	.4
8. Children's programs (not otherwise listed)	25.8	12.8	6.9
9. National, public, and civic affairs	.1	.1	.3
10. News	1.2	2.4	4.3
11. Sports	.4	.9	1.1
12. Adult programs (including educational, labor, agriculture)	2.2	1.8	1.7

Reading interests. One of the most thorough of the studies of reading interests was conducted by Jordan.[16] His study revealed some

[15] W. R. Clark, "Radio Listening Habits of Children," *Journal of Social Psychology,* 1940, Vol. 12, pp. 131–149.
[16] A. M. Jordan, *Educational Psychology.* New York: Henry Holt and Co., 1928, pp. 111–112.

rather striking sex differences in reading interests during adolescence. Table XIX gives the results.

An analysis of the results summarized in the table by the types of books preferred indicates that a considerable amount of overlapping exists for both sexes and also for the different age groups. This is to be expected, since, as has already been pointed out, growth seems to be a continuous rather than a periodic process. The developmental con-

TABLE XIX

PERCENTAGE TABLE INDICATING THE RELATIVE PROPORTION OF BOOKS CHOSEN IN EACH CLASS (*After Jordan*)

AGES	NO. OF SUBJECTS		ADULT FICTION		JUVENILE FICTION		ADVENTURE	
	Boys	Girls	Boys	Girls	Boys	Girls	Boys	Girls
9–11	59	87	4	15	27	67	56	12
12–13	253	336	6	33	19	44	64	17
14–16	846	1,195	18	45	11	30	59	18
17–18	283	414	50	58	9	13	49	12

cept of growth as represented in physical and mental growth will hold true also for growth in behavior units, interests, and intellectual concepts. Various studies of reading interests show a keen interest in fiction among girls, whereas adventure stories are preferred by boys. During later adolescence (postadolescence) there is a natural shift of girls' interest from juvenile to adult fiction, the trend of boys' interests being toward biography, history, travel, information of a general type, and humor; and yet there is considerable overlapping.

In a more recent study by Witty, Coomer, and McBean [17] the books ranking highest in the interests of upper grade boys and girls were determined. The seventeen books ranking highest are listed in Table XX in the order in which they most often appeared. These results are somewhat in harmony with the findings of Harold Saxe Tuttle from his survey of the reading tastes of eighth grade pupils throughout the country.[18] *Tom Sawyer* was first in the list of books these pupils had read and enjoyed, with *Treasure Island* and *Little Women* second and third, respectively. Considerable differences are found in the incidents

[17] Paul Witty, Anne Coomer, and Dilla McBean, "Children's Choices of Favorite Books," *Journal of Educational Psychology,* 1946, Vol. 37, pp. 266–278.

[18] See a report of Professor Tuttle's study in *The New York Times,* October 3, 1943, p. 43.

of reading materials preferred by boys and those preferred by girls. Boys prefer action, such as mischievous pranks, fights, races, moving around, and adventure; whereas girls prefer mystery far more than boys, as well as deaths, accidents, kind acts, and events involving social

TABLE XX

FAVORITE BOOKS LISTED IN ORDER OF FREQUENCY OF CHOICE
BY UPPER GRADE PUPILS

AUTHOR	TITLE
Knight	*Lassie Come Home*
Estes	*Moffat Books*
Boylston	*Sue Barton Books*
Tunis	*Keystone Kids*
O'Brien	*Silver Chief*
Terhune	*Lad*
Tunis	*All American*
Brink	*Caddie Woodlawn*
Spivy	*Call It Courage*
Stevenson	*Treasure Island*
Sewell	*Black Beauty*
Tunis	*Kid from Tomkinsville*
Twain	*Adventures of Tom Sawyer*
Seredy	*The Good Masters*
Beals	*Davey Crockett*
Sture-Vasa	*My Friend Flicka*

and romantic elements. Several studies have been concerned with basic interest pattern rather than with the actual reading preferences of children; Thorndike concludes from his studies that sex is the most important determining factor, and that within a particular school system the bright child's pattern of interest will be most like that of the dull child who is several years older.[19]

The differences noted in the choices of books by boys and girls are also reflected in their choice of magazines. Therefore any consideration of periodicals for the home or school must recognize this gulf. There is some evidence that the magazines to be found in the school libraries have been chosen largely from the woman's world. This may be accounted for, in part at least, by the fact that they are usually chosen by a woman or by a committee made up of school teachers (largely feminine). Three trends in American periodicals were noted

---

[19] R. L. Thorndike and Florence Henry, "Differences in Reading Interest Related to Differences in Sex and Intelligence Level," *Elementary School Journal*, 1940, Vol. 40, pp. 751–763.

by Zanders.[20]   These trends are increased popularity of the digest, the news, and picture types of magazines. This is in harmony with findings by Sterner.[21]

The fifteen magazines most widely read by high school pupils of North Newark, in order of their popularity, were listed as: *Life, Look, Reader's Digest, Movie Mirror, Saturday Evening Post, Movie Story, Modern Screen, Ladies Home Journal, Liberty, Popular Mechanics, Popular Science, Movie-Radio Guide, Silver Screen, Good Housekeeping,* and *Woman's Home Companion.* Such a list would vary considerably from locality to locality. Certainly, the boys and girls from the wheatlands of Kansas or from the cotton fields of Texas would not show the same type of interest in reading magazines as would the boys and girls of Newark, New Jersey. There are many factors that would tend to affect this; yet there are likely to appear some features in common arising out of the nature of adolescent interests as expressed in a democratic land.

The comic book, in its present form, is the most recent and in many cases the most widespread of all the leisure verbal activities pursued by pre-adolescents and adolescents. Even though their history is still a short one, they have been subjected to considerable criticism. They have been looked upon by some as the inspirer of bad deeds; their art has been described as a "hodge-podge of blotched lines and clashing colors"; while the content has been referred to as "sadistic drivel." However, some educators recognize their appeal and have attempted to turn the method of presentation used by the funny books into a channel for developing more worthy concepts and ideals. Case studies show that the bright as well as the dull, and the child from the privileged as well as the underprivileged home read the funny books. As grade increases, however, there is a gradual and continuous decrease of interest in reading funny books. As the individual moves into the post-adolescent stage, pronounced sex differences become evident; although there is still considerable interest manifested by both boys and girls in the comic strips. This is well illustrated in the case of the development of Karl.

At the age of seven Karl showed a great deal of interest in cartoons displaying activities of animals and children. There was a gradual change of

[20] J. Zander, "Modern Magazine Trends," *Chicago School Journal,* 1940, Vol. 22, pp. 63–67.

[21] A. P. Sterner, "Radio, Motion Picture, and Reading Interest," Teachers College, Columbia University, *Contributions to Education,* No. 932, 1947.

interest with increased age, so that by the age of ten he was keenly interested in super activities of men, who were oftentimes made heroes; with the outbreak of World War II, he became interested in constructing comics of his own and developed a series of comics at the age of eleven entitled, *Flying Tommy*.

Karl was always very successful in his school work and had a reading ability two years advanced for his age. He was above average in physical and mental development. At the age of twelve he seemed to lose interest in most comic magazines, but showed an increased interest in a number of comic strips. There was a significant relationship observed between these changes of interest in the comics and the development and change of interest in the radio, movies, and play activities.[22]

THE EXPANSION AND SIGNIFICANCE OF
ADOLESCENT INTERESTS

Expanding interests. The child's general satisfaction with himself and his surroundings gives way during adolescence under the pressure of many problems, difficulties, and maladjustments. Once indifferent to matters not immediately related to pleasure and pain, he now has an intense curiosity and self-consciousness, and a real concern with the social and ethical standards of adults. Curiosity may show itself in a great many different ways, but it is also subject to ready perversion if in unwholesome surroundings. This is true especially of those impulses and interests of the adolescent that are now maturing and becoming more and more important in his life. Satisfaction and complacency in routine is often replaced rather suddenly by a restlessness leading toward idealistic behavior trends or probably into antisocial activities. After years of activities concerned largely with egocentric interests and activity for its own sake, the adolescent is thrown into further contacts with others. With newer interests and contacts, he acquires new purposes and interests in special activities leading to definite results, whether in his play or in his work. But having acquired these expanded interests, he stands in need of further stimulation, inspiration, information, and guidance.

These expanding and maturing interests are clearly revealed in the changes in the topics of conversation among boys as they progress from freshman to senior year in high school.[23] The results of Table XXI are based on the replies of 2,000 boys who mentioned nearly 6,000 topics that they most frequently talked about among themselves.

---

[22] See P. Witty and Anne Coomer, "Reading the Comics in Grades IX–XI," *Educational Administration and Supervision,* 1942, Vol. 28, pp. 344–353.

[23] Urban H. Fleege, *op. cit.,* p. 234.

The most outstanding change noted was the greatly increased interest in girls, dates, and matters relating to sex. However, there is also an increase of interest in current happenings, and in vocational pursuits.

TABLE XXI

Topics High School Boys Most Frequently Talk About Among Themselves
(*After Fleege*)

Rank	Topic	Freshmen	Sophomores	Juniors	Seniors	All Classes
1.	Sports	62.0%	80.2%	75.0%	75.1%	73.3%
2.	Girls	53.6	67.6	70.6	80.0	68.0
3.	School, studies, teachers	27.0	27.2	29.0	27.3	27.6
4.	Social activities, dates, good times	9.6	16.0	20.1	24.2	17.5
5.	Sex, sexual relations, dirty jokes	11.2	12.4	16.4	20.1	15.0
6.	Movies	16.4	10.4	9.0	7.6	10.9
7.	Current happenings	7.6	7.9	7.1	10.6	8.4
8.	Cars, airplanes, machines	5.4	8.2	7.4	7.0	7.0
9.	Generalities	7.4	5.0	4.1	6.4	5.7
10.	Other boys	5.4	4.2	3.3	3.4	4.1
11.	One's experiences	6.0	3.2	2.0	1.6	3.2
12.	Hobbies	2.5	3.2	4.4	2.0	3.0
13.	Job or work	1.0	1.8	2.0	6.1	2.7
14.	Money	1.2	2.8	3.0	3.1	2.5
15.	Things one is going to do	1.9	2.0	3.4	2.2	2.4
16.	Future vocation	1.2	1.4	1.6	4.7	2.2
17.	Miscellaneous: religion, clothes, homes, food, and so forth	4.0	2.2	3.2	4.3	3.4
18.	No answer	3.4	1.6	.8	.4	1.6

Interests and intelligence. There are a number of experiments that have given information about the ways in which the interests of children of superior mental ability differ from those of children of inferior mental ability. Boynton has made a study of the relationship between children's tested intelligence and their hobby participation.[24] The sub-

[24] P. L. Boynton, "The Relationship Between Children's Tested Intelligence and Their Hobby Participation," *Journal of Genetic Psychology,* 1941, Vol. 58, pp. 353–362.

jects of his study consisted of 4,779 boys and girls from the sixth grade of 258 schools. The children were given the *Kuhlmann Anderson Intelligence Tests,* and the teacher arrived at their hobbies from a conference held with each child. Most of the children had from three to six hobbies. Boynton concludes that, "Some hobbies tend to be participated in more frequently by children of high tested intelligence than do other hobbies." Children without a hobby, especially girls, are more likely to be below average in general intelligence. When both sexes were considered together, the hobbies of collecting, playing musical instruments, and reading were found frequently among those of superior ability. No single hobby appeared to be associated with those of lower than average intelligence. The superior children also appeared to have a greater diversification of hobby interests than very inferior ones. This finding is in harmony with certain conclusions arrived at by Bayard from a comparative study of the interests of high and low ability high school students. He concludes the following:

1. The sum of the average likes and dislikes is the same for both groups; but whereas the high ability group likes approximately twice as many activities as it dislikes, the low ability group dislikes slightly more than it likes. . . .

7. A comparison of the profiles shows that in general both groups like and dislike the same activities in the same categories, although the relative liking of one area as compared to another area may be different. The one exception which may be significant is mathematics.[25]

Another study, conducted at Peabody College under the direction of Boynton, revealed that reading interests and intelligence go hand in hand. In this study Lewis and McGehee[26] gathered data from children from 455 schools in 310 communities in 36 states. The interests of those scoring in the top ten per cent on the *Kuhlmann Anderson Intelligence Tests* were compared with those of the lowest ten per cent. According to their study, dramatics, religious activity, studying, scouting, and campfire activity make a greater appeal to children of superior mental ability. More than twelve times as many retarded children as gifted indicated no hobby. The superior children were more interested in both active and quiet games than were the inferior.

According to the genetic case study of interests, conducted by Mac-

[25] B. Bayard, "A Comparison of the Interests of Students of Low Ability Enrolled in Physical Science and of Students of High Ability Enrolled in Physics," *University High School Journal,* 1941, Vol. 20, pp. 15–19.

[26] W. D. Lewis and William McGehee, "A Comparison of the Interests of the Mentally Superior and Retarded Children," *School and Society,* 1940, Vol. 52, pp. 597–600.

kaye[27] with adolescent subjects, early fixation and permanence of interests were most commonly found among those of inferior intelligence. The interests of subjects of higher intelligence were therefore more unstable in nature. Wishful thinking is oftentimes the basis of vocational interests, and the interests may show very little relation to actual ability. This is especially true for interests formulated without experience as a background. Terman's exhaustive studies[28] of gifted children have revealed that there is a preference among the gifted for school subjects demanding abstract thinking. A number of other studies have pointed rather definitely to the same conclusions.

Problems of adjustment in relation to changing interests. There are no difficulties encountered in changing interests by those boys and girls whose interests and values coincide with those of the group with whom they work and play in school and on the playground. However, there are some children whose physiological development is accelerated. For these boys and girls an interest in less mature and less social games is a thing of the past. Such boys and girls may seek connections in the church or in some special neighborhood activities where there are other boys and girls with these more mature interests. A wholesome and friendly home relationship may help the individual during this stage. If there are several others in the grade at school who have more mature interests also, the adjustment may result in the acquisition of close chums or in the formation of small cliques.

Then, there is the boy or girl who is less mature in his or her interests than the other boys and girls of the group. This individual oftentimes develops an attitude of indifference toward the activities of the group as a whole, but may be able to find comfort and the needed friendship in activities with any others of the group who likewise reveal a less mature interest; in which case, such children will not be seriously affected by the time lag in their maturity. Such friendships should be encouraged at this stage.

However, it is for the individual who has advanced at the same pace as the average in his physiological development but who, for some social or cultural reason, is unable to participate in the activities of the group, that the problem is more serious. Racial, religious, or social conflicts between the practices or ideals of the home and those of the group may be responsible for such a condition. An individual who, because

[27] David L. Mackaye, "The Fixation of Vocational Interest," *American Journal of Sociology,* 1927, Vol. 33, pp. 353–370.

[28] Lewis M. Terman, *Genetic Studies of Genius,* Vol. 1. Palo Alto, Cal.: Stanford University Press, 1926.

of some such condition, is unable to change his pattern of interests in harmony with the interests of the growing boys and girls with whom he is thrown, in and out of school, is going to be faced with a difficult adjustment problem, and this problem is likely to affect his school work, his attitude toward his home, and other phases of his personal and social life.

SUMMARY

Many problems of growing up are closely related to changes in interests. The early interests of adolescents are personal in nature; but the social and emotional changes that appear at this stage are reflected in the development of interests in others and especially a changed attitude and feeling toward members of the opposite sex. This changed attitude is discussed at length in later chapters.

The play life of adolescents involve more of the team spirit and group action than was the case during the pre-adolescent stage. Adolescent boys continue to show an interest in adventure, but the adventures are less fantastic and are more closely connected with present day living conditions and problems. The changes in interests are further reflected in the attitude of adolescents toward movies and radio programs. It should not be inferred, however, that these changes are sudden and complete. Much inconsistency will be found in the interests and behavior of adolescent boys and girls. Their interest in a movie of an adventurous nature involving some romance may be followed by interests in make-believe activities resembling those of the ten- or eleven-year-old individual. Differences will be found in the interests of boys and girls, with girls showing a more mature interest; this is in keeping with their advanced physiological development. Also, differences will be found in the interests and activities of adolescents from different class structures. There is much evidence that differences in class structure are very important in affecting differences in interests during childhood and adolescence.

The adolescent's interests lead in many directions and may change considerably in a short time. When they are not expressed in reality they usually appear in his daydreams, wishes, and imagination. It is essential that parents and teachers have a knowledge of adolescent interests, so that they may aid him better to understand himself and direct or guide him toward a more complete fulfillment of his aspirations and possibilities.

THOUGHT QUESTIONS

1. Point out the significance of adolescents' interests in magazines.
2. Study the early life (adolescent period) of some of our leaders of today. What interests dominated their life during the adolescent age?
3. Discuss the range of adolescent interests as compared with the interests of the 8-year-old child.
4. What interests have been somewhat permanent in your own life? Why?
5. Show how a knowledge of the nature of adolescents' interests is of especial value to a school teacher; to a scoutmaster.
6. What changes have you observed among adolescents in their interests in the radio? Compare these with the materials presented in Table XVII.
7. How do you account for the intense interest of pre-adolescents in the comic books? Note the extent to which this appears to hold over into adolescence and post-adolescence.

SELECTED REFERENCES

Averill, L. A., *Adolescence*. Boston: Houghton Mifflin Co., 1936, Chap. VI.

Briggs, Thomas H., Leonard, J. Paul, and Justman, Joseph, *Secondary Education*. New York: The Macmillan Co., 1950, Chaps. XVI and XVII. Some very useful suggestions are presented in these chapters for helping adolescents develop worth-while interests in relation to the educational process.

Cole, Luella, *Psychology of Adolescence* (Third Ed.). New York: Farrar and Rinehart, 1948, Chap. XIV.

Dale, E., *Children's Attendance at Motion Pictures*. New York: The Macmillan Co., 1935.

Eisenberg, A. E., *Children and Radio Programs*. New York: Columbia University Press, 1935.

Fleege, Urban H., *Self-Revelation of the Adolescent Boy*. Milwaukee: The Bruce Publishing Co., 1945, Chaps. XIII and XIV.

Hollingshead, A. B., *Elmtown's Youth*. New York: John Wiley and Sons, 1949. Chaps. XII and XV.

Hurlock, E. B., *Adolescent Development*. New York: McGraw-Hill Book Co., 1949, Chaps. VII, VIII, and IX.

Jordan, A. M., *Children's Interests in Reading*. Chapel Hill, N. C.: University of North Carolina Press, 1926.

Lehman, H. C. and Witty, P. A., *The Psychology of Play Activities*. New York: A. S. Barnes and Co., 1927.

Sterner, A. P., "Radio, Motion Picture, and Reading Interests," Teachers College, Columbia University, *Contribution to Education,* No. 932, 1947.

Terman, L. M. and others, *Genetic Studies in Genius*. Palo Alto, Cal.: Stanford University Press, 1925, Vol. I.

For a fairly complete summary of investigations on attitudes and interests, see Strang, Ruth, *Behavior and Background of Students in College and Secondary School*. New York: Harper and Brothers, 1937, pp. 229–270.

VIII

Social Growth and Development

Growth as a developmental process. It has been observed that individual differences in personality characteristics can be detected from the very beginning of the child's active life. Thus, the responses of children to social situations will vary from early childhood throughout life. However, the principles of development applicable to the mental, physical, and emotional realms are also applicable to the social development of the individual. There is a sequence of behavior patterns that tend to appear at different stages of maturity. However, experience in harmony with the maturational level is essential for wholesome social development. The *Vineland Social Maturity Scale*, developed by Doll, has been found useful in evaluating growth in social competency. (This scale is presented in Appendix D.) A study of the characteristics or features listed for the various stages of maturity indicates the nature of an individual's growth in social competency. It will be noted, for example, at the age of 11–12 the average child does simple creative work, enjoys books, newspapers, and magazines, and assumes increased responsibilities. The extent to which these things are done will depend in a large measure on the habits that have been established during the earlier years of life. From 12 to 15 the adolescent enjoys difficult games, exercises greater care for his clothes, and assumes increased responsibility, especially in connection with his spending money or allowance.

The nature of social development. The child's social development is of a gradual and continuous nature. Early in life he is largely egocentric. A little later he discovers himself as a member of a social unit, in which he stands in certain relations to others. In the home he receives his first social training, and first comes into contact with the routine of life. He establishes regular habits with regard to meals, play

157

hours, sleep, and so forth, and learns that his actions must be conditioned in relation to the behavior of others.   Social habits are formed in harmony with various social forces.   A reasonable amount of conformity is essential, but the problem of individual versus social develop-

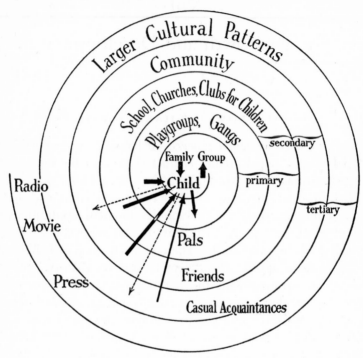

Fig. 28. *The Widening of the Child's Experience and the Resulting Interaction.* (F. J. Brown, *The Sociology of Childhood.* New York: Prentice-Hall, Inc., 1939.)

   The thickness of the arrows represents the probable relative importance of the interaction between the child and the various individual and groups.

ment is not always adequately solved by the various social forces.   The social development of the child is conditioned by various institutions, and the principle of "learning by doing" applied to social participation is important in the development of a social individual.

   The early work of G. Stanley Hall made use of diaries in studying the social development and change of interests that accompany development into adolescence.   A more recent study of adolescent leisure-time activities, made through an analysis of diaries, revealed some interesting changes and conditions.   Study of the leisure-time activities of 535 adolescents between 12 and 21 revealed an increase in heterosexual activities with an increase in age.[1] These later activities of adolescents consisted of dancing, talking, and the like, whereas their earlier activities included much time given to reading, listening to the radio, and to separate boy or girl club and gang affairs. These differences are well summarized in the report by Meek[2] on the personal-social development of boys and girls from the onset of puberty into and through adolescence. (See Table XXII.) It is interesting to note how the

TABLE XXII

THE DEVELOPMENT OF BOYS AND GIRLS FROM THE ONSET OF
PUBERTY INTO AND THROUGH ADOLESCENCE (After Meek)

FROM	TO
1. Variety and instability of interests.	Fewer and deeper interests.
2. Talkative, noisy, daring behavior with a great amount of any kind of activity.	More dignified, controlled masculine and feminine adult behavior.
3. Seeking peer status with a high respect for peer standards.	The reflecting of adult cultural patterns.
4. A desire for identification with the herd, the crowd of boys and girls.	Identification with small select group.
5. Making family status a relatively unimportant factor insofar as it influences the choice of associates.	Making family socio-economic status an increasingly important factor in affecting with whom one associates.
6. Informal social activities such as parties.	Social activities of a more formal nature, such as dances.
7. Rare dating.	Dates and "steadies" the usual thing.
8. Emphasis on building relations with boys and girls.	Increasing concern with preparation for own family life.
9. Temporary friendships.	More lasting friendships.
10. Having many friends.	Fewer but deeper friendships.
11. Willingness to accept activities providing opportunities for social relations.	Desire for activities satisfying to the individual in line with talent development, proposed vocation, academic interest, or hobby.
12. Little insight into own behavior or behavior of others.	Increasing insight into human relations.
13. Accepting the provision of reasonable rules by adults as an important and stabilizing influence.	Making own rules with a definite purpose in view.
14. Ambivalence in accepting adult authority.	Growing independence from adults and dependence on self for decisions and behavior. Seeking relations with adults on an equality basis.

[1] H. E. O. James and F. T. Moore, "Adolescent Leisure in a Working-Class District," Occupational Psychology, 1940, Vol. 14, pp. 132–145.

[2] Lois Hayden Meek (Chairman), The Personal-Social Development of Boys and Girls with Implications for Secondary Education. Committee on Workshops, Progressive Education Association, 1940, p. 121.

gang interests of the adolescent are discarded, during this period, in favor of closer identification with adult culture and interests. The intense desire for status and approval in the society of his peers, together with the low regard for family status, which he manifests during his pre-adolescent years gradually gives way to a recognition of the socio-economic status of the family. It is at this point that his earlier, more democratic nature breaks down, and he begins to seek the friendship and approval of members of some select group.

The maturing sex drive. As a full-fledged drive, sex does not mature until puberty. The dynamic force of this drive comes mainly from the hormones of the gonadal glands. These glands usually begin to function effectively sometime between the ages of 11 and 15 years, generally earlier in girls than in boys. However, many of the aidant mechanisms that later serve the sex appetite are established at an earlier period. Their development is a result of both *biological* and *social* forces. The biological reason is concerned with the fact that certain areas of the body are well supplied with cutaneous sense organs that become the points of stimulation for the development of overt responses. Furthermore, curves of growth of the sex glands reveal a gradual and constant growth, and any strong stimulation applied in the right way under favorable conditions will affect the sex drive during this early period. It has been pointed out that endocrine factors operate in preparing the pre-adolescent for adolescent changes. Concerning this, Shock states:

> With maturation of the sex glands, increased amounts of male or female sex hormones are liberated into the blood stream, stimulating growth and development of accessory sex organs, and resulting in the appearance of secondary sex characters.[3]

Changed interest in personal appearance. The changed interest of adolescents in personal appearance has been noted by students of adolescent psychology, and has been explained in various ways. Most of these have tended toward oversimplification. Some have regarded this as part of the growing up process—a sort of unfolding of natural tendencies; others have regarded it as a means of securing social approval; while others have conceived of it as the result of an awakened social consciousness and awareness of a sex role. This oversimplifica-

---

[3] N. W. Shock, "Physiological Changes in Adolescence," *Forty-Third Yearbook of the National Society for the Study of Education,* Part I. Chicago: Department of Education, University of Chicago, 1944, p. 76.

Social Development as Revealed Through Forms of Recreation.

161

tion, which has appeared in explanations of other aspects of behavior, oftentimes resulted from the nature and purpose of studies conducted on this problem.   As early as 1897 Hall [4] made use of the questionnaire method for studying the relation between clothing and the development of the sense of the self.   More recent investigators have studied the factors motivating teen-age boys and girls to take an increased interest in their personal appearance.   Some students have attacked this from a psychoanalytic viewpoint, and have found evidence that clothes and personal appearance reveal characteristics of individuals.

There is a general agreement among the various studies that adolescence is accompanied by an increased interest in peer approval.   The nature of this interest will be significantly affected by the experiences and social contacts that these boys and girls have had during early childhood, childhood, and pre-adolescence.   Children are directed into their sex role at a very early age and become conscious of sex differences before they enter school; however, their attitudes toward members of the opposite sex will be closely related to their training and experiences.

One aspect of this problem was studied by Silverman.[5]   A study of the actual clothing and grooming practices of adolescent girls was conducted by means of a carefully devised check list and objective questions on subjects related to the use of cosmetics, wearing of ornaments, types of clothing worn, and the like.   Also a questionnaire entitled "What Do You Think About Clothing and Appearance?" and a Personal Data Sheet were administered to the girls.   The study was conducted in a suburban New Jersey high school with an enrollment of 1,100 students representing a fairly good cross section of the population.   The data were studied for age differences and for differences based upon economic circumstances.   The findings revealed the following:

a. Close conformity in the style of dress for daily wear was prevalent not only within the age groups but among the groups, girls at 12 and at all ages through 18 tending to dress in like fashion.   Differences in dress among the age groups were evident in their week-end apparel, when the older girls were wearing higher heeled shoes, stockings instead of socks, and dresses rather than sweaters and skirts. . . .

b. The girls did not tend to go in for heavy use of make-up although certain items were used almost universally.   Age difference rather than conformity to a pattern was the dominating factor in the use of cosmetics.

[4] G. Stanley Hall, "Early Sense of Self," *American Journal of Psychology,* 1897, Vol. 9, pp. 351–395.
[5] Sylvia S. Silverman, "Clothing and Appearance: Their Implications for Teen-Age Girls," Teachers College, Columbia University, *Contributions to Education,* No. 912, 1945.

There was little difference noted in the practices and type of clothing worn by girls in extreme economic groups. The major differences were in the more expensive clothes and greater number of luxury items worn by those in the best economic circumstances; however, girls from all economic levels attempted to conform to the group style of dress and make-up. It appears, therefore, that growth and development through the teen years is accompanied by a changed attitude toward one's self and one's peers. The dynamic forces back of this changed attitude must be sought both in the changing physiological self and in the environmental forces affecting the individual. Some problems related to these changing interests and attitudes toward personal appearance will be discussed in subsequent chapters dealing with home conflicts, adolescent adjustments, mental hygiene, and guidance.

CONDITIONS AFFECTING SOCIAL DEVELOPMENT

In the previous chapters it was emphasized that any given period of development must be viewed in relation to the conditions and events that have preceded it. Thus, if the emotional and social maturation of the adolescent are to be understood, we must study the physiological, social, and educational forces at work during the years of infancy and elementary school life. It is most important that these prior influences affecting the adolescent's attitude toward members of the opposite sex be taken into account; an awareness of them is essential for the proper direction of the social activities of adolescents. That is what Mursell meant when he said:

Even the most dramatic characteristic [of adolescence]—sexual maturity and emotional fixation upon the opposite sex—is known to be prepared for from earliest infancy. A fortunate sexual evolution during these later years depends far more upon what has happened long previously than upon any special measures or special training [adopted] when the event itself takes place . . .[6]

Relation to physiological changes. Furfey's studies of *The Gang Age and The Growing Boy* reveal that boys reach a stage of growth at which they break away from the gang and begin to enter into activities where girls are concerned. This desertion is not a sudden breach, and is seldom complete. Furthermore, it is usually aided by the fact that other boys of the gang have a similar inclination. The number

[6] James L. Mursell, *Education for American Democracy.* New York: W. W. Norton Co., Inc., 1943, p. 207.

with such an inclination increases in time, since physiological maturity is advancing throughout the group.

The relation between social development and physiological changes has been observed by a number of students of adolescent psychology. Sollenberger [7] studied this problem, using as subjects a group of boys between the ages of 13 and 16. These boys were carefully observed over a period of six months and were rated for their maturity of behavior. Also, three questionnaires designed to measure the changing interests and attitudes of growing individuals were given to these boys. After the behavior ratings were made and the questionnaire given, a urine sample from each boy was assayed for androgenic (male sex) hormone activity. It was found that the degree of maturity as revealed through expressed interests and attitudes correlated higher with hormone activity than with the chronological age. The high-hormone boys expressed greater interests in heterosexual activities, personal adornment, and strenuous competitive sports than the low hormone group. Under the headings "Things to Think About" and "Things to See" the low-hormone group was concerned with imaginary phenomena to a much greater extent than the high-hormone groups; while the high-hormone group invariably expressed interests in real people and real things.

Social adjustment and class status. Wherever large groups of individuals come together some sort of class structure appears. The difference in this respect between class structures in various societies depends in a large measure on the factors that operate to produce class groupings. Certain carefully devised procedures have been developed for the evaluation of class status in America. [8] By means of these procedures estimates have been arrived at for different cities and areas. Milner [9] has presented estimates of the percentage of population in the various class groupings arrived at by Carson McGuire for the United States, and by Warner and others for Yankee City and the midwest. These estimates are presented in Table XXIII for further study and consideration.

[7] R. T. Sollenberger, "Some Relationship Between the Urinary Excretion of Male Hormone by Maturing Boys and Their Expressed Interests and Attitudes," *The Journal of Psychology,* 1940, Vol. 9, pp. 179–189.

[8] W. L. Warner, M. Meeker, and K. S. Eells, *Social Class in America: A Manual of Procedure for the Measurement of Social Status.* Chicago: Science Research Associates, 1949.

[9] Esther Milner, "Effects of Sex Role and Social Status on the Early Adolescent Personality," *Genetic Psychology Monographs,* 1949, Vol. 40, pp. 231–325.

The personal and social effects of the class status in which the individual is reared have been studied by a number of investigators. According to Neugarten's study, social class seems to operate differently in affecting the reputation scores at the fifth and sixth grade level and at the high school level.[10]   At the fifth and sixth grade level upper-class status carries with it almost automatically a favorable reputation score, while membership in the lower class usually results in an unfavorable reputation score.   At the high school level upper-class status insures the individual that he will be given attention and consideration, whether his reputation is favorable or unfavorable.   At this age level the lower-class child either drops out of school or takes on the behavior and values of his middle-class associates and thus tends to lose his lower-class characteristics.

TABLE XXIII

PERCENTAGE OF THE POPULATION IN SIX SOCIAL CLASSES

Social class	United States	Yankee City	Midwest
Upper-upper	} 2%	1.44%	} 3%
Lower-upper		1.56%	
Upper-middle	8%	10.22%	11%
Lower-middle	30%	28.12%	31%
Upper-lower	40%	32.60%	41%
Lower-lower	20%	25.22%	14%
Unknown	——	.84%	——

SPECIAL FEATURES OF SOCIAL DEVELOPMENT

Sociability during pre-adolescence.   There are very few research studies available on social functioning at the pre-adolescent level.   The longitudinal studies conducted during the past two or three decades have provided information regarding the general growth trends of the individual.   These studies indicate that there is a general pattern of social behavior that remains somewhat consistent throughout the period of growth.   Psychiatrists, educators, and sociologists have come forth with varying concepts relative to the social characteristics of pre-adolescents.   It has been referred to as the gang age, the critical stage of life, the age of socialization, and age of intense loyalties to peers,

[10] B. L. Neugarten, "Social Class and Friendship Among Children," *American Journal of Sociology,* 1946, Vol. 51, pp. 305–313.

the awkward age of childhood, and so on. Zachry has described the
boy of this age level as follows:

He does not seem open and confiding toward the adult. More often he
seems to be turning away to other children of his own age and sex. During
these years, a transition is made from the young child's intense preoccupa-
tion with himself and the grown persons whom he has needed to interest
in a group of people most like himself; to age mates of his own sex; and the
child now finds freedom and security in the gang, in which strong loyalties
and antagonisms develop. Boys show this tendency to associate with their
own sex more markedly than do girls.[11]

The attitude of boys toward girls was also studied by Sollenberger.
The reactions of 700 boys age nine to eighteen to a question concerning
their attitudes toward girls is presented in Figure 29. Over 40 per cent
of the boys at age nine regard girls as a nuisance, while another 20
per cent neither like nor dislike girls.

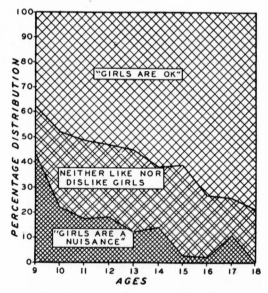

Fig. 29. Reactions of Seven Hundred Boys
Nine to Eighteen Years of Age to a Question
Concerning Their Attitude Toward Girls.
(Based on unpublished data of R. T. Sollen-
berger. From F. K. Shuttleworth, "The Ado-
lescent Period," Monogr. Soc. Res. Child
Developm., 1938, Vol. 3, No. 3.)

[11] C. B. Zachry, "Later Childhood, Some Questions for Research," Progressive
Education, 1938, Vol. 15, p. 522.

Hartley conducted a study of the nature of sociality among boys ten to twelve years of age.[12]   A group of 140 boys from a relatively homogeneous economic and scholastic background was used as subjects for this study.   These boys were given five tests of sociality—a self-rating questionnaire, a schedule for rating classmates, a measure of acquaintance volume and of special friendships, a test of attitude toward activities, and a measure of participation in activities.   Comparisons were made in the test performances of three pairs of groups, representing extremes of sociality in relation to their peers.   The results from the tests are inconclusive.   Only the *Pictorial Extensity Test* and the *Measure of Special Friendships* differentiated the groups somewhat decisively.   Test score profiles generally furnish a better basis for predicting and understanding the nature of sociality than any single measure.

Fourteen boys in the study by Hartley were observed and studied intensely.   One interesting observation was that the boys tended to draw closer together in connection with adverse or displeasing situations or conditions, treating these as a common enemy or difficulty.   Yet, fights, name-calling, and the like were common occurrences.   The socially unsuccessful boy appeared to lack self-control when confronted with his own aggressiveness or that of others, while the socially successful boy showed greater control over his aggressive impulses and was able to modify them to suit the situation with which he was confronted.   Thus, the successful expression of sociality among these boys was bound up with the ability to act aggressively but with self-control, and to accept the aggression with others.   Also, a lack of sensitivity to others was observed among these described as having trouble in their social relationships.

Selection of chums.   There is evidence of selectivity in friendships and attachments even during the preschool years.   Boys are more likely to form attachments for boys; girls, for girls.   The capacity to form friendships and the capacity or tendency to quarrel increases with age.   Children tend to choose their chums from their own neighborhood and their own school grade.   As age increases beyond the preschool period there is a tendency to select friends near one's own age and at the same level of development.[13]   Within these age groupings,

[12] Ruth E. Hartley, "Sociality in Preadolescent Boys," Teachers College, Columbia University, *Contributions to Education,* No. 918, 1946.   Sociality was defined in connection with this study as the degree of acceptance with which an individual reacts to others of his sex-age group.

selections are made on the basis of mental ability, physical make-up, and certain temperamental factors. Bright children tend to associate with bright children, and the dull with the dull. The physically large child more often chooses a child above average in physical development for his associate; the smaller child chooses someone his own size. The bright and large child tends to cross the age lines upward, whereas duller and smaller children tend to cross them downward. The effects of social class, however, are important in this connection at all age levels. The study of social class and friendship among school children by Neugarten[14] revealed that the child in the upper social status occupies an enviable position—many of his classmates from a lower status choose him as their friend. The child from the lower status faces the opposite condition—he is seldom chosen as a friend and then only by children from his class status. It is further observed that chums are usually from the same neighborhood; but here again class status may be operating to a marked degree, since people of a somewhat similar class group usually reside in the same or similar types of neighborhoods. The influence of parents in such choices becomes more indirect and perhaps less forceful as the child grows older and his choices reach beyond certain accidental factors that may be in a large measure under the parents' direct control.

Social development during adolescence.[15] With an increase in age the child enters into social situations of greater complexity. This requires not only a modification of existing attitudes, but also a flexibility for adjustment to varying conditions. Often the period of early adolescence is one of great flexibility, followed by an excessive definiteness, which surprises those most familiar with the child. This cocksureness of the adolescent results in part from his desire to attain greater prestige and increased attention. Positive, inflexible attitudes are one of the means for elevating the self and securing attention. In an attempt to generalize the changes occurring between twelve and fifteen, Tyron states:

[13] J. E. Anderson (Chairman), *The Young Child in the Home: A Survey of Three Thousand American Families.* (Report of the committee on the infant and preschool child, White House Conference on Child Health and Protection.) New York: Appleton-Century-Crofts, Inc., 1936.

[14] *Op. cit.*

[15] Ellis Weitzman, *Growing Up Socially.* Chicago: Science Research Associates. This is one of the Life Adjustment Series publications, and offers many suggestions helpful to the adolescent in his quest for social approval and growth toward social maturity.

During the period between ages twelve and fifteen, values for girls have undergone some revolutionary changes; values for boys have undergone relatively minor changes, mainly in terms of slightly shifted emphases. For the twelve-year-old girl, quiet, sedate, non-aggressive qualities are associated with friendliness, likableness, good humor and attractive appearance. Behavior which conforms to the demands and regulations of the adult world is admired. Tomboyishness is tolerated. At the fifteen-year level, admiration for the demure, docile, rather prim, lady-like prototype has ceased. Instead, many of the criteria for the idealized boy such as extroversion, activity, and good sportsmanship are highly acceptable for the girl. The ability to organize games for parties involving both sexes and the capacity to keep such activities lively and entertaining is admired. In addition, the quality of being fascinating or glamorous to the other sex has become important, but is looked upon as relatively specific or unrelated to other desirable qualities. At the twelve-year level, the idealized boy is skillful and a leader in games; his daring and fearlessness extend beyond his social group to defiance of adult demands and regulations. Any characteristic which might be construed as feminine by one's peers, such as extreme tidiness, or marked conformity in the classroom, is regarded as a weakness. However, some personableness and certain kindly, likable qualities tend to be associated with the more highly prized masculine qualities. At fifteen years, prestige for the boy is still in a large measure determined by physical skill, aggressiveness, and fearlessness. Defiance of adult standards has lost emphasis; though still acceptable and rather amusing to them, it tends to be associated with immaturity. In addition, much greater emphasis is placed on personal acceptability, suggesting the effectiveness of rising heterosexual interests. In fact *Unkempt-Tidy*, related to this constellation, is the only trait among the twenty on which the boys completely reversed their evaluation.[16]

In studying the sex differences revealed by these data, one is impressed by the lack of steadfastness to ideals revealed by the girls, as compared with the boys, over this relatively short period of three years. These data tend to support the theory that the behavior of the female of the species is characterized by expediency, design, irresoluteness, and caprice. A plausible explanation for the phenomenon, which appears early in the social development of boys and girls, is that social activities place a greater demand upon girls than upon boys for flexibility, capacity to readjust ideals, and ability to reorient themselves to new goals.

With the development into adolescence, interests and activities, as we have noted, become related to clubs, fraternities, fashion, the spirit of the times, gangs, and the like. Thus, extracurricular activities of the types present in our high school, in harmony with the interests of ado-

[16] C. M. Tyron, "Evaluations of Adolescent Personality by Adolescents," *Monographs of the Society for Research in Child Development*, 1939, Vol. 4, pp. 77–78.

lescents, are coming to be a more and more potent force in meeting their needs. A good example is the "clubhouse" maintained for members of an adolescent study group at the University of California.[17]   The differences revealed in the sexual maturity of boys and girls were reflected in their social interests. The clubhouse had little appeal for most of the boys in the seventh grade; their interests were rather in individual activities or in playground sports. At this same time, girls were playing dance music and practicing the latest steps with one another. By the latter part of the eighth grade, however, mixed dancing had become the favorite clubhouse activity, and certain other interests of a more individualistic nature were rapidly declining, giving way to activities involving a greater amount of social participation with members of the opposite sex.

There is a gradual growth in the socializing process as the child grows in his general personality pattern, owing to experiences in the home, the school, the playground, and other social agencies. Upon reaching the beginning of adolescence the child tends to become more objective in his thoughts and attitudes; his conversation becomes still more social in nature. Moreover, his conversation comes to have more contiguity, definite conversational trends develop, and thoughts are held over for longer periods of time; thus he has greater reasoning power and more verbal continence. It is only when this stage is reached, incidentally, that it becomes possible to develop personal ideals.

Social consciousness during adolescence.   Cooley[18] was one of the first of the modern sociological writers to emphasize that man is dependent upon his fellows in a large measure for his thoughts, emotions, and modes of behavior. This emphasis was formulated under the term *social consciousness.* According to Cooley and other social psychologists, the consciousness of any single individual is nothing more than the consciousness of the many social groups with which he has come in contact. If we consider the average adolescent girl in the junior year in high school, we will find an individual bound by certain group standards, ideals, and general attitudes. The home and playmates have given her lessons in loyalty, service, cooperation, and interest in others. School studies have brought her, through her imagination, into contact with peoples of other countries and with deeds of men of the past. She

[17] H. R. Stolz, M. C. Jones, and J. Chaffey, "The Junior-High-School Age," *University High School Journal,* 1937, Vol. 15, pp. 63–72.

[18] C. H. Cooley, *Human Nature and the Social Order.* New York: Charles Scribner's Sons, 1902.

thus has a wider and deeper appreciation of direct experience. Her religion, her politics, her pride of family and state, and her respect for the opinion of others have been molded by her social group. However, the adolescent is constantly meeting new social groups, many of which have ideals and attitudes somewhat different from those previously met; and here, Cooley points out, conflicts are likely to develop, since the individual's standards as built up through contact with different social groups may not be harmonious. Thus the adolescent upon meeting such a situation is often referred to as "green" or "nutty," or by some other name which would indicate his failure to understand and thus enter into the behavior of the new social group.

PROBLEMS RELATED TO SOCIAL DEVELOPMENT

The desire for conformity. The normal adolescent, though idealistic in his attitudes, is a slave to group conformity. His ardently poetic and religious interests are seldom carried over into everyday activities. If the group frowns upon noble ideals, he will tend to frown upon such ideals; if the group keeps late hours, he is bent upon keeping late hours; and if the group swears and uses slang, he will again follow its pattern of action. There is at this stage the keen desire to follow the herd and to avoid being marked as "different." This attitude of conformity stands out above almost everything else at this period of life.

It has already been suggested that discrepancies in rate of growth may be a source of psychological tension for the less mature individual. This tension will be reflected in his social attitudes and outlook. Also, the individual maturing early may be faced with various social problems. His peers, because of his physical size, will expect certain things of him, but, since he may not have had the social experiences concomitant with physical development, or since physiological maturity may lag behind skeletal growth, he may not be able to meet their expectations, and this failure may become an important source of psychological tension. Girls who mature early and boys who mature late will be considerably out of line with their associates in development. Problems of social conformity are more prevalent among these "misfits," and may become a source of tension and difficulty for them.

Leadership among adolescents. In the study of leadership, it is essential that we recognize the importance of individual variation, not merely in native or acquired intelligence, but in the whole range of physical, emotional, and social variability. Often a prevailing social status may provide opportunity for leadership otherwise not at hand.

In another social dimension, physical force may be important. The control of others, in some types of activities, demands brute strength. For example, leadership in some forms of athletics will usually be found among those superior in strength, motor coordination, and speed of reaction. It has been observed that where men have to impress other men in face-to-face contact, size and strength count for much in producing prestige and control. Social and emotional characteristics may be distinctly important in the development of adolescent leaders in social situations. It is apparent that with the divisions of life activities and the individual divergencies in life organization, the personality of leaders will vary in accordance with the situations in which they find themselves. Leadership in any field is marked by positive characteristics, such as strength, self-assertion, initiative, willingness to assume responsibility, and the like. Leadership is dynamic, even when it is formalized, as is likely to be the case where large groups are involved. However, leadership is a phase of the entire life organization of the individual, and cannot be explained in terms of a series of special habits or talents.

Family influences that create, on the one hand dominating, or on the other, submissive tendencies may play an important role in the development of leadership characteristics. However, a leader may develop from a family exerting repressive influences, if the boy or girl bears native impulses strong enough to provide a drive to offset or compensate for the sense of inferiority arising from this repression. All leadership, however, does not arise from an act of compensation. Families may be so organized as to provide experiences and conditions that will lead to the development of a personality with actual or potential leadership characteristics.

Seventy-one girls, comprising the junior and senior classes of the Horace Mann High School for Girls, New York City, were used as subjects in a study designed to determine which of a large number of psychological traits, presumably associated with personality, are related to the ability to lead.[19] The teachers of these girls were given lists of forty-six traits and asked to check for each girl the item that could be attributed to that particular girl. In addition, each girl indicated on a scale of ten the intensity of pleasant feeling she subjectively associated with every other girl of her class. Each teacher also indicated on a scale the relative amount of personality possessed by each girl. A defi-

[19] E. G. Fleming, "A Factor Analysis of the Personality of High School Leaders," *Journal of Applied Psychology,* 1935, Vol. 19, pp. 596–605.

nite relationship was found between personality and leadership. Adolescent girl leaders are pleasing to their contemporaries. Traits positively and significantly associated with leadership among these girls, in relative order, are: liveliness, wide interests, intelligence, good sportsmanship, originality, athletic prowess, cleverness, sense of humor, culture, and individuality.

Closely related to this study was one conducted by Reals [20] of the graduating classes of eight high schools in Missouri, Oklahoma, and Illinois. He found that, "The pupil leaders had better school attendance and health records, possessed better general appearance, had had more broadening experiences, and participated and led in extramural activities to a greater extent than non-leaders."

Problems in socialization. It was pointed out in Chapter II that social problems were more frequently experienced by adolescent girls than by adolescent boys. Some of these problems will be discussed further in a later chapter dealing with the adolescent and his peers. Too little social life, lack of friends, wanting to learn how to dance, spending money for social activities, the desire for a new dress and the like are problems experienced by a great many adolescents. The dawning social consciousness appearing during the pre-adolescent stage is an important factor in the development of problems of a social nature at this period. Timidity, moodiness, temper outbursts, and daydreaming tendencies frequently result. The case of Louise shows how a combination of factors may operate to affect the socialization of a young adolescent girl. [21]

Louise, a twelve-year-old girl, was much taller than the other girls of her age and grade. Her intelligence quotient of 90 had made it difficult for her to do satisfactory work in school and consequently she was retarded in her school work. Although she was in a grade where most of the children were one year younger, she was unable to do satisfactory work in school, especially in reading. This was accounted for in part by her inferior cultural home background. The poverty of the home did little to overcome an unhappy home situation and tended to make her still more unattractive to the others of her class, since she was usually poorly and untidily dressed.

The teacher recognized her problems and showed a very sympathetic and understanding attitude toward them. Louise recited from her seat entirely and was never called upon to go to the blackboard for fear that this would

[20] H. Willis Reals, "Leadership in the High School," *The School Review,* 1938, 46, pp. 523–531.
[21] This case was cited by the author in *The Psychology of Exceptional Children,* Revised Edition. Copyright 1950 by The Ronald Press Company.

embarrass her. Furthermore, Louise did not like to march in line with the other members of the class. Although the teacher was rather formal in nature in conducting her class work, she was rather lenient in allowing Louise to remain in the room and complete certain tasks while the other students were going out of the room in the line. However, this attitude on the part of the teacher did not solve Louise's problems. For, in fact, the problem was more than one of self-consciousness. This self-conscious-ness had associated with it certain habits and attitudes toward entering into group activities, finally developing into what might be termed a defiant attitude. Louise came to feel that if she did not wish to do certain things that she should be excused from doing them.

At the end of the school year, Louise, now thirteen years of age, was pro-moted to the seventh grade. The seventh-grade teacher was informed of Louise's problems, and was now in a position to profit from some of the well-meaning mistakes of the former teacher. The teacher set as her goal the bringing about of a better social adjustment on the part of Louise. Through visits to the home she was able to enlist some cooperation from the parents. Fortunately for Louise, one of the neighbors employed her to remain with their children, as a 'sitter.' This provided her with some spending money and gave her needed confidence in her ability and worth. At the end of the year considerable improvement was noted in her socializa-tion. However, much guidance and direction is still needed.

Sex differences. Since boys and girls though of the same age do not develop uniformly, there will be a variation in their desires for social experiences. And, although social participation is essential for healthy social development, boys and girls should not be pushed into such experiences before they have reached the stage of social and emo-tional maturity at which they are ready for them. There is, likewise, a corollary to this proposition, namely, that adolescents should not have the activities in which they desire to participate closed to them when they have reached the stage of development suitable to such participa-tion. If social activities are provided periodically by the home, the church, the school, and other organizations concerned with the social development of boys and girls, there will be opportunities for the latter gradually to take part in them. Some of a particular age-group will be mere onlookers on certain occasions, but gradually the desire to play an active role will emerge, and later on they will begin to take some part in social activities; later still, there will probably be full-fledged and unanimous participation.

SUMMARY AND GENERALIZATION

Concerning the general nature of social growth Catharine Conradi has written as follows:

Developmentally, the child proceeds through a series of stages—from liking to be in the presence of human beings, to experimenting with the qualities and reactions of human beings; from watching a person do something, to trying to do the same thing; from doing the same thing as another person, to understanding that people may have a common objective as they do the same thing; and from this, to the final step of understanding that people may further a common objective by doing different things—i.e., by sharing effort and by specialization of effort.

The child ultimately comes to realize that a group objective can be satisfying to both the individual and the group. But he must always work at the level of his own ability and his own insight, whether he is under the guidance of the nursery-school teacher, the elementary-school teacher, or the high-school teacher.[22]

It has been emphasized throughout this chapter that development into adolescence is accompanied by an increased interest in personal appearance, one's peers, and social activities involving members of the opposite sex. The extent to which the adolescent is able to make satisfactory social adjustments will depend upon a number of factors, including the operation of the developmental process during the earlier years. This is in harmony with the statement quoted from Conradi, which emphasizes the concept that growth and development is gradual in nature. There is, therefore, a necessity for boys and girls to have had opportunities to develop through social participation at the various stages in life. Such participation must always be on the level of the child's maturity and past experiences, if it is to be most effective.

In Chapter VII it was pointed out that with the onset of adolescence there is an increased interest in participating in such activities as clubs, team games, and so forth. If adolescents have had the opportunities for normal development under favorable guidance, they will constantly seek the companionship of members of the opposite sex as well as of their own. Social qualities become quite pronounced in speech, conduct, and common motor expressions. In the development of a social being there must, of course, be contact with others, but some other elements are essential, such as: (1) some important activity in common, for example, a language, symbol, creed, or aim; (2) the effect of suggestion by the activities of others; (3) an acquaintance, unity, or some general interfeeling and intercommunication.

---

[22] Catharine Conradi, "Participating in Shared Child-Adult Activities," *Fostering Mental Health in Our Schools*. 1950 Yearbook of the Association for Supervision and Curriculum Development, pp. 161–162.

THOUGHT PROBLEMS

1. What do you consider the most important factors affecting a child's social development? Just what do you understand the term *social development* to mean?

2. What type of activities characterize the play and recreational lives of ten-year-old boys? Of ten-year-old girls? Of thirteen-year-old boys? Of thirteen-year-old girls?

3. Consider some chum or pal of yours during your early teen years. What factors entered into your selecting the particular person as a pal? Is this in harmony with the materials on this problem presented in this chapter?

4. List in order of seriousness five or six problems connected with socialization.

5. How is language related to social development? Will language training speed up social development or the socializing process? Explain.

6. Just what is meant by social intelligence? What are some evidences of the presence of a high degree of social intelligence?

SELECTED REFERENCES

Anderson, J. E., *The Psychology of Development and Personal Adjustment*. New York: Henry Holt and Co., 1949, Chaps. XIII and XIV.

Havighurst, R. J., and Taba, Hilda, *Adolescent Character and Personality*. New York: John Wiley and Sons, 1949.

Hurlock, E. B., *Adolescent Psychology*. New York: McGraw-Hill Book Co., 1949, Chap. V.

Murphy, L. B., *Social Behavior and Child Personality*. New York: Columbia University Press, 1937.

Partridge, E. De A., *Social Psychology of Adolescence*. New York: Prentice-Hall, Inc., 1938.

Sadler, W. S., *Adolescence Problems*. St. Louis: C. V. Mosby Co., 1948, Chap. XVII.

Wile, I. S., *The Challenge of Adolescence*. New York: Greenberg, 1939, Chap. XII.

Zachry, Caroline B., and Lighty, Margaret, *Emotions and Conduct in Adolescence*. New York: Appleton-Century-Crofts, Inc., 1940, Chaps. V and VI.

For rather complete reviews of studies in this area see, "Growth and Development," *Review of Educational Research,* 1947, Vol. 17, No. 5; *Review of Educational Research,* 1950, Vol. 20, No. 5.

IX

Growth in Attitudes and Religious Beliefs

THE EXTENSION AND MODIFICATION OF ATTITUDES

The development of attitudes. The term *attitude* has been adopted to express a phase of development of a more highly integrated nature than that of factual learning. Thurstone, a number of years ago, defined an attitude as: "The sum total of man's inclinations and feelings, prejudice or bias, preconceived notions, ideas, fears, threats, and convictions about any specified topic. Thus a man's attitude about pacifism means here all that he feels and thinks about peace and war. It is admittedly a subjective and personal affair."[1]

Attitudes always relate to situations around which we have constructed various habit patterns and built up various images and concepts; it has been constantly observed that physical and social contacts result in the establishment of conscious adjustments and reaction tendencies. The child born and reared in a social world is continually subject to ever-changing social stimuli; socially, he becomes what his environment has made him. He develops attitudes toward objects and persons, and through such attitudes brings himself into adjustment with his world.

As boys and girls mature their attitudes and beliefs develop and change, a result of the influence of their families, community mores, religion, and peer influence. In a study of the moral beliefs of the youth of Prairie City, Havighurst and Taba[2] report that class structure effects the adolescent's sense of honesty and responsibility, but is less effective as far as loyalty, moral courage, and friendliness are involved. There is good evidence from this and other studies that social class

---

[1] L. L. Thurstone, "Attitudes Can Be Measured," *American Journal of Sociology,* 1928, Vol. 33, p. 531.

[2] R. J. Havighurst and H. Taba, *Adolescent Character and Personality.* New York: John Wiley & Sons, 1949, p. 95.

status is more important in the formation of attitudes and beliefs than is usually suspected. Attitudes and beliefs are "soaked up" from the milieu in which the child develops. The child of lower-class status is usually faced with the limitation of not being wholly accepted by those of his age level, or if accepted he is given a rather minor role in their activities. Also, he is limited in the type and in the scope of the activities in which he can participate with the group, since most of these involve some expense. Thus, the individual growing up in a lower-class family spends more of his time at home and is less influenced by the activities and by the code of the group than is the adolescent from the upper-lower or middle-class groups.

Change of attitudes. It has already been suggested that attitudes are "determiners of behavior," and that they develop out of social experiences. There is evidence that the deep-seated attitudes acquired early in life are not changed to a marked degree by later experiences, and that, when such changes do occur, they are more temporary in nature than changes in those attitudes not so deeply rooted or even for those acquired at a later period. However, intelligence, educational environment, and years of schooling are positively related to changes in individual attitudes when such changes are in harmony with reason and understanding rather than with emotions and feeling states. A study by Clem and Smith in which a questionnaire involving 15 items was administered to 1,172 secondary school pupils throws further light upon this problem. This study was designed to determine the attitude of high school pupils of different grade levels to certain moral situations. Some conclusions reached from this study are:

1. The attitude of pupils toward such personal habits as swearing, drinking, gambling, and playing cards on Sunday becomes more tolerant in succeeding grades of the six-year secondary school. In general, the reverse is true for cheating, lying, conceit, vulgarity, selfishness, gossip, and extravagance.

2. In terms of "badness," stealing is uniformly considered by all grades the worst of all items studied, and dancing the least "bad."

3. It is evident throughout the study that lower grade pupils are more inclined to make choices on the basis of indoctrination than are upper grade pupils. Upper grade pupils exhibit better social discrimination and judgment, and less ingrained respect for the law.

4. In terms of law observance, upper grade pupils are more inclined to substitute personal judgment for blind obedience: the spirit for the letter of the law.[3]

[3] O. M. Clem and Marcus Smith, "Grade Differences in Attitudinal Reactions of Six-Year Secondary School Pupils," *Journal of Educational Psychology,* 1934, Vol. 25, p. 308.

The effects of puberty. In order to determine the effect of the menarche,[4] Stone and Barker[5] studied the interests and attitudes of 1,000 girls of two large junior high schools of Berkeley, California. It was necessary to include all the children of the age range in which from one and one-half to two years' difference in menarche appeared. These girls were matched with respect to chronological ages and social status, but were significantly different in physiological development—the one group being considered postmenarcheal and the other premenarcheal. From this study it was found that postmenarcheal girls favor the interest and attitude items that are more mature in nature to a greater degree than the premenarcheal. A greater proportion of postmenarcheal than of premenarcheal girls of similar chronological ages favor those responses that indicate an interest and favorable attitude toward the opposite sex. The postmenarcheal girls were more interested in adornment and display of the person than were the premenarcheal. The postmenarcheal girls, according to their responses, engaged in daydreaming and imaginative activities of such types to a greater degree than did the premenarcheal. There was no noticeable difference found in the extent to which the two groups rebelled against or came into conflict with family authority. The postmenarcheal girls indicated less interest in participation in games and activities requiring vigorous activity. These comparisons indicate a growing interest in adult activities, an increased independence, and an increased interest in the opposite sex, as a result of forces associated with the menarche.

Probably the most striking feature of development at this stage is the psychological differentiation of the sexes. Stolz, Jones, and Chaffey have described this as follows:

The girl feels a necessity to prove to herself and to the world that she is essentially feminine; the boy needs to demonstrate that he has those masculine qualities which will require others to recognize him as a man. This characteristic accounts for the girls' spending a large part of their leisure time in shopping and in personal adornment. This is the secret of the manicured nails, painted red to match vivid lips. This is why they must wave and curl their hair, and, having perfected the process, must pin into it ribbon bows, bits of lace, or flowers. This is the reason for the boy's urge to learn to drive a car and for his willingness to move heaven and earth to

[4] Onset of the menstrual period denotes neither the beginning nor the end of the pubescent period. Even prior to the menarche, there are some noticeable changes present in the contour of the girl.

[5] Calvin P. Stone and Roger G. Barker, "The Attitudes and Interests of Premenarcheal and Postmenarcheal Girls," *Journal of Genetic Psychology*, 1939, Vol. 54, pp. 27–72.

borrow or own one. Along with this development, also, we are told by our group that a girl to be popular must be modistly pretty, keep herself clean and neat, be a good mixer. A boy, on the other hand, must be aggressive and must excel at sports. He must have the ability to dance and to talk easily with girls, and in addition he must show that he can compete readily with other boys; that he can achieve and master. This picture of adolescent development is often disturbing to adults, but it should be reassuring to know that, once the girl has arrived at the status in the group to which she has aspired, or has learned to adjust herself to a version of the universal feminine model which suits her own personality, she will be a happier person and a pleasanter one to teach or to have around the house. Likewise, once the boy feels that he is accepted as a man, he can go on with the important business of preparing himself for a job or for college. We have repeatedly noticed that those boys and girls who have acquired some understanding of their personal relations to others and have made a place for themselves in a mixed group, have become more stable and predictable.[6]

The home and attitudes. There is a rather widespread notion that the youth of each generation revolt against the ideas of their parents and of their parents' generation. A study by Remmers and Weltman[7] was undertaken to gather data bearing on such a hypothesis. The *Purdue Opinion Poll for Young People* was available for gathering data in this study. A representative sample of 88 sons, 119 daughters, 207 fathers, 207 mothers, and 89 teachers from ten school communities in Indiana and Illinois were available for the study. Comparisons and interrelations were obtained between the responses of parents and children, daughters and sons, and teachers and pupils. These comparisons showed that a high degree of community of attitudes exists between parents and children. The strength of this relationship varied, however, with the general nature of the attitude object. There was a closer relationship between mother and father than between the parents and children. Likewise, daughter and son attitudes agreed more closely than those of parents and children. This reveals some tendency for those of one generation to agree better than individuals a generation apart in age and outlook. However, the close correlation between the attitudes of parents and those of children and the lower correlation between teachers and pupils indicate that home influences are most important factors affecting the attitudes of adolescent boys

[6] H. R. Stolz, M. C. Jones, and J. Chaffey, "The Junior-High-School Age," *University High School Journal*, 1937, Vol. 15, pp. 63–72.

[7] H. H. Remmers and N. Weltman, "Attitude Inter-Relationship of Youth, Their Parents, and Teachers," *Journal of Social Psychology*, 1947, Vol. 26, pp. 61–67.

and girls. There are many indications that this similarity is greater for pre-adolescents than for adolescents and post-adolescents, although early home attitudes tend to persist to a very marked degree into later years.

A poll of the ideals of 1,526 seventh and eighth graders in the public schools of Massachusetts was conducted by L. A. Averill.[8] The movies and radio, although alluring for many, do not have the appeal sometimes attributed to them. Only 14 per cent chose their ideals from the movies or radio, the radio influencing the boys and the movies influencing the girls more. Historical characters headed the list. About one out of three made a choice from this group. A total of 268 boys expressed the desire to be like some figure in sports; the number of girls listing a figure in sports as their ideal was significantly smaller. Only 12 of the total group questioned named a religious figure as an ideal. The subjects were also asked to name a person whom they wanted to be like 10 years from now. About one out of five named an occupation rather than a person. The most popular occupations named by the boys were airplane pilots, musicians, bankers, physicians, engineers, actors, tradesmen, and writers or reporters. More than one-third of the girls listing an occupation specified teachers or nurses. Other occupations listed in order of frequency were columnists, musicians, writers, and reporters, secretaries and stenographers, and airplane hostesses.

Intelligence and attitudes. Intellectual maturity, as an integral part of the total maturity of a growing child, is accompanied by pronounced changes in attitudes. As the child grows into adolescence, he becomes more discriminating in the choice of friends. At this time prejudices formed earlier as a product of home and neighborhood contacts become more generalized. Attitudes take on a fuller meaning and reveal an increased complexity. Many things of an abstract and nonpersonal nature become more significant and personal. Proof of a close relation between intelligence and the development of social attitudes and habits is demonstrated in studies of this problem.[9] The *Furfey Developmental Age Test* was given to 26 boys, median age 11 years, and to 24 girls, median age 11 years, of superior intelligence. There was a con-

---

[8] L. A. Averill, "Modern Youths' Ideals," *Science News Letter,* April 23, 1940, p. 263. This is a report of a paper presented at the meeting of the Eastern Psychological Association in Springfield, Massachusetts.

[9] R. L. Thorndike, "Performance of Gifted Children on Tests of Developmental Age," *Journal of Psychology,* 1940, Vol. 9, pp. 337–343.

siderable variation in the maturity shown for the different items. The highest maturity was revealed on the items concerned with choice of books to read, future vocations, and things to think about.

The effects of movies and radios on attitudes. The various studies that have been made dealing with the influence of the motion picture on children's attitudes indicate that this form of entertainment may be a potent force in conditioning or reconditioning certain attitudes. These studies indicate that changes brought about in this way are not wholly temporary, but tend to persist.

As children approach adolescence, they become more conscious of characteristics regarded as distinctively feminine or, conversely, distinctively masculine. This new discrimination may be looked upon as one phase of their developing social consciousness, or social development, and Hoban noted that it was reflected in their rating of films involving the cultural status of men and women. He had pupils rate the film *The Truck Farmer* on the scale from 1.00 to 5.00, the highest rating being 1.00 and the lowest 5.00. In the fifth grade class, the boys gave the film an average rating of 1.37; the girls, an average rating of 1.36. The average rating given the film by the eighth grade boys was 1.54; the average by the girls, 1.61. The tenth and eleventh grade boys gave the film an average rating of 1.57; the girls gave it an average rating of 1.83. These differences may be accounted for by the growing awareness of boys and girls in things and activities relating to the cultural and social patterns of men and women. This film related to men's activities involving crating, shipping, farming, and the like. According to Hoban,[10] when the film reveals activities common to both sexes, boys prefer that these activities be presented from the boy's point of view, whereas girls prefer that they be presented in terms of a girl's interests and attitudes. Also, in cases where the film shows machinery, industrial processes, educational and scientific procedures, and activities requiring stamina, strength, and endurance, all ordinarily associated with men, significant sex differences in responses are revealed.

THE ADOLESCENT AND HIS RELIGION

Religion during adolescence. Various attempts have been made to relate the religious activities of man to instinctive tendencies. The religious activities so universally present have apparently developed out of a medley of impulses, such as fear, assertion, sex, and the developed

[10] C. F. Hoban, *Focus on Learning.* Washington: American Council on Education, 1942.

desires and interests of the individual. These impulses, some of which are outgrowths of native impulses, become integrated as drives in the intellectual and social habits of man.

The average adolescent today, when confronted with the popular question "What is religion?" may give any number of strange and incoherent answers. There is little likelihood that any two young people will give an identical definition. Strange though it may seem, this is to be expected. Religion goes beyond a mere definition to be mechanically learned and carried from generation to generation. We find, however, that there are certain fundamental principles and concepts upon which the religious experience of the adolescent is based.

Studies have been made of the religious development of adolescents from adolescent diaries, letters, and poems; and these, together with results from questionnaires, have given valuable materials relative to the development of the religious self.[11] Little can be learned from a study of the religion of childhood, as such, since felt and understood religious experiences do not ordinarily appear until puberty. The religious community and the temperament of the individual determine whether the development shall be continuous or catastrophic and leading to conversion. Factors such as sex, nature, and love influence religious development, but it cannot be said that the development is exclusively determined by them. Some of these forms affecting the religious nature of adolescents are given special consideration in this chapter.

Comparatively little is known about the adolescence of Christ. According to the description presented in the New Testament, Jesus experienced a very early and definite emergence into what we call adolescent independence. At the age of twelve, Christ went with his parents to one of the Jewish festivals at Jerusalem. The account states that He remained behind, engaged in a discussion with the religious teachers in the temple, while His parents assumed He was following with the other children.

This description of the experience of Jesus tarrying behind to discuss religion with the temple teachers is an illustration of how early the mind of youth may turn to nonmaterial phases (perhaps some philosophies) of life. Youth is confronted with the problem of making the transition from the religion of childhood to that of his elders, which

---

[11] For a study of religious development from diaries, letters, and so forth, see O. Kupky, *The Religious Development of Adolescents.* (Tr. William Clark Trow.) New York: The Macmillan Company, 1928.

is to become a personal possession, an internal experience. To the child, God is away off somewhere; to the developing adolescent, He becomes an internal presence.

In one investigation a group of girls between the ages of 15 and 17 were questioned about when and where they first experienced a feeling of reverence.[12] The meaning of reverence was made clear to them before they were presented with this assignment. During this period the project was so developed that the girls freely gave this information and recognized that such experiences did not need to relate to a certain formal religious creed or program. Of the 148 girls questioned, 22 stated that they had never had any emotional experience that could be called "reverence"; 68 girls stated that such feelings arose at a time when they suddenly came to realize the beauty and wonder of nature; only 31 girls, or 21 per cent, reported their first feeling of reverence or awe to be connected with some religious observance. Some of the girls expressed reverence toward some person they had known—in some cases this approached what is usually termed pity—while others reported reverence toward special types of music or some masterpiece of art. The results suggest that reverence tends to be aroused toward anything that is impressive, beautiful, or extremely thought- and emotion-provoking.

The period of conversion. Turning to the transition from childhood to adolescence, one finds some important religious significances. The general development of the child is complicated in nature and is conditioned by many factors, among which are the development of the original tendencies charged with their incoherent energies, and also a constantly growing stock of energy seeking an outlet. The child develops in an environment that perpetually provides material for the formation of complexes of all sorts that are more or less an outgrowth of original tendencies. At the same time the environment establishes a mental conflict between purely egoistic impulses and sex on the one hand, and various growing social habits on the other.

The adolescent period is characterized by various physiological changes that have very definite influences on the individual's psychic development. This period has already been described as one in which there is manifested a marked expression of self-consciousness, as well as a marked development of social consciousness. This development of a social consciousness, during which the child comes to be looked

[12] O. Kupky, *op. cit.*

upon as a social rather than as an egocentric individual, tends to follow naturally the realization of life's purposes and the consciousness of perfected physical and mental powers. "In cases of normal development the religious teaching and impressions of childhood now come to a head, and are invested with a reality and significance they formerly lacked." [13]

This period often represents a crisis—a development from the earlier years in which religious ideas are only half understood and are concrete in nature to a natural and healthy growth into habit patterns involving a more definite religious awakening. This growth, if the individual has been supplied with religious surroundings of a wholesome but nondominant type, will be gradual and become more intensified in feeling and more vital and real in its issues and meaning. This is a process of religious growth by education, and is to be preferred to religious development of a "storm and stress" nature. The latter type of religious experience is accompanied by vivid emotional experiences. The individual has had painted for him a dramatic picture filled with emotional stimuli, and this picture tends to establish morbid fears and extreme shame, as well as a feeling of guilt. A feeling or sense of sin may be established in the individual who has actually lived a normal healthy life, and this sense of sin is often connected with sex development. The individual is given distorted ideas of the relation between the self and God, and comes to feel that he is an outcast and has fallen wholly from the path set by God. Here we find a real contrast with the former case, in which the child has always been given a wholesome yet nondogmatic view of life and God. Young folk who have developed balanced habit systems under proper guidance will often confess that they never have realized that they were true sinners, and see no reason why they should either resort to trembling penitence in order to be saved from their past wrongdoings or give way to morbid fears.

Not always do we have presented a picture of the convert who has passed through a storm and stress period. If the individual is awakened and stimulated to further thought and activity with a positive emphasis on new loyalties, the group welfare, and proper habits of conduct, there will probably be a more healthy and balanced growth in the social, educational, and religious life. It is when the negative emphasis, in which the sins of the past are recounted and the natural

[13] W. H. Selbie, *The Psychology of Religion*. London: Oxford University Press, 1926, p. 176.

sex and the various social tendencies are criticized so vehemently, that we find morbid fears developing and becoming prime factors in the development of emotional instability and perversions. Adolescent boys and girls are susceptible to religious appeals. Statistics of conversion as well as various testimonies, however, show that girls are more affected by the emotional appeal in religious life, whereas boys are more attracted by codes of honor, ethical sanction, and group activity.

Youth and the church.[14] To arrive at sound conclusions about the part the church is playing in the lives of young people is not a simple task. The obvious difficulties are aggravated by the fact that it is impossible to isolate the church as a single factor in the experience and background of youth. It is quite possible, of course, to discover the conditions under which the youth of different church groups are living, and also to find out whatever differences may exist in the ways that they react to current problems. However, to presume to measure the extent to which these differences are due to dissimilarities in religious backgrounds and affiliations is not only unscientific but highly dangerous.

It is one thing to suggest that certain variations in concepts and attitudes are associated with such religious groups as Protestants, Catholics, and Jews, but quite another thing to insist that these dissimilarities are directly the result of different church affiliations. For example, almost 20 per cent of the youth from Protestant homes were Negroes. Thus, what may appear on the surface to be a distinctly religious factor turns out to be influenced by the factor of race. Of 35 per cent of the youth from Catholic homes, either the mother or the father or both were foreign-born. This means, of course, that ethnic as well as religious backgrounds contribute to whatever differences may appear in the Catholic and non-Catholic groups. Similarly the attitudes and the conditions of the Jewish youth are, without doubt, considerably influenced by the facts that 84 per cent of their parents were foreign born (more than half of them came from Russia) and that their

---

[14] Howard M. Bell, *Youth Tell Their Story*. Maryland was chosen for a study of youth because it was believed that it presented in miniature form the major economic and social characteristics of the nation. The study is a forceful analysis of what young people are doing and thinking based on personal interviews with more than 13,500 young people between the ages of 16 and 24. Chapter VI gives some very interesting and valuable information dealing with youth and the church. Most of the materials presented under this topic are taken from the Maryland survey.

median grade attainment was about two grades higher than that of the youth in any other religious group.

Thus it is that differences which, on the surface, may appear to be basically religious in character are, in fact, profoundly affected by such factors as race, nationality, locality of residence, and educational attainment.

It is of considerable interest to note the extent to which youth from homes of different religious backgrounds tend to accept the religion of their parents. Over four-fifths (81.1 per cent) of the youth with some church affiliation had adopted the faith (Protestant, Catholic, or Jewish) of both their parents. When both parents had church affiliations, but when there was a difference between the persuasion of the father and mother, there was more than twice as strong a tendency to accept the faith of the mother. The proportion of youth who had adopted a belief different from that of either parent is quite negligible: 4.2 per cent for the Catholic youth, 2 per cent for the Protestants, and none for the Jewish.

Like many other activities, the matter of church membership seems closely related to the population density of the various areas. Table XXIV indicates that church membership becomes more general as the

TABLE XXIV

CHURCH MEMBERSHIP OF YOUTH ACCORDING TO LOCALITY OF RESIDENCE

LOCALITY OF RESIDENCE	PERCENTAGE WHO CONSIDER THEMSELVES MEMBERS
Farm	59.2
Village	64.7
Town	74.4
City	80.2

population of the area increases. This can hardly be taken to mean that there is something peculiarly devout about young people living in cities, and something peculiarly otherwise about youth living on farms. The smaller proportion of farm youth who said they were members of some church may quite possibly reflect the comparative inaccessibility of churches in certain rural areas. Moreover, the 80.2 per cent of church membership of youth living in cities also reflects the fact that most of the religious group with the highest degree of membership, that is, the Catholics, were city youths.

The *Fortune* survey of 1942[15] shows that there is very little agnos-

[15] "Fortune Survey," *Fortune,* December 1942, p. 18.

ticism among high school students. Only 6.6 per cent of the students stated that they did not believe in a God who punished after death, nor in life after death. Furthermore, these young people were much better churchgoers than their elders. The results at that time may have been beneficially affected by the world war then in progress. These data are presented in Table XXV.

TABLE XXV

RESPONSE OF HIGH SCHOOL STUDENTS TO THE QUESTION "ABOUT HOW OFTEN DO YOU GO TO CHURCH AS A USUAL THING?"

FREQUENCY	ALL STUDENTS	BOYS	GIRLS	FRESHMEN	SENIORS
Weekly or more often	56.5%	49.6%	63.5%	61.4	48.8
Two or three times a month	22.0	24.3	19.5	19.1	24.6
Monthly	7.8	9.2	6.4	5.7	10.1
Less often than monthly	8.6	10.0	7.3	8.4	10.8
Do not attend church	5.1	6.9	3.3	5.4	5.7

RELIGIOUS BELIEFS AND PROBLEMS
OF ADOLESCENTS

Change of religious beliefs during adolescence. Growth has been described throughout the previous chapters as a gradual and continuous process. New experiences and maturity bring about an enlargement and a reorganization of old concepts. Thus, as the child grows, his concept of God takes on an added and to some degree changed meaning; his understanding of his relationship to God becomes more inclusive and less concrete in nature; and his concept of the "brotherhood of man" is enlarged in scope. These changed concepts are a part of the development of the *total self,* and affect changes in emotional, personal, and social behavior patterns. These beliefs are closely related to one's philosophy of life and to standards for evaluating character and conduct.

In the study of the religious beliefs and problems of adolescents by Kuhlen and Arnold [16] a questionnaire was prepared that listed 52 statements representing various religious beliefs. The subjects completing the questionnaire were instructed to mark each statement according to

[16] R. G. Kuhlen and M. Arnold, "Age Differences in Religious Beliefs and Problems during Adolescence," *Pedagogical Seminar and Journal of Genetic Psychology,* 1944, Vol. 65, pp. 291–300.

whether he *believed it, did not believe it,* or was uncertain and *wondered about it.* Responses were secured from 547 (257 boys and 290 girls) sixth, ninth, and twelfth grade pupils. These three groups were chosen since the sixth grade group is largely prepubescent (especially boys, many girls are pubescent or near pubescent), the ninth grade group pubescent, and the twelfth grade group postpubescent. Approximately three-fourths of the pupils were Protestants, 22.85 per cent were Catholic, and several were either of Jewish faith or indicated no church attendance.

The findings were analyzed by determining what proportion of each group checked each statement indicating the nature of their belief regarding that statement. The results from this study are analyzed in Table XXVI. It is evident from the results here presented that significant changes appear in the religious beliefs of boys and girls as they reach adolescence and grow into maturity. Of the 52 statements included in the study statistically significant changes appeared in 36 of them. A pronounced change appears in the attitude toward the scriptures as shown by the responses to the statements, "Every word in the Bible is true," and "It is sinful to doubt the Bible." The responses to a number of the statements provide evidence for the assumption that a greater tolerance toward different religious beliefs and practices appears with increased maturity of the growing individual.

Factors related to adolescent beliefs. It has already been suggested that the pre-adolescent has accepted quite completely the beliefs he has been taught in the home and by religious teachers. However, as early as twelve or thirteen some degree of doubt and oftentimes opposition begins to be manifested, as is revealed in Table XXVII. Certain investigators have sought to determine the factors related to the development of religious beliefs. MacLean found a negative correlation between chronological age and accepted beliefs.[17] The revolt of girls against accepted beliefs comes at a later age than that of boys and is found less frequently and is probably not as inclusive. The general relationship between the educational and economic level of the home and the acceptance of definite religious beliefs is negative, as is also that between intelligence and acceptance of beliefs.

[17] A. H. MacLean, "The Idea of God in Protestant Religious Education," Teachers College, Columbia University, *Contributions to Education,* No. 410, 1930. Also see Luella Cole: *Psychology of Adolescence.* New York: Farrar and Rinehart, 1936, pp. 171-173. This material is used by permission of the Bureau of Publications, Teachers College, Columbia University.

TABLE XXVI

Changes in Specific Religious Beliefs During Adolescence as Shown by
the Percentage of 12, 15, and 18 Year Old Children Who Checked
Various Statements Indicating (a) Belief, (b) Disbelief, or
(c) Uncertainty (Wonder) (*After Kuhlen and Arnold*) *

STATEMENT	BELIEVE			DISBELIEVE			UNCERTAIN		
	12	15	18	12	15	18	12	15	18
God is a strange power working for good, rather than a person	46	49	57	31	33	21	20	14	15
God is someone who watches you to see that you behave yourself and who punishes you if you are not good	70	49	33	18	37	48	11	13	18
I know there is a God	94	80	79	3	5	2	2	14	16
Catholics, Jews, and Protestants are equally good	67	79	86	9	9	7	24	11	7
There is a heaven	72	45	33	15	27	32	13	27	34
Hell is a place where you are punished for your sins on earth	70	49	35	16	21	30	13	27	34
Heaven is here on earth	12	13	14	69	57	52	18	28	32
People who go to church are better than people who do not go to church	46	26	15	37	53	74	17	21	7
Young people should belong to the same church as their parents	77	56	43	13	33	46	10	11	11
The main reason for going to church is to worship God	88	80	79	6	12	15	4	7	6
It is not necessary to attend church to be a Christian	42	62	67	38	23	24	18	15	8
Only our soul lives after death	72	63	61	9	11	6	18	25	31
Good people say prayers regularly	78	57	47	9	29	26	13	13	27
Prayers are answered	76	69	65	3	5	8	21	25	27
Prayers are a source of help in times of trouble	74	80	83	11	8	7	15	10	9
Prayers are to make up for something that you have done that is wrong	47	24	21	35	58	69	18	17	9
Every word in the Bible is true	79	51	34	6	16	23	15	31	43
It is awful to doubt the Bible	62	42	27	18	31	44	20	26	38

* Discrepancies between the total of "Believe," "Not Believe," and "Uncertain" and 100
per cent represent the percentages who did not respond to the statements.

All religions seem to have as their central idea a belief in a supreme
being. For example, Christianity, with which most Americans are
closely identified or at least familiar, teaches belief in God as the

heavenly father of mankind. Religious leaders have held through-out the ages that certain ethical or moral values attend one's concept of God. Thus, the question arises: What relationship exists between

TABLE XXVII

Per Cent of 13-Year-Old Children Marking the Statements True (N–646) (*After MacLean*)

	Per Cent
1. Religion consists of obeying God's laws	70
2. God is simply imagination	21
3. We learn about God through dreams and visions	28
4. God made us, the animals, the stars, and the flowers, and everything in the world	82
5. God knows everything we say or do	78
6. God cares what we do	89
7. God has a good reason for what happens to us, even when we cannot understand it	92
8. God protects from harm those who trust him	70
9. God cares whether we repent of our sins or not	82
10. God hears and answers our prayers	85
11. True prayer consists of thinking of the wonderful ways of God in the world	66
12. It is possible to get things by prayer	31
13. The soul lives on after the body dies	71

one's conception of God and one's behavior? In order to throw some light on this problem, Mathias constructed the *Idea-of-God Test*. The purpose of the test as expressed by Mathias was "(1) to draw out an individual's social attitudes concerning God, and (2) to crystallize the person's viewpoint of God from the angle of available information regarding the universe and its mysteries, as we conceive them." [18]

Correlations were obtained between sixteen factors, referred to as background factors, and composite Idea-of-God scores. Correlations were positive but low, the highest being between moral knowledge and scores on the Idea-of-God test. Correlations were also obtained between fifteen factors, referred to as behavior patterns, and composite Idea-of-God scores. All correlations were again positive but low, the three highest having as second members high motives (.28), self-functioning (.25), and school deportment (.21). It appears, therefore, that certain background factors, as well as personal factors, tend to be

[18] Willis D. Mathias "Ideas of God and Conduct," Teachers College, Columbia University, *Contributions to Education,* No. 874, 1943, p. 43.

associated with high composite concepts of God. However, this does not mean that one is the cause of the other; but rather "that what have been designated as desirable concepts of God tend to be found in those pupils who come from homes with church affiliation and who have a good cultural background, and in individuals of high intelligence, moral knowledge, and social attitudes." [19]   Also, the results support the notion that high motives in conduct, independent action, and church and club participation tend to be associated with desirable concepts of God.

Adolescent doubts. Many adolescents, especially those whose early training has been dogmatic in nature, become very skeptical of all problems not concrete and not specific in nature. As the growing, developing youth increases his realm of knowledge and develops better habits of thinking, he is led to question many of the things he had formerly accepted uncritically. The youth coming into contact with more of life's realities assumes more mental and moral independence. He is thrown upon his own initiative and required to make decisions for himself. He therefore develops habits of thinking and analyzing on the basis of fact. He comes to learn that many of the things he had been taught earlier and had accepted uncritically are not in harmony with the facts presented at school or in his everyday readings. Early faith, so firmly entrenched, thus receives a serious setback when the child learns that the answers to many of his questions are not based upon almost obvious facts.

This critical attitude develops according to the developmental viewpoint as presented throughout this study. It has its beginning with the first observations of the child that the things he has been taught are not wholly in harmony with facts observable in later life. New and broader experiences often aid in destroying faith in other early teachings. This destruction of early faith continues with the acquirement of certain scientific principles which are out of harmony with early learning. Thus the development of doubt continues and finds further support in the behavior and attitude assumed by those who have a powerfully suggestive influence over the life of the subject.

Functional peculiarities of beliefs and attitudes are at this stage of life quite prevalent. The adolescent may desire to stay away from church for some social reason; therefore he comes to doubt the value of the work of the church as well as the general honesty of the leaders. This doubting may serve further to effect the satisfaction of a desire

[19] *Ibid.,* p. 75.

that has been blooming, or justify some need already existing. During adolescence there are usually several elements in the situation that combine to augment doubts extremely.

How should doubts be treated? In the first place, it should be recognized that doubting is not confined to the religious sphere of life. Neither should anyone be misled into believing that doubting is a universal trait and therefore similar in nature to an instinctive form of behavior. The adolescent does not need a dogma or creed to anchor on: his need is to find himself, and to interrelate in his own thinking the processes of the universe with the general plan of life. An anchorage in open sea in a storm is analogous to the type of treatment usually given the individual during this stage. But the first essential in helping the individual to find himself is intellectual honesty. Of course, facts and knowledge should be gathered in harmony with individual needs and interests.

Adolescent problems involving religion. In some cases adolescents develop a peculiar state of hyperconscientiousness. Conscience, instead of being a friendly adviser, turns into an inquisitive persecutor and seems to devote itself to the task of producing an increasing sense of guilt, aggravated by serious doubts. One of the phases of the study by Kuhlen and Arnold dealt with problems involving religion.[20] Each student was asked to respond to 18 problems by encircling an *N,* an *S,* or an *O,* depending upon whether the particular problem *never* bothered him, *sometimes* troubled him, or *often* troubled him. The results for the three age groups are presented in Table XXVIII. There were no differences in the average number of problems checked by the three groups. Thus, these findings do not substantiate the hypothesis commonly presented that adolescence is an age with increased religious problems. However, certain age trends were noted with respect to specific problems. A study of the mean problem scores for the Catholic groups showed that Catholic boys and girls had lower scores in both "wonders" and problems than did the non-Catholics.

Certain age trends may be observed from a further study of Table XXVIII. More than 50 per cent of the 18-year-old group indicated that the following problems troubled them often or sometimes: dislike church service, failing to go to church, getting help on religious problems, wanting communion with God, wanting to know the meaning of religion, Heaven and Hell, sin, conflicts of science and religion, and wondering what becomes of people when they die. A rather

[20] R. G. Kuhlen and M. Arnold, *op. cit.*

significant change was found for those problems that are checked by an asterisk. The most pronounced change with age was in response to the problem of disliking church services. Youth is naturally skeptical, and sometimes its doubts become very disturbing. The apparent conflict

TABLE XXVIII

FREQUENCY WITH WHICH PARTICULAR RELIGIOUS PROBLEMS EXIST AT VARIOUS AGES THROUGH ADOLESCENCE AS SHOWN BY PERCENTAGE OF DIFFERENT AGE GROUPS WHO CHECKED EACH PROBLEM AS SOMETIMES OR OFTEN PRESENT

PROBLEM	Age 12	Age 15	Age 18
Having a different religion from other people	34	25	27
Disliking church service	35	47	60*
Being forced to go to church	30	31	27
Disliking parents' religion	11	8	12
Failing to go to church	67	67	67
Changing my idea of God	27	32	31
Losing faith in religion	37	44	35
Doubting prayer will bring good	37	44	35
Getting help on religious problems	53	54	56
Choosing a religion	21	20	15
Parents objecting to church membership	23	14	11*
Wanting to know the meaning of religion	53	48	60
Wanting communion with God	59	47	57
Heaven and Hell	53	53	66*
Sin	71	62	72
Conflicts of science and religion	42	50	57*
Being teased about my religious feelings	26	22	18
Wondering what becomes of people when they die	67	56	80*
Number of cases	174	243	130

between science and religion may serve as a storm center around which this turmoil rages. In some cases the inner disturbance follows a spectacular conversion which, the youth finds, has failed to solve all his psychic, social, and religious difficulties; in other cases this confusion is produced by the conflict between sexual urges and high spiritual ideals.

The introverted youth naturally tends toward introspection, and introspection plus overmuch religious thinking often leads to psychic depression or even to melancholy, a condition that demands the closest attention of parents, teachers, and psychiatrists. Adolescent melancholy should never be neglected on the assumption that this condition will readjust itself; it often does, but its inherent threat is too great to be taken lightly.

When adolescents become introspective in a religious sense, they should be encouraged immediately to seek help from a religious adviser. Of all the forms of spying on one's self, that of a religious nature is the most dangerous. Introspection can lead a young man to imagine not only that he has some grave physical disease, but also that he is one of the most wretched sinners on the face of the earth. A youth in this dilemma should be put on a proper program of physical hygiene and mental medicine, with suitable guidance in the acquisition of ideals and a more harmonious philosophy of life.

Furthermore, the church has an important function in connection with the sex life and function of adolescents. There has been a gross misinterpretation of the sex drive by many who are probably well-intentioned religious enthusiasts. The sex drive has been looked upon as sinful and a reason for shame, so that many individuals have considered themselves possessed by evil spirits or by unwholesome ideas when the drive appeared. Margaret Mead's studies of the Samoans, a group somewhere between primitive culture and the culture of our present Western civilization, show that when this drive is dealt with more frankly and with less hypocrisy, there are fewer conflicts and also that adolescents do not have to pass through the trying time of life referred to as the "storm and stress" period. If a storm appears, it is because they have not been prepared for a natural manifestation of the sex drive. There is evidence that the youth of today is facing this in a much franker manner than ever before in our civilization. Some of these thoughts as they relate to the problems of adult love, courtship, and marriage will be discussed in the last chapter of this study of adolescents.

By adopting a more enlightened approach to this problem, adolescents will be better prepared for a rational understanding and control of the natural emergence of the sex drive. Reassurance is needed that the various phases through which the boy or the girl as an individual is passing are normal and common to everyone.[21] Guidance and counsel based upon such frank assurance will be far more effective than that too often found in denials and misrepresentations.

Ideals and the adolescent. The integration of behavior units into a general schema or pattern, the development therefrom of a potent force that acts as a drive or tendency toward further activity, has been referred to in connection with habits as drives to behavior. Now it is in this integration of the various units of behavior that ideals arise

[21] "Sex Education: A Re-evaluation," *Child Study,* 1930, Vol. 16, pp. 83–97.

and thus come to control the behavior of the individual. During the early days of life, ideals are passing through an elementary formative stage in harmony with the child's innate tendencies and the environmental forces playing upon him. The individual's experiences are then rather narrow and his ideals very elementary, involving mainly the welfare and pleasure of the ego. (The socializing process at work on the playground, in club activity, in social life, and so forth, has already been discussed in this connection.) But as we look upon the socializing process as a process of growth and development, so must we consider the growth of ideals similarly, especially during this expanding and developing period of life from 12 to 21. Ideals are thus dependent upon maturation and experience, and may be narrow or, in harmony with a wider and fuller life, broad.

Hartshorne [22] has pointed out the need for creative religious activities resulting from an educational process in which the individual is a participant in the development of religious concepts. Since religious growth is a function of culture, he points out further, it should reflect the culture of our society. Our culture is one of socialization and participation, but religion is too often one of authoritative control over would-be passive subjects.

SUMMARY

The development of attitudes during adolescence has been discussed throughout this chapter. It has been pointed out that attitudes become better integrated as a part of the total self with growth and development toward maturity. The attitude of the individual toward current problems and the extent to which they have become an integral part of the total self become measures of the social and mental maturity of the individual. Attitudes toward members of the opposite sex are furthermore profoundly affected by the physiological maturity of the individual. This accounts for some of the changes in attitudes that occur during adolescence.

Ideas of religion, interest in religious problems, religious convictions, and changes in life outlooks appear in the lives of many adolescents as they grow toward maturity. The mental and social developments of adolescents are closely related to a religious awakening and changed attitudes. During adolescence conversion reaches its peak, only to be followed in post-adolescent years by doubts. Doubting grows out of

[22] H. Hartshorne, "Growth in Religion," *Religious Education,* 1939, Vol. 34, pp. 143–151.

wider social and intellectual contacts, and in this the adolescent needs sane, reliable, and honest guidance. It appears that adolescents are eager to find something of value in religion and are often disillusioned. Such a condition may lead to cynicism, doubt, and withdrawal from religious activities. But this is not always true; many a youth in his teens has ascended to heights of religious experience unsurpassed even by adults. When ideals are established and integrated in these religious experiences, there is an increased permanency in the dynamic force in operation. Ideals represent an integration of behavior units into a larger pattern, which comes to be a vital force in determining conduct. With the fuller mental and social growth and the development of ideals come habits of self-control, which form the essential element in the development of a moral nature.

THOUGHT PROBLEMS

1. Study the methods suggested for developing attitudes. Indicate attitudes you now possess that were developed by each of these methods.

2. What are the effects of puberty on the development of attitudes? Did you note any pronounced change in attitudes that occurred in your life during this period?

3. Write out a frank and accurate account of the genesis and development of your own religious attitudes from childhood up to the present time.

4. Is it conceivable that religion will ever be stripped of its contrasting and varying creeds and points of view? Would this be desirable? Give reasons for your answer.

5. Consult available statistics concerning the modal age of conversion. What is the significance of this?

6. What are the dangers inherent in the emotional stress sometimes associated with religious conversions? What are some features that would characterize a desirable form of confirmation?

7. What are the effects of movies on attitudes? Can you cite any change in attitudes observed by you as a result of the movies?

8. Why does religion appear to play an important role in the life of the adolescent?

SELECTED REFERENCES

Anderson, J. E., *The Psychology of Development and Personal Adjustment*. New York: Henry Holt and Co., 1949, Chap. XII.

Cole, Luella, *Psychology of Adolescence* (Third Edition). New York: Farrar and Rinehart, 1948, Chap. XI.

Dimock, H. S., *Rediscovering the Adolescent*. New York: Association Press, 1937, Chap. VIII.

Duffy, E., "Critical Review of Investigations Employing the Allport-Vernon Study of Values and Other Tests of Evaluative Attitude," *Psychological*

Bulletin, 1940, Vol. 37, pp. 597–612. This presents a very complete summary of investigations concerned with measuring attitudes.

Fleege, Urban H., *Self-Revelation of the Adolescent Boy.* Milwaukee: The Bruce Publishing Co., 1945, Chap. XIII. This is an interpretative study of the attitudes of 2,000 boys from representative Catholic high schools.

Hartshorne, H., and Lotz, E., *Case Studies of Present Day Religious Teaching.* New Haven: Yale University Press, 1932, Chap. III.

Hiltner, Seward, *Religion and Health.* New York: The Macmillan Co., 1943. The approach of this volume is mainly from a mental hygiene point of view. The author has served as Executive Secretary of the Commission on Religion and Health of the Federal Council of the Churches of Christ in America.

Hollingshead, A. B., *Elmtown's Youth.* New York: John Wiley and Sons, 1949, Chap. X.

Hurlock, E. B., *Adolescent Development.* New York: McGraw-Hill Book Co., 1949, Chap. X.

Nelson, E., "Student Attitudes Toward Religion," *Genet. Psychol. Monogr.,* 1940, Vol. 22, pp. 323–423.

Weaver, Paul, "Youth and Religion," *The Annals of the American Academy of Political and Social Science,* November, 1944.

Wieman, R. W., *The Modern Family and Religion.* New York: Harper and Bros., 1937.

PART III

ADJUSTMENTS OF ADOLESCENTS

X

The Adolescent and His Peers

ADOLESCENT PEER RELATIONSHIPS

Throughout the earlier chapters the adolescent has been described as if he were composed of a number of separate parts. Although attempts have been made to show that growth is interrelated, each chapter has given special emphasis to some special aspect of the adolescent's growth. Such a treatment should give the student of adolescent psychology a better understanding of the nature of these various phases of growth and development during adolescence. Such a treatment, however, may lead the student to conceive of the adolescent as composed of many rather discrete selfs—the physical self, the emotional self, the mental self, the social self, the spiritual self, and so forth. The chapters that follow will bring together these aspects of growth into a total personality. The adjustments of the adolescent in his peer relations, at home, in the community, and at school will be treated in this and the following chapters. This treatment is then followed by a description of personality and particularly the personality characteristics of adolescents. Special consideration is given in the last chapters of Part III to the personal and social adjustments of adolescents and to problems connected with juvenile delinquency.

Importance of peer relations. We have a tendency to explain a child's behavior on the basis of the family and organized institutions that he has been associated with—thus minimizing the importance of the experiences of boys and girls with each other in their day-by-day activities. The importance of peer relations during adolescence when many personal and social problems appear has been emphasized by Caroline Tryon, when she stated:

"If we were to examine the major developmental tasks which confront boys and girls in late childhood, during pubescence, and in later adolescence, it would become apparent that many of these can only reach a satisfactory solution by boys and girls through the medium of their peer groups. It is in this group that *by doing* they learn about the social processes of our culture. They clarify their sex roles by acting and being responded to, they learn competition, cooperation, social skills, values, and purposes by sharing the common life." [1]

It has been suggested in previous chapters that adolescents are faced with many important developmental tasks. They do not feel that their parents always understand their needs and problems. Likewise, they do not feel that most teachers understand their problems or are sympathetic with them in their effort to solve them. There is considerable evidence that teachers do not usually understand many of the problems faced by adolescents. The fact that the teacher has already met or by-passed these problems in his development oftentimes makes such problems appear as a mere trifle in the life of an individual. Problems of social approval, making friends, being popular, being accepted, and the like are real to most growing boys and girls. Since they feel that their parents either do not understand and appreciate them or are often critical of them in relation to their activities, and since they are unable to secure the needed help and guidance from their teachers, they seek help and sympathy from their peers.

Studies in social psychology and sociology show that in our society a sort of subculture operates among boys and girls. The operation of this is most obvious in our larger towns and cities, where boys and girls are brought together in greater numbers and where family ties are perhaps less binding. The culture operating in these adolescent groups is similar to that found in adult societies, but with the emotions and mental immaturity of younger children commonly manifested. These groups have their own standards, values, purposes, and methods of protecting themselves from too much adult interference. They use the methods found in adult society for securing conformity. Adults are frequently excluded from these groups by indirect means, such as "Oh, this is just for us kids," or "The other kids' parents don't interfere." The group remains somewhat constant for a number of years during the growing life of the adolescent. New members are continuously being admitted from the younger group, while older ones drop out for one reason or another. Materials presented in later chapters will show that such groups are likely to become sources of diffi-

---

[1] Caroline M. Tryon, "The Adolescent Peer Culture," *Forty-third Yearbook of the National Society for the Study of Education,* Part 1, 1944, Chapter XII.

culty and delinquency if they are not given the correct type and amount of guidance and direction.

Attaining a satisfactory role. The attainment of a satisfactory role among peers is a development task faced by the child as he develops. This becomes even more important as he passes from childhood into adolescence. The insecure child and the rejected child find social development a most unpleasant undertaking, and may try out various adjustment techniques in an effort to solve their problems. The failure to attain a satisfactory role presents a critical problem to the adolescent. Studies have revealed several important findings relative to a child's relation to his peers.

1. The child desires the approval of his peers.
2. The relative importance of peer approval increases as the individual grows into adolescence.
3. Pre-adolescents and adolescents like to imitate their peers or those slightly older than they are.
4. Good peer relation during pre-adolescence is perhaps the best assurance available for good peer relations during adolescence and post-adolescence.
5. Early adolescence is accompanied by the formation of cliques. These cliques play an important part in satisfying certain felt needs of adolescents.

There is probably no period in the average individual's life when he does not have the desire to be popular among his peers. This desire is perhaps keener during the teen years than at any other time of life. The results of a survey by the Purdue University Public Opinion Poll, taken of more than 10,000 high school students across the nation, shows the nature and extent of teen-agers' desire for popularity. Girls were somewhat more concerned than boys with being popular. This difference is no doubt closely related to social conventions that place greater restrictions on the initiative of girls in making friends. Sue may admire Jim and want him to date her, but as a rule she must await Jim's move. However, it is well known that girls learn a technique for securing dates and attention from boys, without appearing to make the first move. The results for boys and girls were as follows:[2]

Almost 50 per cent of the boys and 60 per cent of the girls checked the item *I want people to like me more.*

[2] These results of the Purdue University Opinion Poll are adapted from *The Atlanta Journal* for July 4, 1949.

The item *I wish I were more popular* was checked by 30 per cent of the boys and 47 per cent of the girls.

The desire to make new friends was checked by 45 per cent of the boys and 56 per cent of the girls.

Many of the young people felt the need for help in making better social adjustments. High school students writing to the panel requested information about how to overcome shyness, how to carry on a pleasant conversation, how to handle social relations of an embarrassing nature, and the like. On the basis of the results, girls apparently felt more secure than boys in their conversational ability. They were, however, more concerned about gaining self-confidence than were the boys. Specific social skills that help individuals to get along better did not seem to be a significant problem for most of the boys and girls. Sixteen per cent wanted help in introducing people properly. Nineteen per cent felt that they should learn to be more tactful. Twenty per cent wanted information on how to act on formal occasions. Only eight per cent felt that they should be less aggressive in their social behavior.

A nationwide essay contest open to high school seniors gives further data on the aspirations of youth. Almost 2,000 papers were written in response to the following question: "If you could suddenly acquire superiority in *one characteristic* or *ability,* which one would you choose? Describe the characteristic or ability, and give reasons for your choice." [3]

The replies were tabulated and revealed that the largest group aspired most of all to have friends, to be popular, and to succeed in human relations. Closely related to this was the group desiring *musical ability,* especially in playing and singing. The reasons implied for desiring musical ability are as follows: (1) personal popularity; (2) success among contemporaries; and (3) general public recognition. More than half of those aspiring to musical success gave as their primary objective "social success." Girls were much more numerous in aspiring to musical success than boys.

The dominant desire for popularity was revealed in the great variety of aspirations. Some craved special skills as a sort of "show off"; some longed for wit and humor; while others desired a good speaking voice, with the ability to speak in public.

Adolescent friendships. Availability of social contacts and mutual satisfaction of needs were found by Reader and English to be the most

[3] Elizabeth Tate, "What Youth Wants Most," *Ladies Home Journal,* August, 1948, p. 31.

important variables in adolescent female friendships.[4] Girls with similar interests and tastes would thus appear more likely to be able to satisfy these mutual needs. This is in harmony with results obtained from the California growth studies of adolescents.[5] Responses were obtained to the question "What kind of people do you like to be with best?" At all levels studied the majority of boys and girls checked the response that they preferred to be with their own age groups; although the results presented in Table XXIX show that many adolescent girls (girls in the ninth, tenth, eleventh, and twelfth grades) indicate a desire to be with people a few years older. This no doubt stems from the greater physiological maturity of a large per cent of girls at these grade levels (in comparison with the boys at the same levels).

A study of the friendship fluctuations of rural adolescent boys and girls was conducted by Thompson and Horrocks.[6] In this study 421 boys and 484 girls living in rural areas were studied over a two-week period. They found an increase in the stability of friendships from age ten to seventeen. No significant difference in friendship fluctuations was observed between boys and girls. This increased stability of friendship during the adolescent years does not provide support for the hypothesis early advanced by Hall and others that adolescence is a period characterized as one of "storm and stress," and instability.

A second study of friendship fluctuations, conducted by these same investigators, compared urban and rural adolescent girls.[7] The 969 subjects used in this study were obtained from two cities in New York state and from one city in Pennsylvania. These individuals were selected from six to twelve. An attempt was made to select girls from families of approximately average socio-economic status; a similar attempt was made in the case of the rural adolescents in the study made at an earlier date. A comparison of the rural and urban boys in their fluctuations indicates a slightly greater stability in friendships

[4] Natalie Reader and H. B. English, "Personality Factors in Adolescent Friendships," *Journal of Consulting Psychology,* 1947, Vol. 11, pp. 212–220.

[5] Caroline McCann Tryon, *U. C. Inventory I: Social and Emotional Adjustment.* Revised form for presentation of cumulative record of an individual with group norms for a seven-year period, 1939. Forms are presented for both boys and girls.

[6] G. G. Thompson and J. E. Horrocks, "A Study of the Friendship Fluctuations of Urban Boys and Girls," *Journal of Genetic Psychology,* 1946, Vol. 69, pp. 189–198.

[7] J. E. Horrocks and G. G. Thompson, "A Study of the Friendship Fluctuations of Urban Boys and Girls," *Journal of Genetic Psychology,* 1947, Vol. 70, pp. 53–63.

among urban than among rural adolescents, although the difference is not statistically significant. Figure 30 shows the relationship between age and the percentage of boys and girls choosing the same person as their best friends. For both boys and girls there is a decided tendency toward an increased stability of friendship, with the girls showing the greater increase.

TABLE XXIX

RESPONSE OF ADOLESCENTS TO THE QUESTION *"What kind of people do you like to be with best?"*

	5HL6	6HL7	7HL8	8HL9	9HL10	10HL11	11HL12
				BOYS			
Grown people (grown-ups)	1	1	1	3	0	0	0
People younger than I am	0	3	7	1	1	1	1
People about my age	86	80	89	87	89	85	95
People a few years older than I am	7	14	4	10	10	17	17
I would rather be by myself	6	1	0	1	1	0	0
				GIRLS			
Grown people (grown-ups)	4	6	1	1	1	0	0
People younger than I am	3	1	1	1	0	0	0
People about my age	82	85	94	78	69	62	56
People a few years older than I am	3	8	6	25	36	42	50
I would rather be by myself	8	3	1	1	1	0	1

This is no doubt closely related to the more democratic nature of boys and the greater tendency on the part of girls to form small "cliques" and have a single friend. These data further indicate that one of the characteristics of growing up is that of maintaining more stable friendships.

The formation of cliques. Gangs and cliques characterize the adolescent age. When the fourteen-year-old adolescent daughter is asked where she has been, she may reply: "Oh, I have just been down street with the gang." In this case she refers to her small group, a sort of self-sufficient unit. The study of friendship formation among Elmtown's youth furnishes worth-while information on this problem.[8] This study was designed to test the hypothesis that the social behavior

[8] A. B. Hollingshead, *Elmtown's Youth*. New York: John Wiley and Sons, 1949.

of adolescents is related to the place their family occupies in the social structure of the community. This midwestern community consisting of some 10,000 inhabitants was found to be stratified into five classes. The group studied consisted of 369 boys and girls between the ages of 13 and 19 inclusive. This provided a good cross section of the teen-age group.

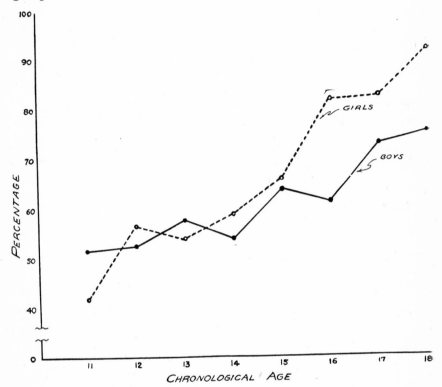

FIG. 30. *The Relationship Between Chronological Age and Percentage of Boys and Girls Choosing the Same Person as Their Best Friend on Two Occasions Separated By a Two-Week Interval.* (After Horrocks and Thompson)

Observations at school, at church, and on the streets revealed the existence of many cliques. After school two or three boys in a group might be seen together in the pool hall, or several girls might be found at the corner drug store. Within these cliques confidences are exchanged, personal matters are discussed, and sentimental ideas of great important to the participants are expressed. The clique has an important impact on the members, adding members as well as dropping members on the basis of informal controls within the group. Thus, these cliques, made up of one's peers, have an important influence on the activities of their members. Hollingshead states:

"Social pressure in the adolescent groups operates far more effectively, and with greater subtlety to channelize friendships within limits permitted by the social system of both the adult and the adolescent social worlds than the hopes, fears, and admonitions of anxious parents." [9]

The dual operation of the parents and peers in the clique is well illustrated in the case of Joyce Jenson's (class III) relationship with her clique and especially with her friend Gladys Johnson, class III.

We influence each other a lot. She influences me almost as much as my parents do. I listen to them, especially when it comes to choosing friends, but I don't agree with everything they tell me. I've had them really give me the dickens about going around with some girls I wanted to go with or maybe Gladys did. Most parents don't want their kids running around with certain other kids, and they'll give them advice and they'll follow it or they won't, but when my folks put the foot down on me I listen.

I know that the folks give me good advice, but sometimes they just don't understand what kids want to do, and they think we ought to act like they acted twenty years ago. My parents, especially my mother, influence me in what I do, but Gladys probably influences me as much or more.

I don't want to run any of the kids down, but there are certain girls here who are just not my type and they're not Gladys's type; they'd like to run around with us, but we don't let them.

Pauline Tryon (class IV) and her bunch would like to run around with us, but we turn our backs on them because they run around all night, cut school, and hang out down at Blue Triangle.

There are some kids we'd like to run around with, but they don't want us to go with them. Gladys and I would like to go around with "Cookie" Barnett (class II) and her bunch, or the G. W. G.'s, but they snub us if we try to get in on their parties, or dances, or date the boys they go with.

An analysis of the 1,258 clique ties observed in the study of Elmtown's youth revealed that approximately three out of five are between boys or girls of the same class position, two out of five are between adolescents who belong to adjacent classes, and one out of twenty-five involves individuals who belong to classes twice removed from one another. The detailed study of close ties disclosed that from 49 to 70 per cent of all clique ties are with class equals. It also reveals that when a class I boy or girl crosses a class line, and about one-third do, a member of class II is likely to be involved. Likewise, when a class

[9] Reprinted by permission from *Elmtown's Youth* by A. B. Hollingshead, published by John Wiley & Sons, Inc., 1949, p. 208. The case of Joyce Jenson is also reprinted from the Elmtown study.

II boy or girl crosses the class line, he moves into class I or class III. Thus, we note that the polar classes are largely isolated one from another in so far as intimate, personal, face-to-face relations are concerned.

HETEROSEXUAL INTERESTS AND ACTIVITIES

Heterosexuality. By heterosexuality is meant the focus of interest upon members of the opposite sex. The study by Kuhlen and Lee [10] reveals an increase of heterosexual relationships with an increase in age. Their findings, based upon the choice of companions for a number of social activities and situations made by pupils at several grade levels, are set forth in Table XXX. Of the sixth graders, less than one-third choose companions of the opposite sex, but of the twelfth graders, almost two-thirds do so. There is also a significant trend for both boys and girls to choose as companions members of the opposite sex as they advance from the sixth to the twelfth grade—the greatest change occurring between the sixth and ninth grades.

Now in adolescents a wide range of reaction patterns relative to the opposite sex exists, and it is very difficult to generalize concerning the

TABLE XXX

CHANGES IN SOCIAL RELATIONSHIPS AT ADOLESCENCE AS SHOWN BY THE PERCENTAGE OF BOYS AND GIRLS AT DIFFERENT GRADE LEVELS WHO CHOOSE MEMBERS OF THE OPPOSITE SEX, AND ARE CHOSEN BY MEMBERS OF THE OPPOSITE SEX, AS COMPANIONS FOR VARIOUS ACTIVITIES (*After Kuhlen and Lee*)

	GRADE		
	VI	IX	XII
Percentage of boys choosing girls	45.0	72.5	75.0
Percentage of girls choosing boys	39.2	59.7	63.0
Percentage of boys chosen by girls	31.2	49.1	65.8
Percentage of girls chosen by boys	30.8	52.4	59.7

reactions of the group as a whole. However, since the sexual urge is present in every individual and probably begins to function influentially, if indirectly, quite early in life, it is evident that the differences between the reactions of various adolescents result from the direction that this urge has been given, rather than from its mere existence.

[10] Raymond G. Kuhlen and Beatrice J. Lee, "Personality Characteristics and Social Acceptability in Adolescence," *Journal of Educational Psychology,* 1943, Vol. 34, pp. 321–340.

Heterosexuality itself can be properly established only by social contacts with members of the opposite sex, and in these contacts two environmental conditions are essential: first, members of the opposite sex must be of sufficient numbers, of appropriate age, and of attractive personal qualities; second, an intelligently encouraging attitude is necessary on the part of parents and others concerned with the individual's guidance and welfare. If these essentials are absent, the child may emerge from adolescence with warped and shameful attitudes toward sex matters that may encumber him permanently.

It has been observed that, in some species of animals, characteristic patterns of behavior appear *de novo* when pubertal changes in the primary and secondary sexual characters and accessory organs of sex are most in evidence. Among the primates in particular, a limited amount of sexual play is said to appear prior to the pubertal changes. In either case it can be said that the sexual drive is greatly augmented as the time of somatic puberty approaches, and that it continues to grow in strength for some time thereafter by virtue of factors of maturation and of sexual contacts and experiences. Since the sexual drive is at heightened strength as a result of the development processes at work, the manifestation of increased sexual activities and sexual play by adolescents is to be expected. The savage youth was prepared to gain his living by the time the sex drive ripened. In contrast, it has already been revealed that in a civilized community most adolescents are in school when this happens, and that economic security and independence are still a dream. It is therefore not possible for the fourteen- or fifteen-year-old boy or girl to enter into economic pursuits in order to support a family; moreover, customs as well as laws do not permit him to do so; and yet, there has been no significant change in the period of the onset of the sex drive.

Dating during adolescence. The extent of dating by high school students will depend upon the customs, living conditions, social backgrounds, and special interests of the particular age-group concerned. Seniors date more than freshmen, and report chaperonage less frequently. This, no doubt, reflects their greater social and physical maturity, and the increased willingness of their parents to allow them, as they grow older, to associate freely with members of the opposite sex. Harold H. Punke[11] reports a study of youth from nine states,

[11] Harold H. Punke, "Dating Practices of High-School Youth," *The Bulletin of the National Association of Secondary School Principals,* January 1944, Vol. 28, No. 119.

ranging from North Carolina and Pennsylvania in the east to California and Washington in the west. Materials relating to frequency of dating are presented in Table XXXI. According to these data, no sig-

TABLE XXXI

FREQUENCY OF DATING OF HIGH SCHOOL PUPILS, AND SINGLE VERSUS DOUBLE DATING, ACCORDING TO GRADE AND SEX OF PUPIL (*After Punke*)

GRADE AND SEX OF PUPIL	FREQUENCY IN NUMBER OF DATES PER MONTH					SINGLE VS. DOUBLE DATING	
	Number Reporting	Percentage Distribution, according to Number of Dates per Month				Number Reporting	Percentage of Dates That Are Double Dates
		None	*1–4*	*5–10*	*over 10*		
Freshman: Boys	1276	53.8	20.6	16.0	9.6	623	45.4
Girls	1490	53.0	21.3	17.0	8.7	814	81.0
Senior: Boys	1408	21.6	29.5	28.9	20.0	1094	55.6
Girls	1454	13.6	19.4	33.1	33.9	1412	80.6

nificant sex differences appear among the freshmen; this, even though girls mature earlier than boys, is a circumstance that suggests greater parental restriction of girl's social life. Senior girls, however, according to these data, do more dating than senior boys; for this there are several possible explanations, but it is quite likely that many of the former are dating older boys who have already finished high school, a supposition in harmony with findings relative to the earlier marriage of girls than of boys.

In Punke's study the students were asked to indicate, in order, the three types of activities in which they most frequently engaged, and the three types they preferred. The answers to these questions were tabulated for Georgia and California. There was very little difference between the activities listed as most commonly engaged in and those preferred. The three items that led both of these lists were: dancing, attending movies, and riding in automobiles. Punke concludes further:

In California both freshmen and seniors of both sexes ranked movies highest among things done, whereas in Georgia boys of both grades placed carriding first among things done, and girls gave first place to movies. For both grades and sexes in California, dancing was intermediate between car-riding and movies so far as activities engaged in were concerned, whereas in Georgia, both sexes combined, the freshmen placed dancing first among those three activities and seniors placed it last. In both states and for both grade levels, boys reported that they engaged in some type of athletic activity when on dates more typically than did girls—*i.e.,* tennis, swimming, bowling, skating, hiking. It is interesting that members of both sexes, for both grades and both states, reported that when on dates they *engaged* in athletic activity much more commonly than they observed athletic events. In Georgia youth engaged in religious activity when on dates to a substantially greater extent than was true in California, and in Georgia seniors of both sexes engaged in such activities more extensively than did freshmen.[12]

The intra- and interclass dating patterns closely parallel the clique patterns.[13] No dates were observed among Elmtown's youth between members of class II and members of class V. In this case the social distance between the classes is too great. On the other hand 61 per cent of the dates belong to the same class; 35 per cent to an adjacent class; and 4 per cent to a class separated by one intervening class. The association of class with class is clearly illustrated in Figure 31. This chart shows that the boy is more willing than the girl to date someone in a lower class structure than himself, or has more opportunity to do so since he is the one that takes the initiative in the dating. Many factors may be introduced to show why the boy dates in a class position below his more often than a girl does. The fact remains that when an Elmtown boy dates outside his class position the chances are two to one that he dates a girl in a class below his; while when the girl dates a member of a different social class, the chances are two to one that she dates in a class above her own.

Petting is the current slang expression for a form of behavior that has been variously named by each succeeding generation—in all cases it refers to the variety of acts of a more or less sexual import. Such acts are not confined to any one generation or to any particular race or tribe; their manifestations vary as a result of manners, customs, traditions, and education. Concerning such activities Louttit says:

This is a problem of adolescence, the seriousness of which depends largely upon one's point of view. The greatest harm comes from the excessive emotional stimulation which frequently is uncontrolled. A sane sex education

---

[12] *Ibid.,* pp. 51–52.

[13] *Op. cit.,* pp. 231–232. Figure 31 is reprinted by permission from *Elmtown's Youth* by A. B. Hollingshead, published by John Wiley & Sons, Inc., 1949.

which gives the child some sense of values is of much greater usefulness than endless preachments or attempts at strict deprivations.[14]

Desire for social approval. The desire for social approval has often been referred to as instinctive, but there is ample evidence from various

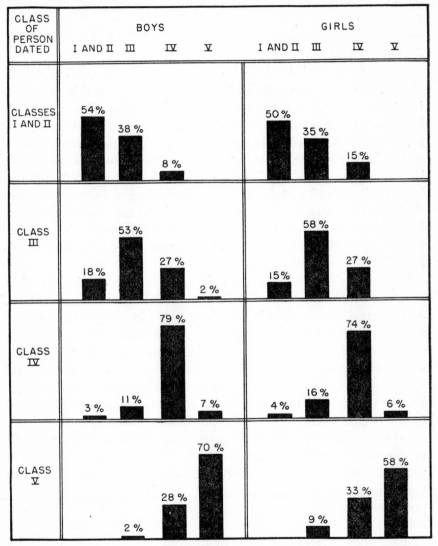

CLASS OF PERSON DATED	BOYS				GIRLS			
	I AND II	III	IV	V	I AND II	III	IV	V
CLASSES I AND II	54%	38%	8%		50%	35%	15%	
CLASS III	18%	53%	27%	2%	15%	58%	27%	
CLASS IV	3%	11%	79%	7%	4%	16%	74%	6%
CLASS V		2%	28%	70%		9%	33%	58%

FIG. 31. *Intra- and Interclass Dating Patterns of Boys and Girls of Elmtown.*
(After Hollingshead)

[14] C. M. Louttit, *Clinical Psychology.* New York: Harper and Brothers, 1936, p. 330. (Reprinted by permission of the publishers.)

observations and studies to show that it is chiefly due to the experience of the individual. The fact that we find social consciousness so clearly revealed in the life of the adolescent lends support to the contention. However, regardless of the extent to which it is instinctive, it does operate as a powerful motivating force during adolescence. The desire for social approval no doubt has as its basis certain major and subtle emotions, and furthermore it may be found existing in many forms during this period.

Sex, notions of self, and the like play a prominent part in the individual's growth and development. It is through these that the group is able to establish and maintain uniformity in manners, styles, and interests. The force of public opinion tends to cause the adolescent to accept readily the standards and customs of the social group; because of public opinion the individual endeavors to further his position in life, and takes pride in his success. The desire for social approval becomes integrated early with the major biological forms of motivation of sex and hunger, the natural tendencies of the individual becoming so modified as to gain it. The very fact that this desire is operating in the life of the individual is evidence that he is becoming a full-fledged member of the social group.

"In the higher forms of social integration, the dominance often goes out of the hands of a single man and is crystallized into law, customs, traditions, and social sanctions. . . . In most social organizations there is a limit to the powers of the dominant person, idea, custom, or force." [15] Now if we begin to study these limitations, we shall probably find homogeneity to be the main force. As the child reaches maturity and becomes more and more a social rather than an individual creature, the force of the role and opinion of the group grows stronger, and is especially prominent in the development of social consciousness. But if the adolescents of the group are homogeneous, the customs, rules, and so forth will play a still more important role than they would otherwise. Homogeneity itself depends upon communicability, similarity of interests and beliefs, and—especially—similarity of general racial features. When this homogeneity exists, control and social integration are more easily effected—a fact that should be carefully observed by those in charge of our educational

[15] Mark A. May, "The Adult in the Community," *The Foundations of Experimental Psychology* (C. Murchison, Ed.). Worcester, Mass.: Clark University Press, 1929, p. 782.

practice and by those dealing with clubs and group programs designed for adolescents.

Again, the desire for social approval might be thought of in connection with more complex adjustments in the life of the adolescent. Let us consider the "sweet girl graduate" from high school just prior to her graduation, and assume that she desires a certain graduation dress and other novelties that will blend with each other and with her general make-up. The images of these articles as they would appear on her constantly run through her mind. She imagines her friends' approval of this attractive outfit; she imagines herself winning Jack's attention, which she desires greatly. But the economic conditions of her family are such that she cannot purchase the clothes, and she therefore must either do without the costume or find some means as yet unknown to purchase them. Thus one will find adolescents and post-adolescents often willing to resort to questionable devices in order to win the approval of their friends. Here we find the girl resorting to various devices in order to appear sexually attractive to the boy she admires. The beautiful wearing apparel will help her to become more attractive, and she recognizes that Jack is quite fond of such a type of costume; she may therefore deprive herself of the movies, other amusements, and even food in order that she may be able to buy what she considers necessities. Again, even petty crimes or misrepresentations may be resorted to in order to win social approval. The average high school girl's ego complex is well developed around certain erotic tendencies, and these become more powerful as they involve the approval or disapproval of the male sex.

This is shown in the study of adolescent motivation by F. H. Lund.[16] Data collected in two junior and two senior high schools, having an enrollment of 8,200 students, showed that the incidence of pupils requesting to be excused from gymnasium classes on the basis of physical disabilities increased markedly during the adolescent period. This increase was most pronounced for girls—being 400 per cent between the seventh and twelfth grades. The fact that the increase was confined to the girls poses a problem in motivation. Table XXXII shows the difference between the boys and girls in the type of disability for which the pupils were excused. The cause for the excessive increase of cardiac, glandular, and miscellaneous disabilities

[16] F. H. Lund, "Adolescent Motivation: Sex Differences," *Pedagogical Seminar and Journal of Genetic Psychology*, 1944, Vol. 64, pp. 99–103.

among the girls (the increase is much higher among white girls and among those best able to pay for medical services) is likely to be found among social factors, such as fear of developing large muscles, desire to remain clean at all times, disturbances of hairdo, and the like. Thus, the girls prefer to avoid exercises that do not serve either directly or indirectly a social purpose.

TABLE XXXII

CLASSIFICATION AND DISTRIBUTION OF "MEDICALS"
AMONG SENIOR HIGH SCHOOL PUPILS (*After Lund*)

PATHOLOGY	BOYS PER CENT	GIRLS PER CENT
Nutritional	6.01	5.48
Cardiac	30.62	33.59
Nervous	4.28	2.13
Glandular	6.73	11.89
Respiratory	4.91	3.65
Post Operative	12.36	10.36
Muscular	6.66	1.52
Structural	6.44	5.48
Traumatic	8.72	5.48
Miscellaneous	13.22	20.42

The study by Kuhlen and Lee [17] throws some light on changes in acceptability of traits from the sixth to the twelfth grade. The five traits most highly related to and the five least related to acceptability at these grade levels are presented in Table XXXIII. The traits "enthusiastic," "friendly," and "popular" are found in all categories of high acceptability; whereas the traits "bosses others," "enjoys a fight," and "seeks attention" are found in all categories of low acceptability.

The adolescent age is an acutely self-conscious one, and for this reason there is need for increased tolerance of the adolescent's demands, ideas, and desires as they relate to his personal appearance. If the other girls go in for nail polish; if they wear peculiar-appearing hairdo's— far less becoming in a mother's view than an attractive bob—still, that mother would be wise to permit her daughter these forms of adornment. In so doing, she gives her greater self-confidence in the society of her peers. This is not to say that the mother should not express her preference for the coiffure more attractive by her own standards;

[17] *Op. cit.,* p. 335.

TABLE XXXIII

TRAITS HAVING HIGHEST ASSOCIATION WITH ACCEPTABILITY

Boys		Girls	
Sixth Grade	Twelfth Grade	Sixth Grade	Twelfth Grade
Cheerful (96)	Friendly (81)	Friendly (83)	Popular (73)
Enthusiastic (92)	Initiates games (81)	Enthusiastic (80)	Friendly (72)
Friendly (92)	Enthusiastic (77)	Good-looking (73)	Enthusiastic (69)
Popular (88)	Cheerful (69)	Popular (72)	Sociable (69)
Good-looking (80)	Popular (69)	Initiates games (66)	Enjoys jokes (65)

TRAITS HAVING LOWEST ASSOCIATION WITH ACCEPTABILITY

Enjoys fight (12)	Seeks attention (27)	Bosses others (0)	Bosses others (31)
Acts older (12)	Restless (19)	Talkative (−3)	Talkative (31)
Seeks attention (12)	Bosses others (17)	Seeks attention (−4)	Restless (25)
Bosses others (0)	Enjoys fight (16)	Enjoys fight (−11)	Enjoys fight (21)
Talkative (−4)	Acts older (4)	Restless (−31)	Seeks attention (14)
Restless (−28)			Acts older (−25)

however, parents and teachers should realize that there is a difference between having standards and imposing them.

Self-assertion before the opposite sex. Self-assertion in the form of self-display before members of the opposite sex has been observed among sexually maturing male animals as well as among adolescent boys. During the mating season there seems to be an overstock of energy that is stored up in animals and is released in the various courting acts that are initiated in response to specific stimulations. In certain species, notably among birds of prey, both male and female show this exuberance, and it is quite common to find it expressed through wonderful flying performances, circlings around each other, and calls peculiar to the kind. The male's showing-off before the female is particularly spectacular. Doubtless a feeling of pleasantness arises from these performances, owing to the growth and maturation of physical structures and reflex co-ordinations and the general release of bodily tensions.

Darwin gives a most striking picture[18] of display by male peacocks and pheasants—their gorgeous crests and tails are given the optimum display before the female. Darwin further writes that the Angus pheasant appears to observe carefully the female's responses to his show; and this could be explained adequately not as a result of some instinct of pride but rather as pride that has developed from experi-

[18] Charles Darwin, *Expression of the Emotions in Man and Animals*. New York: Appleton-Century-Crofts, Inc., 1873.

ence and from the structures of the organism that are now coming to fruition. This courting among various animal types involves activities somewhat subsidiary to sexual ends, and playful exercise is a consequence of superfluous energy that becomes in part directed toward members of the same species and of the opposite sex.

In the human race this assertive tendency can also be seen. Witness the young adolescent, with his daring spirit, overexertion, and constant display of strength and skill. His situation is similar to that of Darwin's pheasant. And the same can be said of the female of the human species: her feminine manners, her slyness, and her persistent efforts to outwit her rivals are all manifestations of this same tendency. Bronner writes:

> It is certainly clear that in order to reach normal adult stature the adolescent must pass, during these years that comprise the adolescent period, from early lack of sex consciousness to a stage characterized by the exact opposite, namely, sex consciousness, and then to the stage of attraction to the opposite sex. The preadolescent boy or girl has very little use for those of the opposite sex. The boy considers his sisters and her friend a nuisance; the girl considers her brothers and his friends rough and rather to be avoided. There upon ensues a time when each is shy and self-conscious in the presence of others; this to be followed e'er long by a stage in which each is attracted by, and wishes to be attractive to, the opposite sex. This last is the stage when in ordinary parlance the boy is in varying degrees 'girl crazy' and the girl, in some measure, 'boy crazy.' [19]

Students of physiology and child psychology have shown that the secondary sexual characteristics of both male and female are dependent, in the final analysis, upon certain internal secretions, particularly those of the sex glands. General internal changes prevalent during sexual excitation have an emotional tone and cause a general restlessness that involves the whole of the organism's behavior. Marston offers evidence that during this state there is, in addition, a general lowering of the blood pressure. The sacral division of the autonomic nervous system is, it will appear, operating more than normally, and this unusual operation tends to direct excessive quantities of blood and glandular secretions into channels which—although they are often not so recognized—are directly related to the sex emotion. These changes are a result of profound visceral and glandular changes and, as we have noted, tend to affect all behavior of the organism.

[19] A. F. Bronner, "Emotional Problems of Adolescence," *The Child's Emotions.* Chicago: Chicago University Press, 1930, pp. 228-229.

Not all members of the sexes are attractive to the opposite sex, nor does the same person make an equal appeal to all. Beauty, good manners, "feminine qualities," health, education, and "personality" are but a few characteristics listed by boys as desirable in girls. Feminine good looks are usually listed as most essential to sex attraction, but their evaluation will differ from decade to decade. During the latter part of the nineteenth century curves were deemed the ideal of beauty; but within a period of twenty-five or thirty years thereafter, curves seriously lost vogue. To draw conclusions, one need only to examine the styles of the past, whether of a century or several centuries, and compare them with each other and with those of today. Girls of former times had, indeed, their sex appeal; they were adored, surely. But if a girl were to appear today with their manners and dress, she would at best be viewed as a curiosity.

The generalization that might be made from these facts is that the current vogue in costumes, manners, language, interests, cosmetics, hairdressing, and so forth makes for sex appeal among those contemporarily on the scene, but that if this vogue is revived later, its followers may be considered ridiculous. The modern mother who insists that her daughter imitate her in dress, manners, interests, and so on either fails to recognize this truth or refuses to live according to its principle. When the facts are rightly understood, we may rightly appraise the value of clothes, appearance, manners, and other subjects of controversy.

The response of the adolescent boy or girl to what is strange or forbidden must not be overlooked, for both curiosity and self-assertion are important in the motivation of conduct. Familiarity with an individual will tend to lessen the sex appeal of that individual. Thus, if the "new girl" in the community has a somewhat different sex appeal, she will have an advantage over the others. Fickleness is indeed characteristic of sexual phenomena, especially in adolescence; and, it may be observed, changes of style serve to augment it by renewing elements of "strangeness." On the other hand, the spirit of self-assertion, which has already been noted as related to sexual life and which we shall consider further later on in this study, leads to love-making in the face of great obstacles. Thus *forbiddance* and *self-assertion* are often present in behavior as a combination which should not be ignored—especially by parents. Because of this combination, troubles often develop between parents and children in connection with courtship and marriage.

THE SEX LIFE OF ADOLESCENTS

Developing desirable attitudes toward members of the opposite sex.
The social activities that permeate much of the club work and the
extra-school life of adolescents are important avenues for the development of desirable boy-girl relationships. Through social affairs, boys
and girls are given the opportunity of working together, and opportunities develop for a division of responsibilities on the basis of sex.
Girls make cookies and decorate tables in home economics classes; the
boys aid by making things in the shop. One group of boys and girls in
a newly consolidated school in Tennessee, with a minimum of help
and a maximum of encouragement from the teachers, provided a social-
recreational room that became a real center for the community.

Many communities are alert to the interests and needs of adolescents.
Recreation centers are being opened in which a reasonable amount of
supervision is provided. At these centers boys and girls can meet to
play quiet games, listen to the radio, sing popular songs around a piano,
play cards, dance, read from current magazines, and the like. These
centers may be located in quarters provided by the school, in an empty
store, or in some other convenient place. Care should be taken to
make them attractive in nature, and to provide supervision that is at
once efficient and inoffensive. As much of the responsibility should be
placed upon the boys and girls as they are capable of shouldering, in
seeing that the place is kept in order and that the group takes a part
in setting the standard for its activities. Boys and girls will enjoy
fixing up and running such a place, provided they are given the
opportunity to do so and are made to feel that it is run for and largely
by them.

The provision of a place where boys and girls can meet informally,
without publicity and a great deal of fashion and show, is especially
helpful to a large group of boys and girls who are not financially able
to look for wholesome commercial entertainment, because of the expense involved. The attraction of the juke joint and the corner drugstore is largely the result of their satisfaction of certain needs and
interests that are not satisfied by other means. The community can
well afford to provide conditions for satisfying these needs in still fuller
measure under conditions still more desirable.

Units of study in natural science give worth-while information about
the birth and care of living things; in the same way, the social sciences
should give a richer understanding of man's institutions, customs, and

ways of living. It is not necessary, however, that such a course of instruction be given in order to provide adolescents with worth-while information and guidance, nor does the teacher need to provide even an isolated lesson or discussion, except as a special need may arise among a group of boys or girls.[20] Through ordinary classroom activities and school programs, materials may be presented and problems projected, provided the teachers do not take a "taboo" attitude toward a discussion of any problems that may have a direct bearing on sex education. In addition there is much information and literature that bring the students into closer contact with some of the problems directly related to social relationships between boys and girls. Many family problems are well illustrated in such books as Undset's *Kristin Lavransdatter,* Maugham's *Of Human Bondage,* and Galsworthy's *Forsyte Saga.*

Aberrations. In the earlier years of life the sex impulse appears to act rather vaguely and indirectly, and therefore often assumes a form wholly out of line with the normal course and final outlet. The relationship between conduct during the earlier years of life and later sex life is not clear, but there is evidence that sex is somewhat related to the love behavior of the young child. With the development of the sex glands, and the maturation of the individual both physically and socially in a social world, many factors may operate to cause behavior resulting from the release of certain drives to deviate from a normal or socially acceptable course. Some of these factors are: repression, ignorance, sex phobias, disgust, curiosity, or some other conditions emotionally toned. It is during the stage of the operation of such factors that trial-and-error behavior ensues. The subject will try many methods of adjusting himself sexually, and some of his efforts may result in perversions—habits that are undesirable either because they will bring ultimate personal dissatisfaction or because they interfere with normal social relations.

[20] This will depend upon a number of factors related to the nature and needs of the boys and girls growing up in the community. For materials related to sex education see: H. N. Baker, *Sex Education in High School.* Emerson Books, 1942. This book presents a survey of sex education in the various states, pointing out the techniques used, problems encountered, and outcomes of the work. It should be most helpful to those considering this problem for high school pupils. B. C. Gruenberg, *High School and Sex Education.* U. S. Public Health Service Bulletin No. 75, 1940. This is an excellent discussion of the problems related to sex education, designed for adolescent and post-adolescent boys and girls. It may be secured from the Superintendent of Government Documents.

Sometimes an extreme attachment will develop between members of the same sex rather than, as is normal, between members of opposite sexes. This is referred to as homosexuality and happens especially when there is a complete absence of the members of the opposite sex during recreational and play life, or when there is an excess of teasing or ridicule about members of the opposite sex; or, still further, when there is a rather complete segregation of the members of one sex from those of the other. Crushes are especially prevalent among girls in camp. At this time the girl is separated from her family and from many of her past friends.[21] The setting, perhaps, is unfamiliar, and there is a lack of a feeling of security. From this lonely and unfamiliar situation, crushes are likely to develop, as a result of the closer contacts and intimacies of a twenty-four-hour-a-day program of living with others. On the other hand, an emotional shock due to frightful stories about sex activities may develop a sexual phobia or a feeling of disgust for members of the opposite sex that will tend to lead to aberrations.

Exact data of a reliable nature on the extent and frequency of masturbation among children and adolescents are not available. Recent studies, however, indicate that the practice is more widespread at the various age levels than was earlier believed. Davis found that 64.8 per cent of 1,181 unmarried women who were college graduates admitted having practiced masturbation at some time in their lives.[22] Of these 17.7 per cent reported the first incident occurred when they were two or three years old. The sex lives of 4,600 unmarried men examined in connection with the Selective Service Act in the summer and fall of 1941 were studied by Hohman and Schaffner.[23] Approximately nine-tenths of these men (ages 21 to 28) admitted masturbating at one time or another in their lives. A summary of various studies indicates that, although the frequency is higher among boys than among girls, girls start the habit more frequently during early childhood or late adolescence, while boys begin more frequently during early adolescence.

There is evidence from comparing figures of the sex lives of college

[21] See E. V. Van Dyne, "Personality Traits and Friendship Formation in Adolescent Girls," *Journal of Social Psychology*, 1940, Vol. 12, pp. 291–303.

[22] K. B. Davis, *Factors in the Sex Life of 2200 Women*. New York: Harper and Bros., 1929.

[23] L. B. Hohman and B. Schaffner, "The Sex Lives of Unmarried Men," *The American Journal of Sociology*, 1946, Vol. 52, pp. 501–507.

women reported by Davis in 1929 with those reported about a decade later by Bromley and Britten that a changed attitude and practice toward sex mores is operating in our society.[24]   Terman's study in 1938 of factors making for marital happiness lends still further support to this contention.[25]   In his study the sex histories of the older and younger age groups of 792 married people were studied.   The experience of sex relations before marriage in the older group was 49.4 per cent for the males and 13.5 per cent for the females; while for the younger group the relationship was increased to 86.4 per cent for the males and 68.3 per cent for the females.   These studies, supplemented by other studies as well as by their own data from 613 subjects, led Porterfield and Salley to support the validity of an hypothesis they set forth, which was:

First, the older universals of the sex mores are being replaced by numerous alternatives in the current sexual folkways, and as a result, the control of sexual behavior is much relaxed.   Second, in the light of this change, it is becoming increasingly difficult to define sex delinquency in any special sense, and perhaps meaningless to try to do so.[26]

Alfred C. Kinsey and his group of scientists have taken sex histories of more than 12,000 Americans in all walks of life.[27]   The study was sponsored by the National Research Council and has received wide recognition.   The study reveals that sex attitudes and habits start in infancy and the sex pattern of the average American male is fairly well fixed by the age of sixteen.   Contrary to general opinion, the male's prime of life as determined by sexual vitality reaches its height in the teens rather than in the twenties or thirties.   The average adolescent girl engages in about one-fifth as much sexual activity as the boy, and the frequency of sexual outlet for her in the twenties and thirties is less than that for the boy.

According to this report there is no single "American sex pattern." The sex attitudes and patterns of the average child are formed through

[24] D. B. Bromley and F. Britten, *Youth and Sex*.   New York: Harper and Bros., 1938.

[25] L. M. Terman, *Psychological Factors in Marital Happiness*.   New York: McGraw-Hill Book Co., 1938, p. 323.

[26] A. L. Porterfield and E. Salley, "Current Folkways of Sexual Behavior," *The American Journal of Sociology*, 1946, Vol. 52, p. 209.

[27] George Gallup, "Dr. Kinsey's Study of Sex," *San Francisco News*, February 24, 1948.   See also Alfred C. Kinsey, Wardell B. Pomeroy and Clyde E. Martin, *Sexual Behavior in the Human Male*.   Philadelphia: W. B. Saunders Company, 1948.

contacts and associations with his playmates and friends, with the home playing an indirect rather than a direct role at most stages. The importance of the home in shaping attitudes toward sex during preschool period is emphasized. Mankind seems not to have changed appreciably: masturbation has as high a frequency today as it had a generation ago.

Normal vs. abnormal sexual manifestations. Sexual abnormalities should be interpreted as deviations in the psychosexual development of the child. There appears to be certain aspects of sexual activity that characterize child behavior at different stages of development. These should not be looked upon as abnormal, and emphasis upon these in an effort to inhibit the child's psychosexual developmental manifestations can only be harmful. George Gardner states, "At various chronological stages in the course of the child's development he enjoys many physical sensations which by the time he reaches puberty will be allied or unified under the dominance of heterosexual forces." [28]

One cannot lay down a fundamental principle whereby all aberration can be graded or classed together, nor can the same causative factors be used in explaining all such behavior activities. Observational studies of infants show that body handling appears very early, and that many children persist in playing with their genital organs or in rubbing their thighs during the nursery school age. Local skin irritations, careless handling by others, too tight clothing, and the like are contributing causes. Emotional stimuli related to sex or inadequate information may lead to a turning of the sex drive from members of the opposite sex to those of the same sex. It is in misinformation and in inadequate information that the sources of many sex problems lie. When such a condition has existed and the child is further shielded from contacts of a wholesome nature with members of the opposite sex as he grows into adolescence, he acquires distorted ideas and attitudes toward sex. Many of the present-day sex problems among growing boys would be solved at an early period if conditions and customs provided for early mating as a means of sexual release. However, this is contrary to the customs, morals, philosophy, and institutions of our civilization. Perhaps we will have to conclude that in the present state of society and public opinion there is no plausible solution to the problem. In this connection Louttit has stated relative to masturbation:

[28] Reprinted from "The Community and the Aggressive Child," in *Understanding the Child,* January, 1949, by George E. Gardner, published by the National Committee for Mental Hygiene, 1790 Broadway, New York.

It is the attitude of parents, teachers, and other adults that the habit is bad, unclean, shameful, and the like, as well as the all too frequent belief in the harmfulness of sequelae, that affects the child and causes him to use any means to circumvent discovery. The patient himself may copy the attitude of others and in so doing cause himself much worry and create feelings of inferiority which are more harmful than the act itself.[29]

It is not to be inferred here that guidance is unnecessary. Guidance, necessary at all stages of life, is especially necessary during the plastic years before ideals and attitudes desirable for the welfare of the group have been established. The failures during this period of life can very possibly be traced to one or a combination of the following factors: (1) failure in some form of care and supervision; (2) inadequate or inaccurate information concerning sex life; (3) the development of sexual phobias or disgust, and (4) lack of proper playmates of both sexes.

The problems that have been presented throughout this study are important not only in relation to the present social and sex adjustments of adolescent boys and girls, but also in connection with their adjustments in family life in the years ahead. Guidance and training of youth toward a well adjusted and happy family life has been too often left to chance. This is, no doubt, an important factor affecting the extent of maladjustments in home and family living. Concerning the importance of early childhood experiences and training for family life, Cochrane has stated:

It is not too much to say that the ability to establish meaningful relationships and to find satisfaction in family life through marriage is largely conditioned by childhood experiences and by the acceptance of masculinity or femininity and sex differences by the individual.[30]

SUMMARY AND IMPLICATIONS

The period of adolescence has been described throughout this chapter as one during which adolescents develop a keen desire for peer approval. This is a period of formation of chums and friends, and friendships have been shown to become more stable at this time. The formation of cliques also characterizes this age, and these reflect the operation of subculture social-economic classes.

[29] C. M. Louttit, *Clinical Psychology* (Revised Edition). New York: Harper and Bros., 1947, p. 355.

[30] H. S. Cochrane, "Emotional Aspects of Social Adjustment for the Child," *Mental Hygiene,* 1948, Vol. 32, pp. 586–595.

The ripening of the sex impulses at this stage is accompanied by a changed attitude toward members of the opposite sex. No longer does a boy look upon a girl his own age as someone to be avoided; he now sees her as a personality whose admiration he desires. The development of this changed attitude is a natural concomitant of the ripening sex drive.

Adolescents have, as one of the major problems of their development, that of identifying themselves fully with the role—masculine or feminine—characteristic of their sex, an identification that began during infancy, when parents and friends made simple distinctions between boys and girls. It is necessary for the individual to learn to play his sex role, if he expects to be acceptable to members of the opposite sex. The infant is born without any awareness of sex or knowledge of its functions. Early in life, however, he learns that boys are treated differently from girls; he becomes familiar with the sex characteristics of his own body, and identifies himself with those characteristics; he accepts the attitudes of others concerning what is masculine and what is feminine; and if his sexual life and social contacts are normal, he eventually adjusts to his own and to the opposite sex in a satisfactory manner. The establishment of desirable heterosexual relations is an important part of social maturity. In the maturing process the adolescent faces many problems and tasks, some of which were considered in previous chapters, whereas others are yet to be considered in chapters to come. They may be summarized as follows:

. . . [The adolescent] must: (1) break away from a relatively exclusive dependence on his family and establish broader social contacts; (2) establish satisfying heterosexual relationships; (3) make vocational choices and move in the direction of efficiency; (4) make fundamental choices of allegiance; and (5) find and begin to assume his role in the social-civic life of the larger community. While performing these tasks, he needs to maintain affection and feelings of security in his immediate personal-social relationships, to develop a personal philosophy, and to establish a feeling of worth through achievement that will make him acceptable and successful in adult society. . . .[31]

THOUGHT PROBLEMS

1. What is the general significance of the findings presented in this chapter dealing with the stability of adolescent friendships? How would you account for the sex differences presented?

2. Do your observations and experiences corroborate the findings presented

[31] Commission on the Secondary School Curriculum, *The Social Studies in General Education*. New York: Appleton-Century-Crofts, Inc., 1940, pp. 92–93.

in this chapter relative to the dating practices and activities within special class groups? What bearing does this have on the stability and continuity of class structure?

3. How would you account for the high degree of loyalty manifested by adolescents to their peers?

4. What are some of the barriers that adolescents often set up to exclude adults from their activities?

5. What do you consider the major values to be derived by adolescents from good peer relations? To what extent do such values enter into the lives of adults? How would you account for any differences to be noted here?

6. Observe the behavior of a group of sixth and ninth graders. What differences did you observe in their interest in members of the opposite sex?

7. Show how the sex drive is related to "the desire for social approval."

8. What are some cautions or principles to be followed in order to avoid the development of sexual aberrations?

For information concerning films for use in instructing youth and adult groups in problems of sex, address The American Social Hygiene Association, 1790 Broadway, New York, N. Y. The film *The Gift of Life* deals with the processes of reproduction through all forms of life. Working with the University of Oregon, Producer-Actor Eddie Albert and Dr. Lester F. Beck of Oregon made a movie called *Human Growth*. This has been used successfully and shown to over half a million high school students.

SELECTED REFERENCES

Bromley, D. D., and Britten, F. H., *Youth and Sex*. New York: Harper and Brothers, 1938.

Butterfield, Oliver M., *Love Problems of Adolescence*. New York: Emerson Books, Inc., 1939.

Campbell, Elsie H., "The Social-Sex Development of Children," *Genetic Psychology Monograph,* 1939, Vol. 21, pp. 461–552. A complete review of the developing sex drive and its socialization during the years covered by the study, 5 to 18.

Cole, Luella, *Psychology of Adolescence* (Third Edition). New York: Farrar and Rinehart, 1948, Chap. VII.

Fleming, C. M., *Adolescence*. New York: International Universities Press, 1949, Chaps. XI and XII.

Hollingshead, A. B., *Elmtown's Youth*. New York: John Wiley and Sons, 1949, Chap. IX.

Hurlock, E. B., *Adolescent Psychology*. New York: McGraw-Hill Book Co., 1949, Chaps. XII and XIII.

Sellin, Thorsten (Ed.), *Adolescents in Wartime*. The Annals of the American Academy of Political and Social Science, 1944.

Terman, L. M., *Sex and Personality*. New York: McGraw-Hill Book Co., 1936.

Tyron, Caroline M., "The Adolescent Peer Culture," *Forty-third Yearbook of the National Society for the Study of Education,* Part I. Chicago: Department of Education, University of Chicago, 1944, Chap. XII.

XI

The Adolescent at Home

"All happy families resemble one another; every unhappy family is unhappy in its own way."

Count Lyof Nikolayevitch Tolstoi

THE IMPORTANCE OF HOME INFLUENCES

Throughout the vital period of adolescence, there is a phenomenal growth that alters the physiological pattern, the anatomical pattern, and the psychological pattern, thus transforming the child into an adult. The nature and extent of this growth are described in earlier chapters. Simultaneously, as the individual approaches maturity, there arises an urge, sometimes of relentless physical strength, sometimes characterized by increased and enlarged mental vigor, and still more often of a socialized nature relating to ambitions and ideals. This urge is seeking expression, uninhibited by the home domination so characteristic of the period of childhood. There is an urge to break away from the semipassive family relationship of childhood to a more independent way of doing and thinking and thus directing one's own plans and destiny toward an adult life.[1] This detachment from family ties does not necessarily involve a physical separation; rather, it connotes an emotional severance—in other words, the casting off of those bonds that would hinder the individual from achieving the things he is striving for and wants independently in adult life. Such a process has been referred to as *emancipation from family domination,* or *psychological weaning,* or by a third term, *achieving independence.*

The in-between person is in a naturally conflicting situation. The unwillingness of parents to recognize that their child is growing up

[1] J. M. Murray, "The Conscience During Adolescence," *Mental Hygiene,* July, 1938.

and maturing at a rapid pace, associated with the latter's growing independence at this period of life, complicates the whole problem. On the one hand, there is the adolescent's desire to break loose from the sheltering walls of the home and to get rid of specific restrictions; on the other, there is his desire for the protection and security the home affords. He cherishes adventure and looks forward to excitement; however, as problems arise, he has a felt need for protection and security. The confused situation resulting from these diametrically opposed drives oftentimes reveals the adolescent in dual roles.

Importance of early home influences. The influence of the home on adolescent behavior is almost synonymous with that of habits formed during the preschool years—the period in which the home's influence is greatest. Frequently, habits are formed in later years that may not seem to be of the same lineage as habits acquired earlier; but the influence of the latter must not be underestimated. Habits are built upon habits, and the earlier habits are likely to give something of their form to the later. This tendency is well illustrated by Rosenheim's descriptive analysis of a thirteen-year-old boy who lacked parental affection during the early years of his life;[2] as a result, he had never learned to show affection for others, and was unable to get along with other boys and girls of his age. Remedial treatment and guidance produced some good results; in spite of this, however, the influence of the early home environment remained constant and more or less pervasive. This influence was especially noticeable in the boy's lack of social responsiveness, and thus in his failure to establish desirable social relations with others; he also lacked steadfastness to ideals and good behavior standards.

Studies of early home influences in their relationship to the personal and social adjustments of adolescents have furnished considerable evidence that satisfactory adjustments are related less to the education of the parents, the size of the family, or the socio-economic level than to the extent to which basic human needs for affection, security, status, and belongingness are satisfied. The case of a somewhat introverted fourteen-year-old girl from the junior high school of Hastings-on-Hudson, New York, shows the influence of a desirable home situation in the development of attitudes and interests. The case D, described by Anderson, is in the ninth grade and has an IQ of 122.

[2] F. Rosenheim, "Character Structure of a Rejected Child," *American Journal of Orthopsychiatry,* 1942, Vol. 12, pp. 486–495.

This girl was identified by her classmates as being someone who prefers to do things by herself and doesn't care to have a lot of people around all the time, who is happy and interested in whatever she is doing, who is co-operative and agreeable, who volunteers to recite in class and always remains sure of herself, who never causes the teacher any trouble, who never tries to avoid responsibility, who is able to work quietly, who never loses her temper, who doesn't take chances for fear she will be wrong, who is shy and never has anything to say around strangers.

D appeared (upon interview) to be a girl of unusual poise. She talked easily and well, smiled frequently and seemed self-assured.

D's major interest is music. She prefers reading and her piano lessons to active games. She doesn't like to dance, and her best friends are admired because they are 'studious,' 'serious,' and 'don't go out much.' She feels that she has a good many friends but would like to have more. Her parents approve of all her friends and encourage her to bring them home. She is allowed to go out in the evenings on special occasions.

D likes school 'in a good many ways' but feels that she is usually required to do too much homework. She was elected president of her class last year and is given a good deal of responsibility this year for class activities.

D's family relationships seem to be harmonious. Although she was scolded or her privileges restricted after misbehavior when she was a small child, it has been a long time since anything like that happened, as she doesn't do things now that displease her parents. She prefers her mother to her father because her mother is 'around most of the time' and is 'kinder,' gives her money when she requests it. D has three older sisters at home; they quite often play cards or do other things with the mother. Several years ago the whole family went to Europe. Sometimes D feels that her sisters try to 'tell her what to do' a little too much but her mother usually intervenes and straightens things out since 'she can see both sides.' On the whole she feels that her parents do not have too many restrictions and are usually right in the standards they set forth.[3]

Parent preference. There is some evidence that the theories set forth by the psychoanalysts relative to the mother fixation on the part of children are true. In the study reported by Simpson[4] of the parent preferences of young children, 500 carefully selected children (fifty boys and fifty girls in each of five age groups) were given a battery of tests designed to measure parent preferences. The children were asked about some pictures, were told stories, and asked about their

[3] John Peyton Anderson, "A Study of the Relationships between Certain Aspects of Parental Behavior and Attitudes and the Behavior of Junior High School Pupils," Teachers College, Columbia University, *Contributions to Education,* No. 809, 1940.

[4] M. Simpson, "Parent Preferences of Young Children," Teachers College, Columbia University, *Contributions to Education,* No. 652, 1935.

dreams. The method for conducting this work was sufficiently standardized to make the results valid. Both sexes showed more mother
preference than father preference in all groups except for the five-
year-old girls. There was an increased percentage of mother preferences with increased age, and in all cases the mother appeared more
frequently in the dreams reported. This preference for the mother is
further borne out by the study of Meltzer.[5] The free association interview technique was used in this study. The reactions of the boys
appeared to be more unusual and complex in nature, indicating that
a greater range of factors and conditions affect their preferences. It is
reasonable to expect that the parent who is in a position to administer
to the needs of the child will be at an advantage in winning his love
to the greatest extent.

In connection with the White House Conference (1930) Burgess[6]
conducted a study of the adolescent in the family. This investigation
reached 13,000 children, from whom 8,000 were chosen for further
study. These children represented a sampling of urban, rural, native,
foreign-born, whites, and Negroes. The children were asked to fill
out elaborate questionnaires, and these were supplemented by data
from their teachers. The study yielded information about the extent
to which children confided in their parents. Tables XXXIV and
XXXV give data on this and show that the mother stands out as a
confidante for both boys and girls, although there is a substantial difference between rural and urban families. It would be difficult to assign
the specific reason for the pronounced difference between the extent to
which rural and urban children confide in their fathers. It seems quite
reasonable to state that fathers in the city are at home far less on the
average than those in the rural environment, and that mothers in the
city are likely to be at home more of the time.

Quite frequently parents adopt unwholesome and undesirable attitudes toward their children through sheer ignorance, carelessness,
or lack of foresight. A mother, failing to realize the necessity for the
child's withdrawal from under her wings of protection, may clutch
strongly the determination to dominate the child. Such a parent cannot hope that the youth will be able to assume responsibility and
exercise self-control if he is not permitted to practice either one. To-

[5] H. Meltzer, "Sex Differences in Parental Preference Patterns," *Character and
Personality,* 1941, Vol. 10, pp. 114–128.
[6] E. W. Burgess (Ed.), *The Adolescent in the Family.* New York: Appleton-
Century-Crofts, Inc., 1934.

day, families are considerably smaller and mothers have relatively more time for each child than did their forebears, who not only raised large families but also bore the drudgery of all the household duties.

TABLE XXXIV

THE DEGREE TO WHICH CHILDREN CONFIDE IN THEIR MOTHERS

Type of Child	Almost Never (%)	Some-times (%)	Almost Always (%)
Rural white boys	21.78	55.71	22.50
Rural white girls	5.15	45.06	49.77
Urban white boys of American parentage	10.82	48.21	40.95
Urban white girls of American parentage	7.21	30.72	62.06

TABLE XXXV

THE DEGREE TO WHICH CHILDREN CONFIDE IN THEIR FATHERS

Type of Child	Almost Never (%)	Some-times (%)	Almost Always (%)
Rural white boys	28.78	54.16	17.04
Rural white girls	31.29	55.10	13.60
Urban white boys of American parentage	24.09	48.88	27.01
Urban white girls of American parentage	29.09	47.71	23.18

Much has been said about the mother's intense love for her children. That such an attachment exists, very few would deny; but in many cases this attachment may be colored by a form of selfishness. Too often the mother's eagerness for her daughter to make good grades stems from the effects that this will have upon the mother's reputation in the neighborhood. Again, the father may desire to have his son enter into some vocation that he himself had earlier desired to enter— thus satisfying his own unfulfilled ambition.

A mother may become so attached to the child and so integral a part of the child's life that she will dislike being pushed into the background when the adolescent finally begins to assert himself and display qualities of independence. The father has been omitted from this discussion, since today he usually plays the role of the breadwinner and assumes less responsibility for the control and discipline of children, except in extreme cases.

It is during this period of transition that parents must use insight, for there is a significant difference between coddling, shielding, and anticipating every desire of the child and the supervision and sympathetic guidance of the adolescent into adult ways of doing and thinking.

The effects of the home. The social adjustments of children living in a tenement area were studied by Boder and Beach.[7] The object of their study was to learn to what extent children on a given street in a large city varied in social adjustment, and to trace some of the factors that may have led to differences among them. By selection of one small street with fairly uniform housing conditions, certain of the most undesirable social and economic factors were held constant; and since all the children of the neighborhood were included in the study, some of the difficulties in attributing causal significance to certain factors were avoided. The outstanding conclusions that emerge from this consideration of some of the factors that might account for the variation in social adjustment displayed by the children living on the same street, are the predominating influence of parental attitudes toward the children and the general relation between what are usually considered good parental attitudes and adequate social adjustment on the part of the children. The various types of maladjustment in parental attitudes appeared to produce rather specific types of reactions in the children. Most of the families in which the children were shy, retiring, or generally socially inadequate had mothers that were—by one means or another—in complete control of the household. Some of them achieved dominance by psychosis or neurosis, others by native ability or by providing the family with economic support. For the most part they overprotected their children, either through excessive solicitude or by undue control of their activities. The fathers were either easygoing, quiet, submissive men or were no longer living at home. On the other hand, the children who were unsupervised and neglected through the mother's laxness or were subjected to the father's violent temper escaped the tense, quarrelsome atmosphere of their homes and became the mischief-makers of the neighborhood.

There is evidence from many sources that the family group is very important in establishing patterns and attitudes affecting the individual's personal-social relations; and a fair degree of security within the home aids in the extension of such relations. Entorf has said of this:

[7] D. B. Boder and E. V. Beach, "Wants of Adolescents: I. A Preliminary Study," *Journal of Psychology,* 1937, Vol. 3, pp. 505–511.

Personally satisfying and socially constructive family relationships seem to depend very largely upon the capacity for genuine and sustained affection and the possession, especially on the part of the parents, of a certain sense of personal adequacy which renders domination, dependence, or emotional exploitation unnecessary within the family circle.[8]

In the development of personal adequacy among children Stott found that the city home ranked first; the farm home, second; and the small-town home, third.[9] These findings may be due to forces and conditions existing within the community. In the small town there is constant gossip about trivialities and a strict censorship is established. This accounts for the findings by Stott that

. . . On the average the attitude of small-town fathers toward the question of strict discipline versus freedom for adolescent children was found to be reliably more in the direction of strictness than that of either the city or farm fathers. It was further shown that the 'strict' attitude of fathers was negatively correlated with their children's scores in independence in regard to personal problems.

The community pattern. The community patterns of attitudes affect the attitudes of the home, and indirectly as well as directly affect the personality development of children as they reach adolescence and enter into more numerous out-of-home contacts. Concerning this Stott points out:

A young person's adjustment to life in general, his attitude toward work, or his independence in solving his personal problems are largely determined by the family situation in which he develops, by the wisdom of his parents in letting him do, and assume the responsibility for, his appropriate share of the household work, and by the extent to which he is allowed and encouraged, without blame for mistakes, to choose for himself and make his own decisions, but they are also conditioned and modified by the culturally determined attitudes and mores of the community or the degree of social integration which characterizes the neighborhood. Whether or not every 'date' an adolescent girl has is a matter of some concern and of considerable conversational value to the whole neighborhood, for example, might have much to do with the sort of personal adjustment she is able to make to life. Clearly the home setting, as well as the quality of the home environment as such, is a factor of importance in the development of personality.[10]

[8] M. L. Entorf, "Ends and Means in Teaching Family Relationships," *Parent Education,* April, 1938.

[9] Leland H. Stott, "Personality Development in Farm, Small-Town, and City Children," *Agriculture Experiment Station Research Bulletin,* No. 114. University of Nebraska, 1939, p. 33.

[10] *Ibid.,* p. 33.

HOME CONFLICTS AND ADJUSTMENTS

Parental attitudes. Some parents are prone either to "baby" their children or to dominate them with an iron hand. There is the case of Henry, which came to the writer's attention several years ago.

The boy's father was an attorney of prominence in the city. The mother had evidently come from a good family, but one puritanical in ideas and attitudes. The boy was an only child. He was a likeable lad and of more than average intelligence. During grammar school years he acquired the average number of acquaintances and friends. He played with those friends whenever he was permitted. But those times were all too few. He had to help with the housework, mow the lawn, weed the garden, and do any other odd jobs that could be found about the home. He completed grammar school at an average age and entered high school. He was developing as a normal boy. He was gaining in weight and stature. His father was a well-built man and it seemed that the son would be the same. Naturally, as he grew larger, he wanted to engage in the same sports and interests as other boys of his age. But it seemed that *mother always had something else for him to do.* Conflicts arose. If he could not get leisure in leisure time, he would take it from the hour of duty. He found another boy who was having the same or similar difficulty. That was the beginning of truancy which ended in expulsion from high school. The expulsion put him at odds with his father, and this led to his staying out late at night and associating with questionable companions in questionable places. Upon learning of this behavior, the father ordered the boy out of the home. A few days later the circus came to town and the boy joined the circus. He stayed with the circus for almost a year and then was arrested and found guilty of stealing.

One of the problems with which many parents have been much concerned is the protection of growing boys and girls from undesirable literature and lurid stories. Certainly, parents won't be able to protect their children from the undesirable side of life forever by a constant *Verboten.* Perhaps a better approach would be one based upon positive guidance, in an endeavor to aid adolescents to develop discrimination. Say to the boy about to read some book considered undesirable, "Go ahead and read it, if that is what you wish to do. Remember, however, that there are more accurate books on the subject if you are looking for facts. There are better books on many subjects if you are merely looking for something interesting to read."

The study by Stott [11] showed that older parents were inclined to voice

[11] L. H. Stott, "Parental Attitudes of Farm, Town, and City Parents in Relation to Certain Personality Adjustments in Their Children," *Journal of Social Psychology,* 1940, Vol. 11, pp. 325–339.

a need for greater control of adolescents than did younger parents. Mothers and fathers scored, on the average, about the same. The town mothers had the most favorable and the town fathers the least favorable attitudes of any of the sub-groups studied. The attitudes of the fathers toward self-reliance in children was not found to be significant, except perhaps in its relation to the adolescent's appreciation of home life. It was concluded that, "a home situation, then, in which both parents agree that high school children should be granted considerable freedom from parental domination is favorable to the development of self-reliance (independence of judgment in regard to personal problems and difficulties)."

A nationwide survey of 10,000 high school youngsters has provided information on what adolescents think about the understanding of their problems by their parents.[12] A question asked of the high school youngsters was: Do you or do you not think that most parents these days understand the problem of their teen age sons and daughters? This question was answered as follows:

Do	35 per cent
Do Not	56 per cent
Undecided	9 per cent

Thus, only one-third of the high school students believed that most parents understand the problems of the teen-agers; while over half of them felt that there was a lack of such an understanding. Perhaps every generation feels that it is misunderstood. It is significant that this attitude exists as a barrier to desirable parent-adolescent relationships.

A further analysis of the results showed that a larger percentage of high school juniors than students at any other grade level feel that parents do not understand their problems. No significant sex differences were noted—there being 58 per cent of the boys and 55 per cent of the girls replying "Do Not." Also, no significant differences were found between the responses of students coming from homes of higher income and those coming from homes of average or below-average income.

Conflicts with parents. The reasons for disagreements arising between parents and adolescents give a further understanding of adolescent yearnings and problems related to home restraint. According to

[12] The survey was conducted by the Purdue University Opinion Poll during the year 1948.

the data of Table XXXVI the most frequent source of difficulty for both boys and girls was the issue of going out or staying out late at night.[13] This is a problem of adjustment or conflict between two different standards: the parents' desires on the one hand, and the attitude of friends on the other. The sex differences presented in this study reveal

TABLE XXXVI

SOURCES OF DISAGREEMENT BETWEEN 348 BOYS AND 382 GIRLS
AND THEIR PARENTS

SOURCE OF DISAGREEMENT	Boys (%)	Girls (%)
Use of automobile	35.6	29.6
The boys or girls you choose as friends	25.0	27.0
Your spending money	37.4	28.8
Number of times you go out on school nights during the week	45.1	47.6
Grades at school	40.2	31.2
The hour you get in at night	45.4	42.7
Home duties (tending furnace, cooking, etc.)	19.0	26.4
Clubs or societies you belong to	5.5	10.5
Church and Sunday school attendance	19.0	18.6
Sunday observance, aside from just going to church and Sunday school	15.2	13.9
The way you dress	14.4	24.6
Going to unchaperoned parties	15.8	27.5
Any other source of disagreement	9.5	8.4
"Do not disagree"	2.0	2.1

that boys' problems exceeded those of girls in such things as the use of the automobile, spending money, and grades at school. For the girls, problems related to home duties, clubs or societies, manner of dress, and nature of parties presented a greater source of disagreement than did the problems of the boys.

It should not be concluded that all parents are domineering and possessive, nor that all children go through a struggle in achieving independence. In most homes parents come to realize, because of one reason or another, that their children are growing up and that they must be dealt with differently as they reach a more advanced level of development. The growth process on the part of the individual child, coupled with wider contacts and increased knowledge, brings with it demands that usually find expression and attain results lead-

[13] Reprinted by permission from *The Family,* by J. K. Folsom, published by John Wiley and Sons, Inc.

ing toward the establishment of greater independence. Furthermore, it should be pointed out that there must be a limit to the degree of independence that anyone attains. Many adults are so completely self-sufficient that they never seek suggestions from others regarding anything, and as a result of their extreme independence they often pay dearly for mistakes that could have been avoided. If the parent becomes a counselor to the child, one who offers help rather than criticism in time of need and trouble, the child will come to recognize the importance of seeking help and advice in a rational manner. Although the boy or girl should be encouraged in this, it should be realized that while the parent gives suggestions and guidance, the growing adolescent must be the one actually to solve the problem and adjust the difficulty.

One of the most interesting and far-reaching studies dealing with adolescent conflicts is that conducted by Block.[14] She found that the conflicts adolescents have with their parents (in her study, mothers) were in many cases the basis for most of the disturbances in their lives.

Over a period of five years, 528 junior and senior high school boys and girls were interviewed. By means of a questionnaire, an index of the conflicts that high school students are facing was obtained. A list of fifty problems indicated by the students was then studied. These problems and the percentage of high school students reporting them are as follows:

1. Insists upon nagging me regarding what I wear and how I dress. (B*-26.3; G**-50.9.)
2. Complaints about how I comb my hair. (B-24.3; G-26.0.)
3. Fusses because I use lipstick. (B-0.0; G-64.6.)
4. Refuses to let me buy the clothes I like. (B-12.7; G-55.6.)
5. Complains about my hands or neck or fingernails being dirty. (B-55.7; G-10.5.)
6. Pesters me about my table manners. (B-74.8; G-63.9.)
7. Pesters me about my personal manners and habits. (B-68.5; G-70.0.)
8. Objects to my smoking. (B-0.8; G-13.4.)
9. Objects to my going with boys or girls she doesn't like. (B-19.1; G-40.4.)
10. Makes me go to bed at the same time that my younger brothers and sisters do. (B-30.6; G-45.1.)
11. Objects to the books and magazines I read. (B-17.9; G-32.5.)
12. Objects to my going to dances. (B-0.0; G-58.8.)

* Boys.
** Girls.
[14] Virginia Lee Block, "Conflicts of Adolescents with Their Mothers," *Journal of Abnormal and Social Psychology,* 1937, Vol. 32, pp. 192–206.

13. Insists that I eat foods which I dislike, but which are good for me. (B-82.4; G-83.8.)
14. Won't let me attend the church I want to attend. (B-4.4; G-53.4.)
15. Urges me to make friends with children of important people in town. (B-9.6; G-13.4.)
16. Won't let me take subjects I want in school. (B-32.9; G-56.1.)
17. Won't let me follow a vocation in which I am interested. (B-64.5; G-34.3.)
18. Insists that I go with friends of her choice. (B-20.3; G-69.7.)
19. Won't let me spend the night with any of my friends. (B-15.1; G-42.6.)
20. Nags about any little thing. (B-26.3; G-66.4.)
21. Insists upon interfering in settling any difficulties I may have with friends or teachers. (B-20.3; G-23.1.)
22. Talks baby talk to me. (B-33.4; G-10.5.)
23. Teases me about my girl friends. (B-51.3; G-0.0.)
24. Teases me about my boy friends. (B-0.0; G-65.7.)
25. Brags about me to other people. (B-50.1; G-22.7.)
26. Holds my sister or brother up as a model to me. (B-66.9; G-75.8.)
27. Spends most of her time at bridge parties, etc., and is rarely ever at home. (B-28.7; G-78.0.)
28. Tells her friends things about me that I tell her confidentially. (B-13.5; G-16.2.)
29. Insists that I be a goody-goody. (B-32.2; G-57.8.)
30. Shows favoritism to my brother or sister. (B-30.6; G-44.4.)
31. Embarrasses me by telling my friends what a good son or daughter I am. (B-49.8; G-26.4.)
32. Is cold to friends of mine she doesn't like. (B-19.9; G-45.1.)
33. Makes a huge fuss over friends of mine whom she likes. (B-34.3; G-36.8.)
34. Scolds if my school marks aren't as high as other people's. (B-82.4; G-85.9.)
35. Gets angry if I don't spend most of my time with her. (B-28.3; G-34.7.)
36. Talks against my father and wants me to agree with her. (B-8.4; G-16.6.)
37. Treats me as if I were a child. (B-5.2; G-16.3.)
38. Objects to my going automobile riding at night with boys. (B-65.7; G-87.4.)
39. Objects to my going automobile riding during the days with boys. (B-49.0; G-66.4.)
40. Insists that I tell her for exactly what I spend my money. (B-80.0; G-81.2.)
41. Won't give me a regular allowance. (B-54.1; G-52.3.)
42. Accompanies me to parties, movies, etc. (B-3.2; G-30.3.)
43. Insists that I take my sister or brother wherever I go. (B-50.5; G-82.3.)
44. Investigates places when I go to parties, etc., before I go. (B-15.1; G-44.4.)
45. Worries about my physical health. (B-26.7; G-58.8.)

46. Won't let me use the car. (B-85.7; G-70.8.)
47. Urges me to beat the next fellow in school work. (B-3.6; G-13.0.)
48. Urges me to outdo others socially, which I hate to do. (B-0.0; G-28.2.)
49. Won't ever let me go to the movies or dancing. (B-7.6; G-13.4.)
50. Won't let me entertain at home. (B-9.2; G-53.1.)

The results from the study by Block pointed to the fact that more conflicts were due to differences in thinking regarding personal appearances, habits, and manners than any other thing. Differences of opinion over vocational, social, recreational, and educational choices also caused some contention. Problems that appeared to have caused disturbances for the largest percentage of girls were in most cases the cause of least disturbances for boys, and vice versa. Girls in the seventh grade had the largest percentage of conflicts, while boys in the eighth grade had the largest percentage. When parents are cognizant of the sources of such conditions, they are in a better position to substitute guidance and understanding for conflict and contention.

Younger adolescents, especially, have difficulty in seeing any reason for many of the protective conventions of society. To insist upon obedience merely for the sake of obedience to some authority will have no value in the development of moral courage, but will, on the other hand, invite conflict and deception. As Butterfield points out:

When adolescents are reaching out to establish and enlarge their prestige with boy and girl friends they are likely to resent anything which restricts their efforts to win favor with such persons. The friendships of youth are precious and when apparently senseless social customs threaten to limit their enjoyment, youth readily adopts a defiant attitude.[15]

Conflicts regarding the proper night hours appear to be among the most common sources of friction between parents and adolescents. The Lynds[16] report that 45 per cent of 348 boys in the upper grades of the high school and 43 per cent of 382 girls who replied to their questionnaire admitted they were having difficulties with their parents about the question of late hours. The causes usually have as their bases the differences in standards between the parents and the social group in which their children are moving. Faced with this difficulty parents all too often resort to scolding and complaining; they either fail to give plausible reasons why they want their children to come in

---

[15] Oliver M. Butterfield, *Love Problems of Adolescence*. New York: Emerson Books, Inc., 1939, p. 33.

[16] Robert Lynd and Helen Lynd, *Middletown in Transition*. New York: Harcourt Brace and Co., 1937.

earlier, or neglect to set up incentives for obedience and to provide a workable plan whereby the children may be able to satisfy their needs for social life and still come in at a more reasonable hour at night. Most young people will be pleased to cooperate when they realize that a plan proposed is a fair one and for their own best interests. The National Tuberculosis Association, for example, has distributed an excellent pamphlet explaining the importance of obtaining the amount of sleep necessary to good health. The use of constructive and noncritical materials of that nature will yield far better results than will parental authority, and go a long way toward developing greater independence and self-control among the adolescent boys and girls.

Home adjustments. An item analysis of the Home Adjustment of the Bell Inventory brought out the following significant differences between boys and girls:

The high school boys had experienced a desire to run away from home more often than the high school girls. The high school girls were more irritated than were the boys by the following home conditions: their parents' personal habits, favoritism among parents, feeling of fear toward their parents, conflicting love and hate for parents, parents with violent tempers, and parents criticizing their appearance.[17]

The study of Stott[18] dealt with rural boys and girls (adolescents) from high schools in Nebraska. By means of personality scales and a home-life questionnaire he gathered data on the effects of certain factors in home life on personality adjustments.

The differences in home-life conditions between well-adjusted and poorly adjusted boys and girls (in terms of test scores on an adjustment test) are presented in Figure 32. According to the home items rated positively, the characteristics of successful farm family life measured by the personal adjustment of boys were, in order of their significance, as follows: (1) an attitude of welcome on the part of parents toward the boy's friends in the home, (2) no recent punishment, (3) a minimum of nervousness manifested in mother, (4) frequently having enjoyable times together in the home as a family group, engaging in

[17] H. M. Bell, *The Theory and Practice of Personal Counseling.* Stanford University, Calif.: The Stanford University Press, 1939.

[18] Leland H. Stott, "The Relation of Certain Factors in Farm Family Life to Personality Development in Adolescents," *Agriculture Experiment Station Research Bulletin,* No. 106, 1938, pp. 40–41. The critical ratios listed in Figure 32 represent the difference between the two scores divided by the standard error of that difference. Critical ratios of less than 2.00 are regarded as statistically insignificant.

such activities as playing games, telling stories or singing, and playing
instruments, (5) relatively little illness of mother.   The items characteristic of successful family life as viewed by girls' adjustments were as
follows: (1) no recent punishment, (2) a confidential relationship
with the father, (3) an attitude of welcome in parents toward the girl's

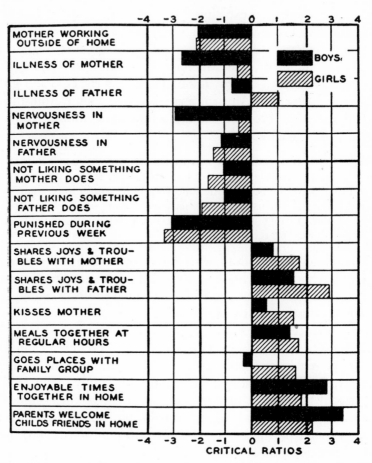

FIG. 32.   *Differences in home life conditions of well adjusted and poorly adjusted
rural boys and girls.*   (After Stott)

friends in the home, and (4) a minimum of participation of the mother
in the work outside the home.

Parental conflicts are not new, but when vast social and economic
changes are being wrought, the conflicts between the older and younger
generations will be much greater.   It was pointed out earlier that the

causes of home maladjustments are many and varied. When these have been studied and efforts directed to eradicate them through a well-planned cooperative program, much has been accomplished. The study by Brown [19] dealt with the problem of determining whether children who lived in the more stable environment of an orphans' home manifest more neurotic tendencies than those who reside with their parents. The Brown Personality Inventory for Children was administered to 200 orphans' home children of both sexes. A comparison group consisting of 200 boys and girls selected at random were likewise given the inventory, and comparisons were made between these results and those obtained from the children of the orphans' home. The results indicated a greater neuroticism among institution children. However, a comparison "with children of low socio-economic status whose parents contribute heavily to the population of institutions reveals a similarity between the two. Socio-economically inferior children living at home resemble the institutional children when compared for neurotic traits." It appears likely, therefore, that the greater presence of neuroticism among institutional children is a result of the environmental conditions from whence they came rather than of the institutional environment.

EMANCIPATION: GROWTH TOWARD MATURITY

Evaluating adolescent emancipation. The question we are confronted with in this topic involves an evaluation of the degree or extent to which emancipation has been accomplished in a given adolescent. Dimock [20] has made several interesting studies on the subject of achieving independence and has furnished a measure or technique for estimating the degree of emancipation. He first compiled a list of several hundred items of conduct and activities that were characteristic of dependence and independence. After the completion of the list, it was submitted to about one hundred judges—psychologists, educators, sociologists, and parents. These judges evaluated each item and the one hundred and twenty most important ones were included in the final test. A sample of this E. F. P. Scale by which the degree of independence can be estimated is presented in Table XXXVII.

From the scale, which is self-explanatory, Dimock turned next to a

[19] Fred Brown, "Neuroticism of Institution Versus Non-Institution Children," *Journal of Applied Psychology,* 1937, Vol. 21, pp. 379–383.

[20] H. S. Dimock, *Rediscovering the Adolescent.* New York: Association Press, 1937, p. 145.

study of the factors that condition emancipation. Chronological age was found relatively unimportant with a correlation of .14 between emancipation scores and age. Physical characteristics such as height and weight, on the other hand, were quite significant.

TABLE XXXVII

ILLUSTRATIVE ITEMS OF THE EMANCIPATION SCALE

BOY'S E. F. P. SCALE

ITEM	What I Do	What I Want to Do	What My Parents Want Me to Do
Decide things for myself	Yes No ?	Yes No ?	Yes No ?
Do what my father or mother decides on every question	Yes No ?	Yes No ?	Yes No ?
Depend on my parents to buy all my things for me	Yes No ?	Yes No ?	Yes No ?
Spend my allowance as I choose	Yes No ?	Yes No ?	Yes No ?
Pick out and buy my own clothes....	Yes No ?	Yes No ?	Yes No ?

Emancipation is evidenced from an analysis of movie attendance of children. As the child grows older there is a decline of movie attendance with members of the family but an increased attendance with friends and others, as is shown in Figure 27, page 143. The greater independence of the boys at all age levels is shown here.

If Dimock's E. F. P. Scale were employed in helping to determine John Jones's degree of emancipation and we found that he is still psychologically unweaned, what would be some characteristics of his behavior? First, John would constantly be seeking the advice and help of others simply because he cannot act or think independently. Mother has always been near to shield the youngster in her own inimitable way from burdensome tasks and difficult decisions. Help in school and supervised study are both necessary for John even to keep up with his classmates. His teachers wonder if he is capable of following printed instruction without having someone there to explain each step. Again, if John is forced to leave home for a visit, he suffers nostalgia to the extent that he loses his appetite and is unable to sleep. Perhaps this lad profoundly desires to become independent but is ignorant about the means of achieving this state. As a shield for his attachment to his parents he indulges in dramatic overcompensations such as getting drunk or using profanity. These radical behavior patterns are his outlets to show his independence. But looking into the future, we see the instability and unhappiness of an unweaned individual. He is not able to get along with the employer because he expects extra sympathy and "giving in" to his whims. Many a marriage has been wrecked owing to this same condition. The case may be that of an only child who constantly seeks the advice of an over-anxious mother. It is not necessarily the mother who spoils the child. A case called to the attention of the writer illustrates this quite adequately.

Jane, an orphan child brought up outside the orphanage, was cared for by an older sister. The older sister accepted full responsibility for Jane's clothes, education, and late love affairs. This was so complete that, even after marriage, Jane still consulted with her about things. Owing to varying circumstances, Jane finally came to make her home in an adjoining town near the older sister. She called her older sister almost daily over long-distance. Jane tried to see her at least each week. As a result of various social problems arising, she eventually found herself under the complete control of the older sister and finally wrecked her own home due to this complete *infantilism* accentuated by the ever-present dominance of the sister.

Principles of establishing independence. Learning to let go means for the adolescent the art of relinquishment. He is confronted with the task of throwing off childhood habits of almost blind obedience, dependence, and desire for protection. His emancipation from almost complete supervision to independence cannot and should not take place in too short a period. Rather, this should be a gradual process, begun during childhood by the parents and developed through carefully planned education for *initiative* and *responsibility*. With the adolescent caught between new urges and old habits, one cannot help but realize the deep need for sympathetic understanding and wise handling on the part of the parents.

What, then, are some desirable procedures to follow in the development of a growing child into a socially adequate and responsible youth? This is not a simple question; neither is there a single key that will answer it. That habits of independence should begin in childhood has already been suggested. With further development, responsibilities and privileges should be increased. The growing child will need more spending money and this can well be increased with advancing age. Again, the adolescent should be given greater freedom in the selection of his friends. The parents can function very effectively here through early training in ideals; for the present situation they can provide encouragement and an adequate setting for desirable friends whom the child has chosen. The adolescent wants greater freedom, for example, in buying his clothes or in doing his shopping for Christmas.

The study by Esther Prevey [21] showed that boys were provided with valuable experiences and training in the handling of money by their parents more often than girls were given such training. This difference appeared in the different parental practices studied, but was most pronounced in connection with experiences in earning money and in being a party in the discussions of the family financial status, expenses, and plans involving money. Follow-up studies of later money habits of the subjects studied revealed a positive relationship between parental practices in training children in the use of money and the ability to handle their financial activities successfully in early adulthood.

Not only are parents prone to thrust their ideals and manners of life upon their children literally in the form of a blueprint, but they may lay out certain vocational plans and try to make their children conform to them. Sometimes such plans are conceived of in terms of the parents' own weaknesses, their rationalizations, or still some other element in their make-up that is without a logical basis. The vocational plans of the adolescent should be made by the adolescent himself, with the aid, of course, of suggestions and information that may be obtained from the wisdom and understanding of those with whom he is in contact. Parents may—and in many cases do—have their own notions about what studies should be pursued in school, and many wellnigh force their child (a developing adolescent) to study particular school subjects without his understanding the reasons for such demands. It is in matters of such choices that parents can best serve as advisers; their advice becomes valid to the extent that emotions and feelings are controlled and reason and understanding, based upon fairness and truth, are used. Consider the following case:

Morris, a boy of fifteen, managed to play truant from school for two full months before being discovered. His feat involved considerable lying, interception of mail, forging a report card, and general deception. Previously Morris had been an unusually satisfactory son and pupil. An only child, he was reared in a household consisting of parents, grandparents, uncles, and aunts. He was an affectionate, obedient child, thoughtful of the adults, and

[21] E. E. Prevey, "A Quantitative Study of Family Practices in the Use of Money," *Journal of Educational Psychology*, 1945, Vol. 36, pp. 411–428.

especially close to his parents, who were deeply attached to each other and to him. He had friends, was reasonably well liked by other boys throughout his childhood, but was more sober-minded than most of his companions, and of his own choice spent much of his free time reading or in recreational activity with his adult relatives. His parents had thought that they understood him thoroughly and had his full confidence. Actually, a small issue had, before the truancy, unconsciously become the symbol of the increasing dilemma of this boy and his parents.

At about fourteen Morris had begun to be interested in the music of name bands, and soon afterward wanted to learn to play the trumpet. Though his parents recognized that he had musical talent and though there was money for instrument and lessons, they feared that Morris would want to form or join a band and that such a band would be the center of a whole section of his life that they could not share with him. Accordingly, they refused permission, rationalizing their refusal by claiming that he needed all his spare time for study. Later they weakened that argument to some extent by encouraging him to take a part-time job in the neighborhood drugstore.

When Morris was just fifteen, he was thrown into a mild depression by the sudden death of a favorite uncle who had represented support of those individual interests that he was unable to affirm in the face of his parents' opposition. Nothing seemed worth doing, and when in the fall a school companion promised an excuse and suggested that they cut school to hear a famous band, he agreed. When, later, the excuse was not forthcoming, Morris continued to play truant, listening to records and attending theaters, all the while in such great conflict over what he was doing that it was eventually a relief to have his deception detected.[22]

In the choice of sweethearts and finally of a mate, parents often find themselves in disagreement with their children. Though well-meaning and eager for the boy or girl to choose wisely, the parent cannot make the choice for the youth. Again, the role of the parent should be that of a counselor; his counsel will be effective insofar as he has been willing to serve as an impartial and ever helpful adviser in the various difficulties and problems that the adolescent faces. Adolescents will welcome suggestions and help, even in matters relating to the choice of a mate, when such help is given in a spirit of sincerity and fairness, motivated by a desire to aid them in finding the greatest harmony and happiness as a result of the choice made.

There is, therefore, a need for a carefully planned program integrated

[22] Norman R. Ingraham, "Health Problems of the Adolescent Period," *The Annals of the American Academy of Political and Social Science,* November 1944, p. 131.

by the schools, churches, and homes in guiding the developing adolescent boys and girls. Many parents are unaware that conflicts exist, and when they are aware of them, they do not in most cases understand the sources of such conflicts. Home situations that take their toll in the form of parental nervousness, family discord, and childhood unhappiness can best be dispelled by studying the underlying sources of such troubles. This was the aim of the study by Block. As a result of this study a program was formulated and its effectiveness proved. Some important characteristics of this program as presented by Block are as follows:

A comparative study of the interviews with children and their mothers demonstrated that many situations producing apparently similar problems were very different in their causal elements. A careful investigation of the total clinical picture of 69.3 per cent of the children in the seventh grade complaining about their mothers nagging them about what they wear and how they dress, showed that the basic cause of the nagging was different for different children. Since no two problems are identical, the home and school must realize that the methods of treating one child exhibiting a definite behavior pattern may be opposite from the method applied to another child exhibiting the same behavior patterns. Each child must be studied by his parents and teachers as an entity in relation to his peculiar physical, mental and emotional make-up and his environmental influences.

An analysis of the interviews revealed the need and desire on the part of parents for a better understanding of the problems of adolescents and for cooperative effort to help boys and girls solve these problems. Parent discussion groups, parent-teacher meetings, personal interviews between parents and advisers, interviews with parents, children and advisers helped to bring the school and home into a very close and cooperative relationship. Teachers were able to obtain clearer understandings of pupils and adjust their methods to the needs of each child. Administrators and supervisors were better able to distribute children intelligently to curricula and extra-curricula offerings that were interesting and challenging to them and to adjust the curriculum in the light of the felt needs of the group. Many children were better able to take advantage of the opportunities offered in the high school; others who had exhibited undesirable tendencies were recognized earlier and were so guided that their attitudes in many cases were modified into socially acceptable behavior. Parents and teachers worked together in defining, interpreting, and planning experiences for children which would be most conducive to well-balanced, satisfying, and challenging experiences for the child. As a result, children were less disintegrated by varying philosophies of treatment as is so often the case when the home, school, and community fail to define mutually a philosophy.[23]

[23] Virginia Lee Block, *op. cit.*, pp. 204–205.

SUMMARY

According to the concepts of growth presented throughout the discussions of adolescence, the adolescent is a product of all that has gone on before; no one ever outgrows his childhood. He develops physically, mentally, and emotionally, but he never escapes the influence of his earlier years. This is very fortunate, since these early years become preparatory periods for adult living. They are fundamental as a stabilizing force in molding the individual into an adaptable member of the society in which he is to live. Sometimes the process of training is undesirable or deficient and the child carries infantile traits into adulthood. "The immature adult is seen to be selfish, wilful, petulant, impulsive, and in other ways objectionable to society." [24]

The causes of conflict between the child and the parents are many, but the failure of parents to realize that the child is growing up stands out as a common observation. The tendency of parents to thrust their exact pattern of conduct and ways of behaving on the child is also generally present. The child may become selfish and wilful under the protection of wealth. The daughter grows up without any sense of responsibility under the dominance of a very strict father. The only child may be pampered and spoiled by an adoring aunt or grandmother. The social pattern in the home will do much to affect the child's social and emotional development. A domineering and ill-tempered father keeps the child ill at ease and repressed. Vacillating and inconsistent authority and punishment will present a condition of bewilderment for the growing child. These childhood patterns become fixed and tend to persist into adult life. Bluemel gives the following illustrations of the operation of such childhood emotional patterns:

Little Hettie frequently quarreled with her sister, and because she was the younger and smaller of the two children, she could do little to help herself in the situation except indulge in cutting remarks. This became an accomplishment, but unfortunately she carried the patterns of response into adult life, and she now frequently offends people with her snippy and flippant comments. Her acquaintances regard her as snobbish and superficial, and she has made no lasting friendships.

When Jeffrey was a child, he and his mother had many encounters at mealtimes. As he was underweight his mother insisted that he eat everything that was set before him. In this situation her attitude was one of

[24] C. S. Bluemel, *The Troubled Mind*. Baltimore: The Williams and Wilkins Company, 1938, p. 468.

stubborn insistence maintained with complete silence. The boy responded with the same stubborn attitude and thus an hour or two would pass in which each would contend against the other's will. The boy has carried much of this taciturn resistance into adult life.

Benny has always been the spoiled child of his widowed mother, and it has been her desire to smooth his way in life and indulge his every wish. When he encountered trouble as a boy he could always run to his mother, knowing that she would take his part and protect him in his difficulties. If she could meet the situation in no other way, she would buy him candy or a new toy, or give him extra spending money and thus divert him from his troubles. Benny still regards his mother as a refuge now that he has reached manhood. He brings his marital troubles to his mother, and she takes him into her home. He brings his financial problems to her, and she pays his bills. Now that his debts have become too large for her to meet, she urges that he take bankruptcy and thus continue his evasion of responsibility.[25]

THOUGHT PROBLEMS

1. Just what do you understand the term "emancipation," as used in this chapter, to mean? What is its significance in relation to adjustment problems?

2. Present a descriptive case of personality maladjustment due to unfortunate or undesirable family conditions.

3. Show how difficulties between brothers and sisters arise. Illustrate this by some case with which you may be familiar.

4. How do you account for the results from the Purdue University Poll presented on page 236?

5. Consider some adolescents that are happy and apparently well adjusted. What are some of the special characteristics of their home life?

6. It has often been stated that the most important single factor making for satisfactory adjustments among adolescents is the *attitude of the parents*. What do you regard as desirable attitudes? Be specific.

7. How do you account for the *parent preference* results presented in this chapter? What factors would tend to affect such a preference?

SELECTED REFERENCES

Arlitt, A. H., *Family Relationships*. New York: McGraw-Hill Book Co., 1942.

Brown, F. J., *The Sociology of Childhood*. New York: Prentice-Hall, Inc., 1939, Chap. VIII.

Cole, Luella, *Psychology of Adolescence* (Third Edition). New York: Farrar and Rinehart, 1948, Chap. IX.

Crow, Lester D., and Crow, Alice, *Our Teen-Age Boys and Girls*. New York: McGraw-Hill Book Co., 1945, Chaps. III and IV.

Dimock, H. S., *Rediscovering the Adolescent*. New York: Association Press, 1937, Chap. VII.

[25] C. S. Bluemel, *op. cit.,* p. 470.

Fleege, Urban H., *Self-Revelation of the Adolescent Boy.* Milwaukee: The Bruce Publishing Co., 1945, Chaps. III, IV, V, and VI.

Fleming, C. M., *Adolescence.* New York: International Universities Press, Inc., 1949, Chap. VIII.

Frank, Lawrence K., "The Adolescent and the Family," *Forty-third Yearbook of the National Society for the Study of Education,* Part I. Chicago: Department of Education, University of Chicago, 1944, Chap. XIII.

Hollingworth, Leta S., *The Psychology of Adolescence.* New York: Appleton-Century-Crofts, Inc., 1928, Chap. III.

Levy, D. M., *Maternal Overprotection.* New York: Columbia University Press, 1943.

Myers, Charles E., "Emancipation of the Adolescent from Parental Control," *The Nervous Child,* 1946, Vol. 5, pp. 251–262. The meaning of emancipation is first presented. This is followed by a review of studies bearing on the effects of different parental attitudes and controls on personality development during and following adolescence.

Pierce, W. C., *Youth Comes of Age.* New York: McGraw-Hill Book Co., 1948.

Symonds, P. M., *The Dynamics of Parent Child Relationships.* New York: Bureau of Publications, Teachers College, Columbia University, 1949.

Train, F. B., *Your Child, His Family and Friends.* New York: Appleton-Century-Crofts, Inc., 1943.

"Youth Education Today," *Sixteenth Yearbook of the American Association of School Administration.* Washington, D. C.: 1938, Chap. IV.

Zucker, Henry L., "Working Parents and Latchkey Children," *The Annals of the American Academy of Political and Social Science,* Vol. 236, 1944.

XII

The Adolescent at School

INTRODUCTION: PROBLEMS AND PURPOSES

Increased school attendance. The institution of the school has evolved in all highly civilized societies. In a democratic society, it holds an outstanding and dominant place. The perpetuity of a democracy is predominantly dependent upon an enlightened citizenry, and, therefore, upon educational opportunities for all. According to the school census for 1947, 91.6 per cent of youths fourteen and fifteen years of age were enrolled in school, and 67.6 per cent of those sixteen and seventeen years of age were in school.[1] This age level corresponds with the last two years of high school, and in most states is beyond the compulsory-attendance age level, which is usually the end of the fifteenth year.

The comment made by Ralph B. Spence relative to the school program in the state of New York is applicable to many places. He says:

We have done an excellent job in building up a school program that will take care of a large percentage of children. We cannot rest satisfied, however, until we have acquired the skill necessary to meet the needs of 100 per cent of the children. It is necessary, therefore, that each school carefully check its curriculum, its guidance program, its recreation program, and its other services, to see if these are taking care of the special needs of the vulnerable group.[2]

Our notion of what constitutes child labor has changed enormously during the course of the past century. Slightly over a century ago (1842) the state of Massachusetts specified that children under twelve

---

[1] Data secured from the Current Population Reports, Population Characteristics Series P-20, No. 19, July 30, 1948. *School Enrollment of the Civilian Population, October 1947.*

[2] Ralph B. Spence, "New York State's Program for Preventing Delinquency," *Journal of Educational Sociology,* 1945, Vol. 18, p. 442.

years of age should not work more than ten hours per day. However, the first minimum-wage law, passed by Pennsylvania in 1848, established a twelve-year-minimum age for workers in textile mills. This was a higher minimum than that provided a few years later in Connecticut and Massachusetts. The concern during this period was over the control of the hours of labor for children rather than adolescents. There was little concern about child-labor legislation for adolescence until a fairly recent date. Thus, child labor in industry was at its peak in 1910. At that time approximately thirty per cent of the fourteen- and fifteen-year-old boys and girls were listed as gainfully employed, while many others worked on the farms and at other tasks and did not attend school regularly. Furthermore almost one million youngsters under fourteen years of age were gainfully employed. The steady decrease in the number of children employed each year since 1910, together with the corresponding increase in the number in school, is a reflection of the growth of the concept that universal education is needed for intelligent and efficient citizenship.

The problem of the school. Materials relative to school survival rates reveal a high mortality rate with advancing age and contacts with more complex school materials. The curriculum of our secondary schools was not designed for children of subnormal or even average abilities; consequently it has little significance to many pupils. It has already been stated that the child is an active organism, functioning according to the organismic concept in a unitary manner. The principle of "learning by doing," which was early recognized as an essential element in all learning developments, applies to social and emotional phases of life as well as to those classified as mental.

The problem of providing for the needs of *all the pupils* when they reach the high school age is becoming more acute with the increase of the secondary school enrollment and the development of a universal recognition of the necessity of a high school education. As Witty observes:

Elasticity in school demand and freedom to develop individuality are essential in the new school, if maximum growth is to take place. Uniformity and excessive conformity are foes to growth. Outmoded codes must be abandoned, and we should aim, through cooperative reconstruction of experience, to develop more adequate values. In this process of creation, cooperative endeavor is a most important determiner of growth.[3]

[3] P. A. Witty, "Enriching the Life Experience of Exceptional Children," *School and Society,* 1934, Vol. 39, p. 106.

Probably at no period in life is some sort of social adaptation unnecessary. The adolescent entering high school has received much training in cooperation, punctuality, and other obligations; but with increasing maturity, social contacts are widened and new adaptations must be

Courtesy, School Life.

Adolescents at school.

made. So the school in a number of ways attempts to further the socialization of its pupils. Recognizing that direct suggestion through lectures and formal study is of little value in establishing desirable social habits, it must provide social situations in which the young may develop according to the laws of learning. Thus extracurricular activities have been encouraged. However, there is one danger in the administration of such activities that should be clearly seen and guarded against. It results from the carrying over of the teacher's classroom attitude to nonclassroom activity; it is illustrated in athletic teams whose aim is to win rather than to improve their social life.

School success is not sufficient. The child who is successful in school is quite often looked upon as a model; such success is regarded as a crowning achievement of such qualities as will power, tenacity, desirable drives, and good mental habits and powers. That this is true in a great many cases is not questioned here; but when such an achievement is attained at the expense of a well-balanced personality, it is fraught with danger and should be looked upon with suspicion. The writer is reminded of a case that came within his observation a few years ago:

A girl, referred to here as Josephine, had always been a good student. She enjoyed her work at school and spent most of her time working with the assigned lessons. She was third from the top in a class of more than thirty pupils in the sixth grade. Since she was about average in intelligence (the Stanford Revision of the Binet Test gave her an IQ of 107), she had to spend most of her time at work on her lessons in order to make the mark and hold the position in her class toward which her aims were always pointed. Her parents as well as friends and kinspeople commented favorably to her about her school work, and this was an added urge to keep trying. From observations of this girl for a period of five years following this first general observation, it has become apparent that the girl is not developing her social qualities as she should. She has a very narrow range of interests; though a leader in her school work she is not a leader among the group of girls of her class, nor has she given just consideration to her health and general appearance. It is unlikely that she will be able to go to college (unless her continued academic drive operates in this connection). Although she is not a problem case of any kind, she has not developed the various phases of herself that would serve her well in difficult situations or enable her to adapt herself to the groups with which she will come into contact during the course of her life.

Anne G. Beck[4] suggests that among individuals completely immersed in their studies, either one or both of the following conditions are often found: There may be a withdrawal mechanism established in which the individual finds more satisfactions and pleasures from an introverted, introspective type of experience than from a more active, extroverted type of life, constantly in contact with others. Secondly, there may be a definite defense mechanism established in which the individual is trying to overcompensate for some inadequacy of which he is conscious. By such a technique he is able to obtain esteem and prestige, at least in his own way of thinking.

Social development is important for adequate adjustment in our social order today. Among some of the things essential for an adequate adolescent personality, aside from school success, are participation in group activities, the development of a desirable range of interests and abilities, habits of self-control, and a reasonable amount of psychological weaning. Students with high personality quotients usually assume positions involving leadership and are more popular with other students. Students with low personality quotients are not necessarily problem cases, but they do not always show marked sociable habit patterns. Students with high personality quotients appear to be more dependent on environmental than hereditary factors, and are probably influenced by a wider range of environmental factors.

[4] Anne G. Beck, "School Success as a Withdrawal Mechanism in Two Adolescents," *Journal of Abnormal and Social Psychology,* 1934, Vol. 29, pp. 87–94.

ATTITUDES AND INTERESTS

The adolescent brings with him to the classroom certain attitudes and interests formulated as a result of his activities and experiences at home, on the playground, in school, at church, and elsewhere. The teacher who fails to take into account the various cultural forces that have operated in the development of a particular child's attitudes and interests will be unable to appreciate and understand him and the problems he encounters.

Attitudes toward classmates. Among the attitudes formulated by junior and senior high school students are certain notions about the activities of their classmates. There is a strong tendency for these boys and girls to assign special roles to each other, based upon their special characteristics or abilities. The results of the California studies show some of the special likes and dislikes of adolescents in this connection. Items from the inventory used in the California growth study bearing on attitudes toward classmates that were most frequently checked are presented in Table XXXVIII along with the percentage of boys and girls checking each of the items at three age levels.[5] There is a continuous decline in the per cent of both boys and girls that checked those items connected with the tendency for adolescents to form into cliques as they develop from the pre-adolescent to the adolescent and thence to the post-adolescent period. The High Fifth and Low Sixth grade groups checked a significantly larger percentage of the items than did the High Eleventh and Low Twelfth grade groups. No consistent sex differences were observed in the responses to these items.

Attitudes toward the teacher. The attitude assumed by adolescents oftentimes represents a group and sometimes a community attitude. When the leaders of certain gangs or cliques assume some special attitude toward a teacher and assign him or her a special role, this may become the model for most other adolescents to follow. Sometimes a community may place a teacher in a favorable or unfavorable role and thus influence the actions of adolescents. However, there is evidence that adolescents are more honest and forthright in their appraisal than are the parents. Adolescents respond favorably to fair treatment, and have in many cases taken a fair attitude toward a teacher who was being

---

[5] For a further discussion of these items and a method for using the inventory for studying the development of an individual during the adolescent years see H. E. Jones, *Development in Adolescence.* New York: Appleton-Century-Crofts, Inc., 1948.

maligned by members of the community because of some belief or practice not wholly in harmony with that held to and practiced by leading
members of the community.

TABLE XXXVIII

Aspects of School Life Disliked by Adolescents: Attitudes
Toward Classmates

ASPECT	H5 L6		H8 L9		H11 L12	
	Boys	Girls	Boys	Girls	Boys	Girls
Classmates who plan games or hikes or parties and then won't let others in on the fun	45	32	20	25	15	7
Groups or gangs or crowds that won't have anything to do with pupils outside of these groups	42	31	21	26	11	26
Having some of the pupils start a club which they won't let others into	39	31	17	14	8	6
Having the classmates you like most turn out to be stuck up	46	44	24	22	30	11
One's classmates are snobbish and stuck-up	39	44	27	37	30	21
Having certain pupils run everything in the school	61	49	38	43	27	46
Classmates whispering and making fun of one behind one's back	55	47	20	26	23	10
Having a few pupils in the school make fun of some of the other pupils	46	42	18	25	10	15
Being laughed at when one recites in class	39	35	35	35	23	21
Being called nicknames	18	11	3	8	0	0

In the California growth study 71 boys and 72 girls completed an
inventory consisting of fifty items entitled "Things You Do Not Like
About School." [6] This inventory was checked annually by these boys

[6] *U. C. Inventory I: Social and Emotional Adjustment.* Revised form for presentation of the cumulative record of an individual, with group norms by items for
a seven-year period. University of California. This consists of two forms, one
for girls and one for boys.

and girls for seven consecutive years.   Items from the school situation involving teacher-pupil relations are presented in Table XXXIX.   The numbers of pupils checking the items at the first checking (H5 L6), the fourth checking (H8 L9), and the seventh checking (H11 L12) are presented as a basis for comparing these boys and girls at the pre-adolescent or early adolescent years, the adolescent stage, and the post-adolescent stage.   At all of these stages of development both boys and girls showed a strong dislike for unfair practices on the part of teachers. Boys were slightly more inclined to check the various items than girls, except for the one relating to being embarrassed by the teacher before the class.

TABLE XXXIX

Aspects of the School Situation Disliked by Adolescents:
Teachers and Discipline

Aspect	H5 L6		H8 L9		H11 L12	
	Boys	Girls	Boys	Girls	Boys	Girls
Being punished for things you do not do	69	58	69	56	59	36
Teachers who are not interested in their pupils	30	26	38	33	34	44
Teachers who make one feel embarrassed before the class	51	46	54	67	46	60
Teachers who mark you down because they do not like you	49	39	68	65	58	54
Examinations that are unfair	35	25	65	51	62	49
Too many teachers' pets	61	42	49	39	38	22
Teachers who have the wrong opinion about you	30	28	52	51	51	36

Interests in school subjects.   The general conclusions of the various studies pertaining to interest and ability do not reflect so much the individual's capacity as compared with that of others, as they do his hierarchy of abilities.   Thus the individual is likely to be most interested in those things he can do best; but this "best" does not of necessity mean superiority over others in the specified task.

According to the data presented from the *Fortune* Survey, conducted in 1942, English and mathematics rank first and second, respectively, as the most liked and least liked subjects. (See Table XL.) [7] It might be stated that these are the subjects most widely given, and that they generally run throughout the high school curriculum, although there

TABLE XL

High School Subjects Liked Best and the Ones Liked Least by High School Seniors

Subject	Best	Least
Mathematics—algebra, geometry, trigonometry	20.0%	26.7%
English—grammar, composition, literature, etc.	17.7	22.2
Sciences—general science, biology, chemistry, other (except social)	14.8	18.7
History	11.0	12.4
Vocational courses—home economics, typing, and other business	10.3	3.9
Languages—French, Spanish, etc.	6.8	13.4
Civics, government, social science, etc.	1.8	3.2
Don't know	2.4	5.3

Courtesy, Los Angeles Public Schools

A group exploring the secrets of the radio.

is a tendency to make more of the advanced high school work in mathematics elective in nature. Cross tabulations from the Survey show fur-

[7] "Fortune Survey," *Fortune*, December 1942, p. 14. The materials of Table XL are based upon answers to the questions: "Of all the subjects you have taken so far in high school, which one have you liked best? Which least?"

ther that certain patterns of preferences there are frequently found. Those students disliking English, languages, and history often offer as their preferences mathematics and the laboratory sciences; and vice versa.

The study of children's interests reported by Jersild and Tasch showed that, at all grade levels, items in the broad category that includes academic subject matter areas were mentioned most frequently when children told what they liked best in school.[8] Nature study and natural science were mentioned infrequently in the earlier grades but showed a gain in popularity at the junior and senior high school levels—grades 7–9 and 10–12. The results of this study (presented in Table XLI) show, however, a decline of interest in the academic and educational features of school life, and an increased interest in sports, games, discussion clubs, student council, and the category of people. These activities are in many cases more meaningful and significant to the adolescent than to the pre-adolescent, and are in harmony with the physiological and social changes occurring at this stage.

TABLE XLI

WHAT I LIKE BEST IN SCHOOL

CATEGORY	AGES 9–12		AGES 12–15		AGES 15–18	
	Boys	Girls	Boys	Girls	Boys	Girls
Sports, games, physical education	13.3	9.5	30.6	33.2	34.8	34.4
Areas of study, subject matter	69.7	76.3	44.4	60.1	41.3	45.5
Art activity or appreciation: music, painting, drawing	11.1	14.8	10.0	15.9	16.2	13.8
Crafts, mechanical arts	.3	0	19.8	0	15.5	.4
Discussion clubs, student council	1.3	.8	1.0	.5	3.6	6.4
People: both pupils and teachers	2.5	6.1	4.1	5.6	6.0	11.4

Interests and abilities. Several studies have been made of the relationship between interests and abilities during different periods of

[8] A. T. Jersild, and Ruth J. Tasch, *Children's Interests*. New York: Bureau of Publications, Teachers College, Columbia University, 1949.

life. Thorndike[9] was one of the first to investigate this general subject. He had a group of 344 college students rank their interests in the elementary school, the high school, and the college period in seven different school abilities. Correlations were computed between the individual's order of interest and his order of abilities, and were found to be .89 each for the elementary school, for the high school, and for college. Bernard O. Nemoitin[10] investigated the relation between interest and achievement. The data of his study were gathered by means of a questionnaire and the use of school records of the students. He found that the degrees of relationship between ability in high school courses "liked best," "liked second best," "disliked most," "disliked next as much," and average ability for high school courses are expressed by the correlation coefficients, $.60 \pm .04$, $.49 \pm .04$, $.58 \pm .04$, and $.57 \pm .04$, when the data obtained from 150 high school seniors are considered. The relationship between interest and ability was found to become more variable and hence less reliable as the degree of interest considered moved from the extremes.

One of the ultimate measures of the vitality of the experiences gained in school is the extent to which the experiences lead to desirable interests and habits that endure into maturity. Interest and motivation are very closely related. It is well recognized by successful teachers that when work is properly motivated and based upon the interests of the subjects it appears easier to the student. When he is interested in a task, his attention remains more nearly in the marginal context and does not fluctuate far from the general pattern. Interest tends to focus the attention within a marginal field and thus should be considered as selective in nature as well as a driving force. Since learning is so largely dependent upon the attentive response of the subject, one will find a direct relation existing between interest and amount of learning. Attitude, which is closely related to learning, has been studied by various investigators as to its effect on both amount and duration of learning. It has been shown that, when different attitudes are set up by different purposes, the same subject will exhibit marked differences in amount learned. It might be laid down as a fundamental proposition that "interest breeds ability and ability breeds interest." It does not

[9] E. L. Thorndike, "Early Interests: Their Permanence and Relation to Abilities," *School and Society,* 1917, Vol. 5, pp. 178–179.

[10] B. O. Nemoitin, "Relation between Interest and Achievement," *Journal of Applied Psychology,* 1932, Vol. 16, pp. 59–73.

appear likely that one could be interested in a task if one knew nothing whatsoever about it.

Despite the fact that interest is related to ability, it cannot be concluded that a student is of especially high ability in some special line merely because of his interest in that line. In the first place, there is the question of individual variation: the student might be more interested in this special line of endeavor than in any other activity, and have better ability in it than in most other fields, but still have very little ability because of a general deficiency. A boy is observed as displaying a keen interest in baseball, but this does not mean that he will be able to make the high school team. It will be much safer to predict that he will succeed better in baseball than in any other form of athletics; that is, he is probably more able to compete with a fair degree of success in this sport than in any other. There has indeed been some confusion in the drawing of conclusions concerning the relation between interest and ability. It is safer to consider the ability of the individual in the field of his intense interest in relation to his ability in other kindred activities, than to compare this ability with that of others displaying a less intense interest in this line.

Parents and teachers can do a great deal by showing appreciation of the desirable qualities, potentialities, and abilities of their children. However, they should also recognize their children's limitations, and not expect things of them beyond their possibilities. Gibby has given a detailed case history of a scholastically retarded boy that is typical of a great number of cases:

James was a poorly dressed boy of 14 years and 7 months. His face and hands were dirty and he was always in need of a haircut. He was 5 feet and 2 inches tall and weighed approximately 128 pounds; his muscular co-ordination, however, was not good. He is left-handed, and no efforts have ever been made to change this condition. He has no speech defects. His vision is poor and he wears glasses. According to the Stanford Revision of the Binet Test, he has a mental age of 7 years and 4 months. This would give him an IQ of 52. Upon the Ohio Literacy Test, James had a raw score of 0, which would give him a mental age of less than 5. From his reactions to other tests, it was quite obvious that he was sufficiently subnormal to have very great trouble understanding directions, and that he had not developed the ability to read.

James thought that he was clever, and would usually answer questions that he didn't know by saying: 'I haven't learned that at school yet.' He would oftentimes look at the examiner and merely smile in response to questions. His ranking on Schedule A of the Haggerty-Olson-Wickman Behavior Rating Schedule was in the 74th percentile, and his ranking for

Schedule B was in the 99.9th, indications of a very serious maladjustment on his part. James has been a problem child to the school officials ever since he entered the school system. That is, he has been a problem in that the school officials have not known what to do with him. He did not manifest any serious behavior problem traits. He was ranked, at the time of this study, in the sixth grade; however, he had never actually passed the work equivalent to that of the first grade. In the school room he would remain quiet most of the time, but two or three times a day would burst into an uncontrollable laughter and clap his hands together several times. He attended school regularly, and was awarded a certificate of honor for this perfect attendance.

His arithmetic age was about 7 years, and he showed no ability to read. This is revealed in the following episode: "The letters A, B, C, D, and E were placed on the blackboard. James was told the name of each letter, and was drilled on them again and again. As long as the letters were in their proper order, he could repeat them correctly, pointing to each letter as it came; but just as soon as the sequence was changed, or he began with any other letter than the first, he was unable to distinguish the one from the other." His experiences with a pre-primer reader containing stories reveals further the nature of his reading: 'He was delighted to get the book, and told everyone that he was going to learn how to read. He came to school the next day, said that he had finished the book, and that he wanted another. Upon being asked to read the first story, James rattled it off correctly. He was praised for this and then asked to read the next two stories, which he also did correctly. However, upon being asked what the word "run" was, he responded with the word "baby." Questioning brought out the information that his grandmother had read the stories to him the evening before, and that she had read them to him so often that he had remembered them word for word.'

James is clearly a mentally retarded individual. He presents no serious behavior problems, but accepts blindly all rules and authority. His lack of muscular co-ordination makes it well-nigh impossible to train him for usefulness in most manual activities. Any work activities that he is ever able to do will have to be carefully planned for him, and each stage supervised. He has not benefited from his enforced period of attendance of school for nine years by mastering reading and the other 'tool' subjects. He has thus far been kept from becoming a ward of the state at the expense of the inconvenience and trouble that it has cost the school to care for him. Because of the nature of his home conditions, which are at a very low level, he will likely become a ward of the state once he leaves school.[11]

The school and the expansion of interest. The schools offer the adolescent the opportunity, among other things, to expand his social contacts, to achieve status, and to prepare himself for a normal adult

[11] R. G. Gibby, "A Clinical Study of Thirty-Two Scholastically Retarded Special-Class Boys," Master's Thesis, Ohio State University, 1939.

life.   Not all adolescents can master academic subjects, but the majority can learn to live a reasonably normal life according to the training and influence the school implants within them.

The friendships made in school during this period have a marked bearing upon the shaping of character and personality and the stabilizing of adulthood.   Records and charts show that such friendships are based primarily on common interests.   This fact is illustrated in the activities of any adolescent one knows: the adolescent is obviously drawn toward a fellow athlete, musician, or craftsman.   Moreover, the frequency of such associations will determine the duration of these interests, and here the school promotes recurring association by means of its many avenues of approach to these varied interests, in the form of extra activities not in the regular curriculum.

Loyalty, too, has its foundation in the school.   At first this quality is directed toward the adolescent's classmates and teachers, but soon it embraces the entire school in the form of school spirit.   The school does not complete its function when it merely *teaches* loyalty, honesty, and democracy; the real value of these lessons is realized only when they are put into practice in all the organizations of the school. The result proves to be a stepping-stone to good citizenship, through loyalty to community and, ultimately, loyalty to country.   It is at this point that many activities taking place outside the classroom function most effectively.   These activities have often been referred to as extra-curricular in nature; there is, however, a distinct tendency to regard them as an important part of the school's program.   Such activities have been classified in various ways.   Hausle suggests a four-division classification as follows: "(1) Athletics—interscholastic and intramural; (2) Clubs—subject, hobby, welfare, honorary; (3) Semi-curricular activities—those for which a school may grant subject-credit; (4) Citizenship activities—service." [12]

The disregard for individual variation by our school system is the chief reason why so many students are constant failures, academically and emotionally.   The mentally and physically inadequate (by school or society's standards) are all victims of constant failure in their ambitions.   An adolescent is more sensitive than an adult to the inability to win some sort of acclaim in his actions.   An adult can take pleasure in bowling or golf even if he is not too skillful; he soon realizes his limitations.   The adolescent, on the other hand, always has a hope,

[12] Eugenie C. Hausle, "Objectives of a Program of Extracurricular Activities in High School," *Recreation,* 1940, Vol. 34, p. 361.

more or less intense, of becoming a champion or a leader. It is more difficult for him, therefore, to accept his role and status in society. Competition and some measure of success temper the adolescent, since they give him facts and ideas that enable him to resolve and interpret his role.

PROBLEMS AND GOALS

The significance of the teacher. It is evident that today's educational standards require the teacher to be an informed, well-integrated, and far-seeing adult member of a community. It is no longer enough that the teacher be the possessor of knowledge. Today's teacher must be capable of setting up a desirable environment for learning. He is responsible for teaching pupils to respect one another's personalities, and to work and play cooperatively with others under restrictions and privileges established and maintained by majority will.

The classroom must be considered a social laboratory in which children learn to live with others cooperatively and harmoniously. It must be a place in which control evolves from within the group and is exercised for the welfare of the majority. The general atmosphere must be characterized by mutual understanding and mutual respect of pupil for pupil, pupil for teacher, and teacher for pupil.

Children have emotional needs that require particular attention and sympathetic understanding. Under the pressure of group action these needs become intensified and more complex in nature. Every child needs to feel his own worth and developing power. He needs recognition and encouragement. Only as he accepts and understands himself does he function at his best. Fears and inhibitions concerning himself and inability to direct his attention to external conditions detract from his well-being and happiness. The teacher must understand his need for recognition and so shape events that every pupil has sufficient successful experience to insure in him a sense of his own security and worth. The teacher must have a sympathetic understanding of the behavior tendencies of different levels of ability and of what constitutes appropriate experiences for each level.

Again, it should be pointed out that human nature is not basically bad. Sometimes a group of individuals are badly trained and therefore act badly; that misbehavior is the result of the training imposed rather than of the nature of man. This thought further suggests that the responsibility of the teacher is not only a great but is also a very important one. Western civilization is characterized by its *dynamic* quality. This fact should constantly be kept in mind by those who are concerned with the organization, administration, and operation of our schools. It is well to note the trends that have been underway in administration and operation. By so doing one is in a better position to see the directions in which our schools and other institutions are now moving. If education is to function effectively and efficiently in producing citizens for a better world order, it is essential for those concerned with and responsible for the educational program to have certain fairly well-defined goals. As we look forward to the future and see the task that lies ahead, we are confronted with the general question as to whether we are holding a funeral over the age that has been, or a christening for the one to come. Certainly it would appear that we are passing into a new era—one that will make new and additional demands upon education.

School problems of adolescents. It was pointed out in Chapter II that educational problems loom large in the lives of adolescents. Problems relating to failure in school, how to study, pupil-teacher relationships, and the like, apparently appear in the lives of many high school boys and girls. Materials from the California adolescent growth studies, presented earlier in this chapter, show that a large percentage of pre-adolescents, adolescents, and post-adolescents disliked elements in the school situation that indicated unfair practices on the part of teachers and snobbish as well as overly aggressive and dominating attitudes and practices on the part of their classmates. Some features of the curriculum and program disliked by a large per cent of boys and girls in the California study are listed in Table XLII.[13] In general the boys checked many more items related to the curriculum and program than did the girls. Dullness and lack of interest as well as lack of value (as viewed by the students) in the subject matter were checked in many cases at all stages of development by approximately fifty per cent or more of the boys and slightly fewer girls.

Norton's study[14] of the student problems met by the teacher indicated that such problems fall into three large groups, which are listed as: (1) school-related problems, (2) non-school-related problems, and (3) post-school-related problems. The number and per cent of guidance problems in these areas met by a group of high school teachers is given in Table XLIII.

[13] *Op. cit.*
[14] S. K. Norton, "Student Problems Met by the Teacher," *School Review,* 1948, Vol. 51, p. 404.

TABLE XLII

Aspects of the School Curriculum and Program Disliked by Adolescents

Aspect	H5L6		H8L9		H11L12	
	Boys	Girls	Boys	Girls	Boys	Girls
There is too much homework	18	4	45	29	45	36
Assignments are too long	32	18	51	32	42	39
Many of the subjects are dull and uninteresting	55	32	63	56	63	53
No chance to pick out the subjects that one likes	48	25	44	25	20	6
Having to take subjects that one dislikes	63	31	61	57	48	42
School work is too monotonous	28	10	46	28	28	21
Having to take subjects which will be of no use to one when grown up	52	28	68	60	55	53

The problems were broken down in the following categories:

I. School-related problems—Those arising directly out of, and chiefly pertaining to, school situations such as choice of study, difficulties with subject matter, extracurricular activities, and school citizenship.

II. Non-school-related problems—Those dealing primarily with the pupils' lives away from school and not directly traceable to school situations, such as homes, their families, their friends, their financial conditions, and their health.

III. Post-school-related problems—Those concerning the pupils' choice of vocations and of educational institutions beyond high school.

TABLE XLIII

Major Classification of 4,682 Guidance Problems
Met by 1,586 High School Pupils

Category	Number	Per Cent
I. School-Related Problems	1,676	35.8
II. Non-School-Related Problems	1,822	38.9
III. Post-School-Related Problems	1,184	25.3
Total	4,682	100.00

Of the school-related problems, choice of study ranks first for over half the group in 54 Michigan communities. This problem should be taken seriously by the teacher, more seriously than routine, because in many cases it determines whether a student can enter the college of his choice. This, in turn, will affect his entire life. The next highest item was academic difficulties. These include such items as low achievement, poor study habits, make-up work, reading difficulties, lack of interest, and the like. School citizenship is very important from the standpoint of both the teacher and the student. Earlier in this chapter items relating to school citizenship were discussed in connection with school situations annoying to adolescents. (See Table XXXIX.) In general, most of these problems of school citizenship are not of the sort that can be completely disregarded. Even though they may not seriously disturb the activities of the classroom, the library, or wherever they may occur, they are symptoms of emotional disturbances resulting from maladjustments.

Non-school-related problems have taken on an added importance within recent years. The high school student's behavior will be largely affected by the conditions and forces he meets outside the school environment. It makes a great deal of difference whether the boy or girl comes from a home where the social climate is a happy and wholesome one, or whether he comes from a home where the social climate is not a happy and wholesome one. Although it is not easy for the school authorities to investigate problems arising from home and community conditions, an awareness of the nature of these conditions will enable them better to understand adolescent problems originating in situations and conditions outside the school.

Vocational choice and employment opportunities top the list of problems encountered by students leaving high school. The need for vocational guidance and follow-up work is recognized by students as well as teachers. Some of the major problems relating to this will be discussed in Chapter XX.

Growth through participation. One of the greatest values of the school lies in its provision of a place where boys and girls are brought together and given the opportunity to participate in wholesome activities under the general guidance of teachers and counselors. There is evidence, on the basis of studies of this problem, that participation in the varied high school activities tends to promote understanding, cooperation, respect for others, and ability to work with them. In addition, such participation supplies certain felt needs of the adolescent and thus leads him to acquire a better adjusted personality.

In the first place, adolescence is characterized by excess energy. This energy should be directed into wholesome activities, for its proper direction will supply certain needs of the adolescent and promote his growth and development. Among the needs satisfied through participation in athletics, clubs, social organizations, specialized groups, and the like are:

(1) *The need to succeed.* A wide range of activities will provide for the expression of a wide range of talents and special abilities, and thus provide nearly all adolescents with the possibility of a reasonable amount of success.

(2) *The need for belongingness.* By becoming a member of a special group and participating in their activities, the adolescent identifies himself with the group.

(3) *The desire for social approval.* This is closely related to the satisfaction of the need for belongingness.

(4) *The need for security.* The old adage, "In unity there is strength," applies here. As a member of a group who takes an active role in group activities, the individual comes to feel more secure and develops an increased confidence in himself and in his ability to do his part and play his role.

It is difficult to estimate the value of clubs, athletics, and socialized programs in our high schools, since the results derived from these are less tangible in nature than are results from textbooks. However, these activities take care of needs not satisfied by other agencies and conditions, and thanks to them the school offers adolescents the best opportunity for active participation in wholesome activities of a meaningful and satisfying nature.

A SUMMARY OF PRINCIPLES

The solution of the economic, social, and civic problems of tomorrow will surely be affected by the program of education adopted today. The problem of individual variation in our secondary schools has become more acute as a result of (1) the increased enrollment in our schools, (2) the lengthened period of school life, and (3) the enlarged program of the schools. Diversified systems of education should provide opportunities for each child to develop his abilities and potentialities. The dull child, the neuropathic child, and the gifted child all need individual consideration.

As he grows through different school activities, the child will tend to find himself competing with others for success and recognition in the various tasks of the school. It is at this point that a number of maladjustments and potential maladjustments, referred to in Chapter XIV, have their onset. This conflict becomes more pronounced as the child grows into adolescence and adulthood, since more is expected of the individual and increased social pressures operate in connection with the requirements of society.

Those concerned with the guidance of adolescents should keep in mind that they are training the whole child, not just some intellectual phase of him. The whole child goes to school and is involved in everything he does while at school. The unitary concept of the growing adolescent, stressed in earlier chapters, must be held if the school is to perform its function. The importance of the school is well stated by Lois Meek:

Although the kind of home and community environment of boys and girls is a major cause of maladjustment, the school must remember that its own influence plays a large part in determining the child's development. Principals should recognize the fact that the school itself may be responsible for maladjustments in the youth it is supposed to serve. Among the school practices to be avoided are the following: mechanically applied systems of merits and demerits which finally engulf the wayward; course requirements designed for the twenty per cent who are not interested; teachers who are utterly uncompromising with the very human nature of youth; lack of opportunity for students to participate in running the school community; subject matter for which pupils can see no use either now or later.[15]

THOUGHT PROBLEMS

1. List in order the factors you consider most important in the increased school enrollment.

2. Show from some case of your acquaintance how school success is not sufficient.

3. How would you account for the interests in school subjects presented in Table XL?

4. Study the materials presented in Table XXXVII showing the percentages of boys and girls disliking certain aspects of school life. List several of the most significant conclusions you would draw from these data.

5. How does the school aid in the expansion of interests? Illustrate.

6. Show how the teacher is the greatest asset and the greatest liability of the school.

7. What are some of the barriers to a desirable pupil-teacher relationship in school? (See especially the reference to Baxter's *Teacher-Pupil Relationships*.)

[15] Lois Hayden Meek (Chairman), Committee on Workshops of the Progressive Education Association, *The Personal-Social Development of Boys and Girls with Implications for Secondary Education*. New York: Progressive Education Association, 1940, p. 141.

8. List a number of problems which you have observed that would be classed as non-school-related problems. What do you consider the school's function should be in connection with such problems?

SELECTED REFERENCES

Appy, Nellie, *Pupils Are People*. New York: Appleton-Century-Crofts, Inc., 1941.

Baxter, Bernice, *Teacher-Pupil Relationships*. New York: The Macmillan Co., 1941.

Cole, Luella, *Psychology of Adolescence* (Third Edition). New York: Farrar and Rinehart, 1948, Chap. VIII.

Crow, Lester D., and Crow, Alice, *Our Teen-Age Boys and Girls*. New York: McGraw-Hill Book Co., 1945, Chaps. V and VI.

Dimock, H. S., *Rediscovering the Adolescent*. New York: Association Press, 1937, Chaps. IX and X.

Fleming, C. M., *Adolescence*. New York: International Universities Press, Inc., 1949, Chaps. IX and X.

Hollingshead, A. B., *Elmtown's Youth*. New York: John Wiley and Sons, Inc., 1949, Chap. VIII.

Landis, Paul H., *Adolescence and Youth*. New York: McGraw-Hill Book Co., Inc., 1945. Chaps. XVIII and XIX.

Meek, Lois H. *et al., The Personal-Social Growth of Boys and Girls*. New York: Progressive Education Association, 1940.

Partridge, E. D., *Social Psychology of Adolescence*. New York: Prentice-Hall, Inc., Chap. XII.

Sadler, W. S., *Adolescence Problems*. St. Louis: C. V. Mosby Co., 1949, Chaps. XI, XV, XVI, and XVII.

XIII

The Adolescent in the Community

SOCIAL AND RECREATIONAL ACTIVITIES

The importance of community forces and conditions in the development of teen-age boys and girls is hard to evaluate. There is much evidence that, with the decline in size and function of the family unit, forces within the community have assumed a more important role; consequently, at some point, the growing individual comes face to face with problems that are not solved on the basis of authority or of sentiment, as are problems arising at home. The importance of the home and community in the development of character was well stated several years ago by John Dewey when he wrote:

In its deepest and richest sense a community must always remain a matter of face-to-face intercourse. This is why the family and neighborhood, with all their deficiencies, have always been the chief agencies of nurture, the means by which the dispositions are stably formed and ideas acquired which lay hold of the roots of character. The Great Community, in the sense of free and full intercommunication, is conceivable. But it can never possess all the qualities which mark a local community.[1]

Social activities outside the home. When the child first attempts to make adjustments outside the home situation, he is confronted with many new conditions that cause him to feel insecure and somewhat inadequate. In the home he is constantly sheltered by older persons, who accept many responsibilities for his needs and provide him with affection and sympathetic treatment; whereas in the environment outside the home, he is confronted with others of his age level and with adult leaders who are responsible for guiding a group of individuals in their activities. There is a felt need on the part of each member of the group in the community for self-expression, and this can only be

[1] John Dewey, *The Public and Its Problems.* New York: Henry Holt, 1927, pp. 211–212.

secured through cooperative activity. Thus, the pattern of group activity outside the home calls for cooperation and understanding on the part of the members of the group.

Within the community culture are definite class distinctions. There are a number of factors contributing to the formation of these distinctions, among which are race, economic resources, family background, national origin, and educational attainment. Social classes develop their own culture patterns, and there is a certain amount of unity found within a class.

The family transmits to the child not only its own culture, but the culture of the class to which it belongs as well. Also, the child is given ideas about other classes which affect his attitude and behavior toward them. This attitude and this behavior sometimes become emotionalized and firmly fixed, and affect the growing individual's approach to all problems involving human relations.

There is evidence that the family culture and status influences the adolescent's interests, activities, and choice of friends. A number of investigators have concerned themselves with the problems connected with the activities and interests of boys and girls. Results from some of these studies were presented in Chapter VIII. Data on children's interests outside of school show that at all age levels those activities falling in the category of sports, games, and play lead in popularity. This is emphasized in the study by Jersild and Tasch. Results bearing on this are presented in Table XLIV.[2]

Consistent with social development during adolescence, there is an increased interest manifested in places of recreation, in going to the theater, and in other activities involving the social element. In the study by Jersild and Tasch there was relatively little emphasis on general cultural activities, such as going to an art exhibit or a concert. Likewise, there is relatively slight mention of hobbies. When the children described what they disliked most outside school, a large proportion listed activities relating to chores, duties, and everyday routines. The impact of urban culture is evident in these community interests of children and adolescents. Those institutions which best provide for the needs of adolescents will offer the greatest appeal. If the youth organization connected with a church accepts the challenge and attempts to provide a program in line with the needs and aspirations of adolescents, much interest will be manifested by the youth in

[2] A. T. Jersild and Ruth J. Tasch, *Children's Interests*. New York: Bureau of Publications, Teachers College, Columbia University, 1949.

the doings of the organization. Almost every community in America
is so patterned that various forces are operating through organized and
unorganized institutions to provide for the outlets of the needs of
adolescents. In many cases, however, the outlets provided are not
what might be regarded as highly desirable.

TABLE XLIV

WHAT I LIKE BEST OUTSIDE SCHOOL

CATEGORY	AGES 9–12		AGES 12–15		AGES 15–18	
	Boys	Girls	Boys	Girls	Boys	Girls
Material things, specific objects, toys, food, shelter, pets, dress	1.9	3.4	3.1	3.2	3.6	2.3
Sports, play, games, outdoor activities, driving car	73.5	68.0	56.4	51.9	56.9	41.8
Miscellaneous places of recreation, parks; travel, camp, resort	2.4	5.7	5.2	8.5	7.4	9.4
Radio, movies, theater, comics	5.5	9.5	9.8	16.7	8.9	18.7
Social activities, organizations, parties, Scouts, DeMolay	1.5	1.0	1.3	8.0	3.2	13.0
Areas of study, reading, school subjects	4.3	3.9	5.3	13.3	8.5	18.3
Art activity or appreciation: music, painting	.6	3.1	2.4	5.5	4.6	8.7
Crafts, mechanical arts	3.5	1.7	4.5	.7	11.3	0
Self-improvement, understanding, including vocational placement or competence	3.3	3.6	.8	5.0	18.1	6.0
Chores, duties, everyday routines	1.2	5.0	2.2	3.9	.6	.8
People: both relatives and non-relatives	3.9	4.8	3.7	8.9	9.3	20.9

Playmates. Just prior to adolescence both boys and girls choose
playmates or some particular chum and build close friendships, interests and attachments. The reason for the choice of a particular
chum and the effect of the chum on the formation of character in the

life of the individual have been carefully studied by Furfey.[3] In a
study of 62 pairs of boys in a group of 296, he found that 45 per cent
were from the same neighborhood and that 89 per cent were in the
same room in school. Correlations were obtained between the chums
and certain variables, these variables being mainly physical measurements.

Table XLV gives the coefficients of correlation with respect to various

TABLE XLV

CORRELATION OF CHUMS WITH EACH OTHER WITH RESPECT
TO CERTAIN VARIABLES

Chronological age	$.39 \pm .07$
Mental age	$.24 \pm .08$
Developmental age (maturity)	$.37 \pm .07$
Height	$.34 \pm .08$
Weight	$.22 \pm .08$

physical measurements. The study did not take into account such
factors as tastes, interests, moral standards, temperament, social status,
and economic conditions that in some cases are probably more important than some of the measurements given. Chronological age, physiological maturity, and height, respectively, were the three measurements that correlated highest, while weight gave the lowest correlation.

A study by Neugarten[4] was concerned with the general question:
To what extent and in what observable ways does the factor of social
status affect the friendship among school children? The subjects of
the study were all children enrolled in grades 5, 6, 10, and 11 of the
public school. The median age for the younger group was eleven
years and three months; for the older group sixteen years and three
months. A sociometry and a guess-who test were administered to
these subjects. The results were studied in the light of the social
status of the different children. He found that, with the exception
of the group of lowest status, children tend to select as friends, first,
children of higher status than their own and, second, children of their
own status level. Children of families of high status received the
favorable votes and children of low status received the unfavorable ones.

The child from a family of upper status occupies an enviable
position—many of his classmates consider him their friend. The

---

[3] Paul H. Furfey, "Some Factors Influencing the Selection of Boys' Chums,"
Journal of Applied Psychology, 1927, Vol. 11, pp. 47–51.
[4] Bernice Neugarten, "Social Class and Friendship among School Children,"
American Journal of Sociology, 1946, Vol. 51, pp. 305–313.

child from a family of lower status faces the opposite situation. He is seldom mentioned as a friend and oftentimes mentioned as a person his classmates do not like. A child, consciously or unconsciously selecting his friends, is probably reflecting the class stereotypes as he has learned them from his parents. In the high school level, upper status is a sure indication that the adolescent will at least be the center of attention in his group, whether his reputation is favorable or unfavorable.

Physical activity is quite important in drawing adolescent boys together, although it is not likely to operate to such an extent among girls. Physical activity and ability are so very often looked upon as masculine traits that they are conceived of as more essential for the boy than for the girl; hence they tend to influence his choice of friends and companions. The home emphasis on social standing is especially influential among adolescent girls in their choice of companions.

That "birds of a feather flock together" has been long recognized and is borne out by evidence in the field of psychology and education. During adolescence playmates or companions are much more likely to be chosen according to individual likings than during earlier childhood, or even during the period following, when business and social standing play so prominent a part for most people in their choice of associates. When the adolescent tends to choose undesirable companions, it is usually of little use to admonish or reproach him. The trouble in most cases is due to early training or environmental surroundings, and much pressure brought to bear during adolescence will, as a rule, serve only to aggravate the general situation and cause the individual to assume an antagonistic frame of mind. It is during the earlier years of life that tastes for good friendships should be established. Ideals of conduct directed toward some desirable goal develop gradually, according to the developmental concept emphasized throughout this study of adolescents. A new environmental setting for the adolescent, a new interpretation of life's value in harmony with certain interests and desires, or a change in general vocational activity may function effectively in the eradication of undesirable chum selection.

Formation of groups and gangs. At the age of adolescence boys and girls become highly interested in forming groups, societies, gangs, and clubs; and these are indeed truly representative of the "gang" stage of life. Scientific investigations show that as a rule the members of a gang are likely to be of about the same level of intelligence. The members usually come from within a certain limited geographical area, as is the case in the selection of chums among adolescent boys.   The gang is very apt to be in the main a neighborhood affair.   Through it individuals are affected by the behavior patterns of others and tend to influence the formation of behavior patterns in others by their own activities.   The group is generally homogeneous in its desires, likes, and dislikes; social uniformity in ideals and attitudes tends to develop in accordance with general activities.   Loyalty to different members of the group reaches a high pitch and may even surpass the loyalty earlier established to such ideals as honesty and truthfulness.

The structure and behavior of a gang is molded in part through its accommodation to its life conditions.   The groups in the ghetto, in a suburb, along a business street, in the residential district, in a Midwestern town, or in a lumber community vary in their interests and activities not only according to the social patterns of their respective milieus but also according to the layout of the buildings, streets, alleys, and public works, and the general topography of their environments.   These various conditioning factors within which the gang lives, thrives, and develops may be regarded as the "situation complex," within which the human nature elements interact to produce gang phenomena.   So marked is the influence of such factors as bodies of water, prairies, hills, and ravines in determining the location and character of gang activities that in Cleveland juvenile delinquents have been classified on this basis.

Gangs represent the spontaneous effort of boys to create a society for themselves where none exists adequate to their needs.   Boys derive from such association experiences that they do not get otherwise under the conditions that adult society imposes—the thrill and zest of participation in common interests, more especially in corporate action, in hunting, capture, conflict, flight, and escape.   Conflict with other gangs and the world about them furnishes the occasion for many of their exciting group activities.

The gang functions with reference to these conditions in two ways: it offers a substitute for what society fails to give; and it provides a relief from suppression and distasteful behavior.   It fills a gap and affords an escape.   Thus the gang, itself a natural, spontaneous type of organization arising through conflict, is a symptom of disorganization in the larger social framework.

As individuals become affiliated with different groups in the school, the church, and the community in general, there may be conflicting loyalties.   This is especially true for those who are members of minority groups, for the larger and more inclusive community organizations and agencies are likely to foster ideals and attitudes dominated by the majority element. The problem of adjustment is more difficult for minority groups—since it is fraught with more chances for conflicts—than for members of the majority group or groups. For example, the behavior, attitudes, and beliefs of the child of Greek-born parents, living in a family culture that is largely Greek but is located in a second- and third-generation Polish or German neighborhood, would be in conflict with that of the children of the community. In adolescence students are keenly aware of loyalties, especially of loyalty to members of their groups. The problems encountered in this connection are sometimes very difficult, as suggested by Lois Meek.

How one can be loyal to one's family, loyal to a small organized group of peers, and loyal to the school becomes a vital question. Boys and girls need help in analyzing these loyalties and in discussing loyalties appropriate to various group affiliations. They need help through which to build a constructive basis for guiding their behavior.[5]

COMMUNITY PROGRAMS FOR ADOLESCENTS

Social recreational programs. The various studies of the effects of lack of guidance on the behavior of adolescent gangs have caused the community to focus its attention more and more upon the need for desirable recreational activities for adolescent boys and girls. According to the studies by Thrasher,[6] gang life thrives in those areas where there is a lack of wholesome and well-directed group activities, and where boys and girls are faced with difficulties in adjustment to persisting problems.

Many cities realize that boys and girls need a place where they can meet together, laugh, talk, and amuse themselves at wholesome activities. The lure of the "juke joint" and similar places is contrived to appeal to boys and girls attempting to satisfy their desire for recreation. Although the community projects carried out by city authorities cannot take the place of homes or parental supervision, they can provide hangout rooms where adolescents may enjoy wholesome recreation under desirable conditions. These rooms may be equipped with a radio, a piano, table tennis, magazines, and the like, and provided with super-

---

[5] Lois Hayden Meek (Chairman, Committee on Workshop), *The Personal-Social Development of Boys and Girls with Implications for Secondary Education.* New York: Progressive Education Association, 1940, p. 128.

[6] F. M. Thrasher, *The Gang.* Chicago: University of Chicago Press, 1927.

vision sufficient to satisfy the need without being obtrusive. Recreation departments have found that for the older high school group the most popular programs are those offering social activities of a rather informal nature, an observation confirmed by the conclusions of the *Fortune* survey: "Ahead of any specific sport came dancing and movies for both boys and girls. After these the favorite pastimes are running around with friends, gab sessions, and the like." [7]

The importance of recreation as a stabilizing force during a transition period is well exemplified in the case of Henry Smith.

Henry left high school at the end of the ninth grade. He was then almost sixteen years of age. Although he was not a failure in school, he didn't find the work too interesting. Furthermore, he came from a home in the lower economic scale. He tried to get a job, first in one of the stores of the community and later in a near by furniture factory. His efforts were fruitless, although his name was placed on file for future reference. Back at home and later on the streets, Henry was faced with several possibilities. He might continue a search for a job. In this quest he might or might not succeed. He might give up completely his search for a job and join a gang of ne'er-do-wells at the back lot, or he might remain patient, and find release for his energy in wholesome recreational pursuits. The last named alternative would tend to keep him in a more wholesome state, and better prepare him for breaks that might appear at a later time. Henry resorted to this alternative and spent part of his spare time at the public library, looking through some magazines and books dealing with mechanics. Furthermore, he continued his interest in Sunday school and during the summer played soft ball with a church team. Through his associations and by constant alertness to find work, he obtained a job at a filling station, an adjustment that seemed to be a happy one for him.

Although this is not a story of the poor boy becoming a prince, it does reveal the need for wholesome recreational pursuits as a means for attaining desired adjustments during a transition period. It is during such a period that the boy or girl is forced to make a choice, and any condition involving choice is fraught with the danger that the wrong one will be made.

In order to provide opportunities for boys and girls to enjoy wholesome recreational pursuits, it was suggested by Dorothy Richardson, Director of the Y.W.C.A.–USO work during World War II, a three-way program is needed. This program would include more adult education relative to the nature and needs of adolescents, well-planned high school recreational programs designed to hold youngsters in their

[7] See *Fortune*, December, 1942.

community and perhaps in school, and wholesome social centers or places of meeting sufficiently inviting to attract the youngsters who might otherwise frequent less desirable places.

An illustration of how one town with a population of approximately 15,000 has met this problem reveals the nature and needs of adolescents today. In this Georgia town there were many adolescents who had almost no responsibilities and few things to occupy their interests. There were the movies, of course, some of the homes attempted to welcome a few of the boys and girls who happened to be friends of the family. The Country Club provided for a few. But the situation in general was grave. With no recreational activities open to most of these boys and girls, they began creating their own without guidance and help from adults. They formed a "100 Club." In order to become a member of this club the adolescent was required to drive a car down a certain street—a dangerous street, crossing side streets and a main highway—at approximately 100 miles per hour. They also found and visited the famous "parking" places. In many cases, the primary purposes were to molest the cars of others, to frighten others, or to commit some act of mischief or some offense.

Some alert parents and some community workers, recognizing the gravity of the situation, became interested, worked together, and were instrumental in the formation of what was referred to as "The Teen Tavern." They took an old mill building, made some slight alterations and needed repairs, repainted it, and turned it into a recreational hall. Nearly all of the work was done by the teen agers, many of them members of the 100 Club. The materials were furnished by one of the civic organizations of the town. Various community leaders provided such supervision and special skills as were needed. The boys and girls elected officers, set up committees for carrying on the various activities, and provided a functionally responsible organization. Included in the building are a snack bar, a bowling alley, and games of a number of different types; dances are held from time to time, and in general The Teen Tavern is a wholesome place in which to meet friends and to have fun. An attraction that interested many teen agers was the introduction of lessons in folk dancing. Although this did not eliminate all of the mischief carried on by adolescent boys and girls, it reduced it to a minimum and at the same time did much to develop wholesome attitudes and outlooks on the part of boys and girls toward problems of everyday living.

The camp. It is becoming more widely recognized that camping activities may contribute much to the personality development of boys and girls.   However, one cannot lay down any general rule about who should go to camp, nor can one say just what is the best type of camp, for this will depend upon the nature and needs of the individual.   The widespread development of camps is a result of (1) the increased recognition of the educational value of camp experiences, (2) the mental and physical health value of camp life, and (3) the need for recreation under guidance during the summer months when schools are not in session.

With the urbanization of our society there have come about new demands for these activities, and the utilization of camping as an educational, health, and recreational agency seems destined to become more general in the future.   The form the camp takes depends in a large measure upon the agency sponsoring it, but the general aims of all camps are somewhat similar; to this statement the scout camp, operated primarily in the towns and cities, and the 4-H Club camp, operated among rural groups, both with the same object and with much the same activities, bear witness.   It is quite likely that more camps supported by public funds and appealing to special interest groups will be developed, and that these will be operated primarily as educational and recreational centers.   By bringing together special interest groups from different localities, camps enable adolescents to form new associations, and these associations give them a broader view on life, liberalize their thinking, and humanize their personalities.

Dimock [8] requested parents of boys who had been away at camp to note changes appearing in these boys a month after they had returned. According to the results of these ratings, there was a pronounced improvement in certain character qualities, among which are the following, listed in order of the number of times the increased rating was given them: confidence in self, courtesy, responsiveness to parental suggestion, appreciation of music, consideration for the welfare of others, meeting and mixing with others, readiness to cooperate, volunteering for service tasks, and so forth.

Dimock [9] also requested the boys themselves to indicate "the biggest things a boy gets out of camp life."   The things mentioned most often were as follows:

[8] H. S. Dimock and C. E. Hendry, *Camping and Character*.   New York: Association Press, 1929, pp. 284–288.
[9] *Ibid.*, p. 18.

No. of boys
mentioning

Skill in such activities as swimming, canoeing, campcraft	39
Learning to get along with others, "mixing," working together	35
Better health, physical fitness, strength, posture	33
Attitude of helping the other fellow, unselfishness	32
Self-confidence, reliance, initiative, thinking for self	20
Development of courage and nerve, losing timidity	17
Appreciation of nature, out-of-doors, and music	17
Meeting and making friends, fellowship	16

Organizations such as the Camp Fire Girls, Girl and Boy Scouts, the Y.W.C.A., the Y.M.C.A., and many others have established camps throughout the country. These camps are generally operated on a liberal expense budget. The child's health is given careful consideration and he is put on a balanced diet; he learns the art of living, working, and playing with others. Children—especially those from the urban sections of the country—are given an opportunity to become acquainted with the world of nature and to explore its possibilities. The camp, like the public school, is democratic in nature, and it provides an even greater opportunity for democratic living. Children from different types of homes are enrolled at these camps, yet they are all required to follow the same rules and are given the same privileges; thus they tend to live, work, and play in a democratic manner.

The social setting. There has been much controversy over the relative advantages of the city, small-town, or country environment for the social and character development of children. It has been pointed out by some that a farm in the open, away from the artificiality and restrictions of the city, is the ideal place; whereas others have pointed out that these advantages are more than offset by such disadvantages as lack of educational facilities, opportunities for social participation, and many modern sanitary and labor-saving conveniences. However, as a result of modern means of transportation and communication, coupled with the more widespread use of labor-saving devices, these differences are not so marked as formerly.

On the other hand, it has already been stated that there is a significant difference in the size of the family unit in different areas, with the rural areas having the largest units. Forty-five per cent of the children of this nation live on farms. In an essentially rural cultural pattern, neighborliness, stability of mores, and close family ties with considerable family dependability are present. A child reared in such a cultural pattern has ordinarily found adjustment to his vocational and adult world a fairly simple matter; however, with the technological developments that have been under way for the past century, many boys and girls are continuously leaving this rural cultural environment and entering into the cultural life of the cities. Many of these individuals are ill-prepared for meeting the more complex and artificial problems of an urban environment and consequently become maladjusted, a misfortune that presents a real challenge to the rural communities to develop an educational program that will offer boys and girls an opportunity to study problems related to urban life, and a chance to acquire vocational skills and understanding enabling them to adjust more readily to increasingly complex social living and modern technological procedures.

Demands are continuously being made in the community for cooperation and group action. Truly, we have passed from an individualistic society to a society requiring group action, group thinking, and cooperative effort. The school program should be sufficiently related to community activities and problems to provide meaningful situations for group activity.

The school has become in many rural and small town communities the social and recreational, as well as the educational, site and center. Such a community center provides boys and girls with an opportunity to meet together, to develop the ability to play together, to attain skill in socially acceptable activities, to learn the art of conversation, and to acquire an appreciation of desirable types of entertainment. Future Farmers of America, 4-H Clubs, and like organizations provide splendid opportunities for boys and girls to meet and work together under desirable conditions. Through such organizations they are able to display leadership and initiative, gain information, and acquire useful and wholesome social skills.

Over thirteen million of our population live in small towns of between 250 and 2,500 people. Some of these communities are railroad towns, others are manufacturing towns with the factory as the center, and still others are college towns, capital towns, lumbering towns, and the like. In such a situation one may find a diversity of nationalities, religions, and cultures; the process of assimilation, however, is very rapid under such conditions, since most of the minority groups are very small in number. Such is not the case in the city, where minority groups isolate themselves in special areas. The small towns are made up of more homogeneous groups than are the cities, and there is less diversity in the social and economic strata.

It has been said that the present-day city exhibits civilization at its worst and at its best. Fifty years ago slightly more than one-fourth of our population lived in cities; today over one-half does so, in conditions that range from those of the slums to those reserved for position and wealth. Sociological studies show that there is a very close relationship between unrest and crime on the one hand and living conditions on the other. The range in standard of living is from the lowest, where children are underfed, ill-clothed, and certainly ill-housed, to the other limit of the scale, where there is an abundance of luxuries of all classes and descriptions. The average standard of living in this country is high when compared with that of most other areas of the world, but the median standard of living is several hundred dollars a year per family below that of the average. The wide variety of groupings in a typical American city, and the great disparity between even the average standard of living and the substandard scale by which many are forced to live are important sources of unrest, mischief, and crime.

Then, too, in the large city the neighborliness and moral stability found in a more homogeneous and informal group of rural and small-town people are likely to break down. In small communities, where everyone knows his neighbor and accepts approximately the same standards as guides for his conduct, public opinion acts as a regulating force. But in the large cities, where such conditions do not exist, there is much less public scrutiny, and increased freedom of action results. This, of course, has both advantages and disadvantages. The existence of the latter creates an increased need for the agencies concerned with the guidance of boys and girls to give added attention to the importance of the development within the individual of standards and values for the regulation, direction, and control of his personal-social relations.

Furfey,[10] from a comparison of the "developmental-age" scores of urban and rural boys, found differences favoring the urban boys "equivalent to about a year and one third at age eleven, about two years at twelve, and about one third of a year at thirteen." It has already been pointed out that the family situation provides the setting, the stimulation, and most of the guidance (especially during the earlier years) that will determine, very largely, whether the child will develop

[10] P. H. Furfey, "A Note on the Relative Developmental Age Scores of Urban and Rural Boys," *Child Development*, 1935, Vol. 6, pp. 88–90.

into a well-adjusted and socially useful individual; however, the home environment never functions independently of its social setting. The quality of the social life and of the person-to-person relationships outside the home are influential factors affecting social development. Stott found that in the area of social relationships the farm home was at a disadvantage. He concludes from his studies:

Two of the personality variables, viz., resourcefulness in group situations and ethical judgment, had particularly to do with facility and discrimination in social relationships. The farm group ranked lowest in both of these variables. The city and town groups averaged about equally but both were significantly superior to the farm group. These differences, however, were almost wholly contributed by the girls.[11]

Democratic vs. autocratic leadership. There is evidence from the study of the early social climates of adults that aggressive behavior and instability are related to an early life dominated by authoritarian control. When the father or mother was the dominating (authoritarian) force in the home, the children obeyed, but their lives were filled with tension and frustration. Lewin, Lippitt, and White have conducted a number of investigations on this general problem.[12] Their studies have furnished considerable evidence for the conclusion that the nature of the experimentally created social climates (autocratically or democratically controlled) affects the behavior of children.

In their first experiment two clubs of ten-year-old boys, engaged in theatrical mask-making for a three-month period, were studied. The group leader treated one group in an authoritarian manner, while the other group was handled democratically.[13] The behavior of the boys was carefully studied by four observers. In the club meetings the authoritarian club members developed an increasingly aggressive, domineering attitude toward each other but an attitude of submission toward the leader. The behavior of the democratic club members toward each other was characterized by friendliness and fact-finding. This group was more spontaneous in its responses and assumed a free

[11] Leland H. Stott, "Personality Development in Farm, Small-Town, and City Children," *Agriculture Experiment Station Research Bulletin,* No. 114. University of Nebraska, 1939, p. 32.

[12] K. Lewin, R. Lippitt, R. K. White, "Patterns of Aggressive Behavior in Experimentally Created Social Climates," *Journal of Social Psychology,* 1939, Vol. 10, pp. 271–299.

[13] R. Lippitt, "An Experimental Study of the Effect of Democratic and Authoritarian Group Atmosphere," *University of Iowa Child Welfare,* 1940, Vol. 16, No. 3, pp. 43–195.

and friendly relation with its leaders.   On the one item, *overt hostility,* the authoritarian group was much more aggressive than the other, the ratio being 40 to 1.   The authoritarian group displayed greater hostility toward each other, used more attention-getting devices, showed hostile criticism, and indicated a lack of a sense of fair play.

In the second experiment by Lewin, Lippitt, and White five democratic, five autocratic, and two "laissez-faire" atmospheres were established.   In the "laissez-faire" groups the leader sat around and left things to the club members.   There was less than half as much participation by him as there was by the leaders of the other types of groups.   The influence of the leader's personality was controlled by having each of four leaders play the role of autocratic and of democratic leader at least once.   Relative to tension created in the autocratic group, the investigators state:

An instance where tension was created by *annoying* experiences occurred when the group work was criticized by a stranger (janitor).   There were two cases where fighting broke out immediately afterwards.

In the autocratic atmosphere the behavior of the leader probably annoyed the children considerably (to judge from the interviews. . . .)

In addition there were six times as many directing approaches to an individual by the leader in autocracy than in democracy.   It is probably fair to assume that the bombardment with such frequency ascendant approaches is equivalent to higher *pressure* and that this pressure created a higher tension.[14]

The value of cooperation was well stated by a classroom teacher in the St. Louis public schools:

This faith in cooperation—belief that people working together to manage their own affairs is the best kind of control—has its roots in the social philosophy of democracy and is the essence of our American heritage.   Yet, even though our society accepts this philosophy verbally, many of its institutions reveal that force and not the "will of all" is the guiding principle.[15]

Need for adult insight.   As children progress from early childhood through later childhood and into adolescence, the adults who deal with them show less and less insight into the role or position of a particular child in his social group.   This is reflected in statements made by adults, such as: "I don't understand why Sue is not more popular with

[14] *Op. cit.,* pp. 291–292.
[15] Dorothy C. Bohn, "Teachers Share in Administration," *Educational Leadership,* 1948, Vol. 5, p. 429.

the other girls; she seems to be such a nice, sweet girl." In a study bearing on this problem, Moreno[16] asked children from kindergarten through the eighth grade to choose two classmates whom they would like to have sit on each side of them. He also asked the teachers to list which children conceivably would receive many choices and which would receive few or none. The teachers' judgments coincided with the choices made by the kindergarten and first grade children about two-thirds of the time. On the other hand, they agreed with the seventh grade pupils only in about one-fourth of the cases.

SUMMARY

Changed social and economic conditions have resulted in profound changes in the organization and activities of the community. The increased leisure time of boys and girls has presented a definite challenge to the communities to provide recreational opportunities and better guidance of boys and girls, especially during the adolescent years. Many of these problems will be studied further in later chapters. Materials have been presented in this chapter showing the activities of adolescents in the community and some of the needs for the community to provide better educational and recreational facilities for boys and girls as they grow through adolescence toward complete maturity.

A few pieces of lumber, some glue, and nails is not a table. Likewise, a collection of boys, girls, and adults is not a community. There must be some common interests and needs, mutual confidence and understanding, association and sharing a common lot, if there is to be a true community. Morgan has said in this connection: "In a true community many activities are shared by the same people. This unified living results in deeper social roots and more unified personalities."[17]

The summer camp has come into use as a means for providing for the recreational, health, and educational needs of adolescents. Increased leisure has presented a problem and a challenge. It is important that this time be not wasted, but used as a road to health, efficiency, and morality. Without a purpose or goal, free time may bring the adolescent in contact with vice, crime, and unconventional practices. But if his community offers libraries, museums, school activities, sports, hobby groups, church groups, "Y" settlement houses, playgrounds,

[16] J. L. Moreno, *Who Shall Survive?* Washington: Nervous and Mental Disease Publishing Co., 1934.

[17] A. E. Morgan, "The Community," *Journal of the National Education Association,* 1945, Vol. 34, p. 55.

movies, and parks, there is less chance that he will divert his energy into undesirable channels.

THOUGHT PROBLEMS

1. What are the main features of a community? Describe some community of your acquaintance, showing the presence of these features.
2. What are some of the new conditions and problems faced by the child in his community adjustments?
3. Show how conflicting loyalties sometimes develop as the individual becomes affiliated with various groups of the community.
4. What are some class distinctions found in a community of your acquaintance? Show how individuals may move from one class to another. What are the major barriers to such mobility?
5. Do your observations verify the findings from the study by Neugarten of social class and friendship among school children? Elaborate.
6. How would you account for the increased tension usually found among boys in autocratically controlled groups?
7. What is the fundamental reason for the gang? What needs does the gang supply?
8. Describe the community facilities for recreation in some community with which you are especially acquainted.
9. Analyze the nature and function of one of the teen-age recreation centers discussed in some fairly recent number of *Recreation* magazine.

SELECTED REFERENCES

Butler, G. D., *Introduction to Community Recreation*. New York: The Macmillan Co., 1940.

Centers, R., "The American Class Structure," in Newcomb, T. M.; Hartley, E. L.; *et al., Readings in Social Psychology*. New York: Henry Holt & Co., 1947.

Cole, Luella, *Psychology of Adolescence*. New York: Rinehart and Co., Inc., 1948, Chap. XII.

Hollingshead, A. B., *Elmtown's Youth*. New York: John Wiley and Sons, 1949, Chap. VII.

Olsen, Edward G., *School and Community Programs*. New York: Prentice-Hall, Inc., 1949.

Partridge, E. D., *Social Psychology of Adolescence*. New York: Prentice-Hall, Inc., Chaps. X and XI.

Smith, Samuel; Cressman, George R.; and Speer, Robert K., *Education and Society*. New York: The Dryden Press, 1942.

Strain, Frances Bruce, *Your Child, His Family and Friends*. New York: Appleton-Century-Crofts, Inc., 1943.

Warner, L. L., and Lunt, P. S., *The Social Life of a Modern Community*. New Haven: Yale University Press, 1941.

Wrenn, C. G., and Harley, D. L., *Time on Their Hands*. Washington: American Council on Education, 1941, Chaps. I, II, and VI. This is perhaps the best single reference available on the general problem of leisure.

XIV

The Adolescent Personality

PERSONALITY: ITS NATURE AND CHARACTERISTICS

Although this chapter is entitled *The Adolescent Personality,* it should not be inferred that adolescence ushers in *a self* that is separate and distinct from that appearing during the earlier years of life. The only distinct changes that adolescence brings with it are those associated with sexual development, and even here the experiences of earlier years are of utmost importance. The earlier chapters dealing with growth indicate that growth is a continuous process and cannot be broken down into special periods, except for certain specific changes that may appear at the various stages in life. During adolescence, however, a better coordination of experiences is made possible because of the mental and educational growth over the preceding years. There is a continued correlation of the physiological self with the demands of society. Attitudes are formed and organized in harmony with these demands. The previous chapters have presented materials relative to the influence of peers, the home, the community, and the school on the formation and extension of the adolescent personality. It is the province of this chapter to describe the adolescent personality, present certain means found useful for evaluating it, and suggest some of the major problems encountered in its development.

Personality defined. The term *personality* is frequently used in our present-day terminology to refer to man's behavior and characteristics. It has been used widely and loosely by the layman, the personality expert, the orator, and the psychologist. The layman looks upon it in terms of qualifying adjectives such as "good," "pleasing," and "queer," whereas the personality expert considers it somewhat like a pair of gloves or a stylish hat—something that can be bought for five dollars or more and worn effectively with a few hints on how to wear it. Orators—and some psychologists—have clothed the term in a sort of mysticism and abstraction similar to that which surrounds the terms *ego, soul,* and *spirit.* In such a case it does not yield readily to definition or even to adequate description.

The concept of personality has a definite value in psychological terminology, and for the man-in-the-street it has not only a general theoretical value but a practical one as well. In order that students of adolescent psychology may have a more exact picture of personality, they should first realize that it is not something that can be imposed from without and thus put on or taken off through some formal teaching-learning procedure. Conforming to conventional practices one may (1) give a list of traits or attributes that constitute personality and let these provide the basis for a definition or description; (2) define it in terms of its general function; or (3) omit all efforts at defining it but give a generalized treatment of it in the life and growth of growing individuals.

Although a few students of personality use the term to signify a group of personal qualities or traits, the majority of authorities use the term somewhat as it was early conceived, that is, to signify the whole person, body, mental qualities, emotions, character, voice, and habits. A definition that avoids a listing of traits and also steers clear of conceiving personality in terms of *uniqueness* has been presented by Katz and Schanck.[1] This is as follows: *"Personality is the concept under which we subsume the individual's characteristic, ideational, emotional, and motor reactions and the characteristic organization of these responses."* The authors of this definition point out that such *characteristics* are more a function of the individual than of the immediate stimulating situation.

Davis and Havighurst formulated a concept of personality in terms of two basic interacting systems of behavior. Concerning these they state:

One system of actions, feelings, and thoughts is (1) cultural. It is learned by the individual from his social groups; his family, his age-groups, his sex-group, his social-class group, and so on. The other system of responses is (2) individual, or "idiosyncratic," or "private." It derives in part from (a) genetic factors and in part from (b) learning. These learned individual traits are responses to (a) organic, (b) affectional, and (c) chance factors, and likewise to (d) the particular deviations of a child's training from the standard cultural training for his group.[2]

[1] Reprinted by permission from *Social Psychology,* by Daniel Katz and R. L. Schanck, published by John Wiley and Sons, Inc.

[2] A. Davis and R. J. Havighurst, "Social Class and Color Differences in Child-rearing," *American Sociological Review,* 1946, Vol. 11, p. 698.

Personality as an integration of traits. The personality of an individual depends not only upon the traits that he possesses, but also upon the integration of such traits. By integration is meant the general organization of traits into a larger unit of behavior, and with some traits becoming subordinate to others in such an organization. Personality, therefore, cannot be considered as so many separate traits; rather, the individual's personality is made up of a totality and pattern of such traits. Many people lose sight of the integrative nature of personality in their study of the individual; this is especially in evidence in the classification of all individuals with the same educational achievement as similar in personality. The same error is made with regard to criminals, professional classes, people of the same intelligence, and so forth. It is only when two individuals have alsolutely identical heredity, identical training, and identical organic conditions that one could expect various personality elements to be integrated into identical personality patterns.

How broad organic traits or behavior trends become is a problem both of physiology and of sociology. The complexity of habit patterns involving the higher levels of behavior will depend upon the integration of these various patterns into a general behavior pattern; the *Gestalt* school in Germany has emphasized certain aspects of behavior allied to this general problem. Few psychologists would affirm the complete isolation of behavior patterns from physiological and social relations; that habits become integrated into larger units of behavior has been emphasized throughout our discussion of the adolescent. More complex behavior patterns involving the social and biological life of the organism are to be regarded as constituting a higher level of organization more complex in nature than the simple behavior habits out of which they grow.

The growing nature of personality. Since adolescence is a period especially marked by physical, mental, and emotional changes, one can expect corresponding changes in the personality of the adolescent subject. Mental maturity is reached during adolescence. Physical growth, which was discussed in Chapter III, is rather rapid early in this period, but there are some rather abrupt organic changes involved. The thymus gland ceases to function; the sex glands begin to function; and thus a new endocrine balance is established. The child's egocentric nature thus takes on a social form, correlated with the changed endocrine self. The child is now held responsible for acts committed by the self; society looks upon the personality as a growing social force, and now sees not Smith's child but Mr. Smith's young daughter.   The impression the growing individual makes upon others is therefore changing with the growing elements that contribute as a general configuration to personality.

Again, it is interesting to note the personality of an individual as we observe it in different situations.   The writer has in mind a 14-year-old girl, whom for convenience we shall call Edna.   She is very disobedient at home, especially in response to her mother's requests, and the mother thinks of her as "a little smarty."   In the presence of her older sister in social situations Edna is quite submissive and timid, but with the boys and girls in the eighth grade at school Edna is quite sociable, and is liked by all.   Not only do we notice different behavior patterns when Edna is in three different situations, but even when she is "performing" in any one of these situations we are likely to notice an at least partial exhibition of these other personality characteristics. Thus, personality cannot be considered apart from the situation in which the various traits are exhibited.   Some situations will call forth some traits, while another situation may call forth a very different pattern of traits.   The combination of traits present in a particular situation will depend upon many variables, such as maturity, sex, habit systems, health, present attitude, general social pattern, and so forth.

Personality types.   Man has ever been interested in dividing individuals into special types.   Individuals so divided can be catalogued and more readily described.   This simplification of individual differences in personality has furnished two-way classifications of the following groupings:

> introverts — extroverts
> dominant — submissive
> theoretical — practical

However, careful studies show that most people represent a mixture of components and cannot be divided according to a two-way grouping. Furthermore, variations are of a continuous nature, going from one extreme to the other, rather than of a discontinuous nature represented by types.

In Chapter III the concept of body build was presented.   Various attempts have been made to discover the relationship of body build to temperament and other variations in personality.   Kretschmer's [3]

[3] E. Kretschmer, *Physique and Character*.   New York: Harcourt Brace & Co., 1925.

classification furnishes a basis for dividing personality into the following types, based upon physical structure:

Body Build	Personality Characteristics
Asthenic or slender build	Withdrawal tendencies
Pyknic or broad build	Volatile, outgoing, assertive tendencies

Sheldon's studies [4] followed those of Kretschmer but assumed a different approach. The greatest support of his theories has come from studies conducted with maladjusted and abnormal individuals as subjects. Sheldon brought forth a tripolar classification parallel to his three large groups of body build types. His first group, the *viscerotonia,* is characterized, in the extreme cases, by general relaxation, sociability, love of comfort, extreme liking for food, and enjoyment of people. The second group, the *somatotonia,* is primarily a muscular type, and is characterized by vigorous bodily activity and the exertion of muscular activity and strength. The third, the *cerebrotonia,* is especially characterized by its inhibitory nature. The individual is secretive and tends to hold the self in restraint. The individual may be rated on a seven-point scale for each of the primary body dimensions. In addition to the variables listed, there are others, such as, intelligence and sexuality.

Marked agreements between the types of body structure and these temperaments have been recorded by Sheldon. Other investigators have not, however, confirmed this close relation.

Measurements devised for the study of personality have revealed that certain trait clusters tend to appear together. Perhaps the most functional classification of personality types is that developed by the Committee on Human Development of the University of Chicago and presented by Havighurst and Taba.[5] Groupings of sixteen-year-old were empirically arrived at. The clinical conference methods used in this study consisted of analyzing the data, observing similarities among certain subjects of their study, and grouping together similar subjects. A profile of personality and character factors that characterized each of the groups was then developed. This is shown in Table XLVI for

[4] W. H. Sheldon, S. S. Stevens, and W. B. Tucker, *The Varieties of Human Physique.* New York: Harper and Bros., 1942.
[5] The materials of Table XLVI are reprinted by permission from *Adolescent Character and Personality* by B. J. Havighurst and H. Taba, published by John Wiley & Sons, Inc., 1949, Chap. XI.

TABLE XLVI

Personality Profiles of Adolescents

Personality Types

Area	Self-Directive	Adaptive	Submissive	Defiant	Unadjusted
Social personality	Ambitious Conscientious Orderly Persistent Introspective	Outgoing Confident Positive, favorable reactions to environment	Timid Does not initiate action Avoids conflicts	Openly hostile Self-defensive Blames society for failure	Discontented Complaining Not openly hostile
Character reputation	High Higher on H and R than on F	High Higher on F than on H and R	Average to high Higher on H and R than on F	Very low Higher on MC than on other traits	Low to average
Moral beliefs and principles	Variable High uncertainty	High Little uncertainty	High Some uncertainty	Low	Low to average
Family environment	Strict family training	Permissive family training No conflicts with family	Some family training	Family training inconsistent, provides no basis for constructive character formation Conflict with family	Variable family training Conflict with family

Social adjustment with age mates	Leader Active in school affairs Awkward in social skills	Very popular Active in school affairs Social skills well developed Popular with opposite sex	Follower Nonentity Awkward in social skills	Unpopular Hostile to school activities	Unpopular Hostile or indifferent to school situations
Intellectual ability	Average to high	Average to high	Low to average Seldom high	Low to high	Low to high
School achievement	High or higher than IQ would imply	Fair to high	Fair Seldom high	Low or lower than IQ would imply	Low or lower than IQ would imply
Personal adjustment	Self-doubt Self-critical Some anxiety, but well controlled Concern about moral principles Average aggressiveness Moves away from people Lack of warmth in human relations Gains security through achievement	High on all adjustment measures Self-assured No signs of anxiety Unaggressive Moves toward people	Self-doubt Self-critical Submissive to authority Unaggressive	Hostile to authority Impulsive Inadequately socialized Moves against people	Aggressive impulses Feelings of insecurity

295

F, H, MC, and R refer to the traits friendliness, honesty, moral courage, and responsibility the five types developed in this analysis. The fact that 31 per cent of the group could not be placed in any of the five types is additional evidence for the contention that individuals do not fall into clear-cut types.

The materials of Table XLVI are useful in observing adolescents and in noting the factors associated with certain character and personality qualities. The material of this table provides a sort of summary of the effects of peers, the family social status, community forces, and the school on the personality and character of adolescents.

The persistence of the personality pattern. Not only do children manifest from the beginning of life differences in personality characteristics that set one child apart from another; but there are certain characteristics within each child that tend to persist during the period of growth and into adolescence and adulthood. The studies by Shirley[6] conducted with two infants as subjects provide good evidence for this generalization. Shirley noted that one of the infants consistently displayed a more irritable nature than the other. As the infants developed, she noted that certain forms of behavior would wane, only to be displaced by other forms of behavior somewhat consistent with earlier behavior activities. Profile charts, showing ratings and scores for each baby on a number of characteristics, furnished information that indicated a definite consistency in personality characteristics at the different age levels. This consistency was sufficiently clear to enable one to distinguish a particular child by means of identifying earmarks in his previous activities, without resorting to names or to other identifying procedures.

This tendency on the part of the growing child to remain somewhat consistent in his personality characteristics as he develops has been observed by other investigators. The studies by Gesell and others at Yale University reveal that in the case of twins one might show greater sociability than the other from a rather early stage of life. It has been further pointed out that the more sociable child would perhaps receive greater attention, which would tend to encourage further social development. This would, then, tend to accentuate any differences already existing. These genetic studies of personality consistency have received further support from case studies conducted with more mature subjects. In a study by Roberts and Fleming 25 college women were selected from

[6] N. M. Shirley, *The First Two Years*, Vol. III: *Personality Manifestations.* Minneapolis: University of Minnesota Press, 1933.

a large list of 100 cases.[7] Case studies as well as group data treated statistically indicated that the home relationships were most important in the development of personality traits, and that while there was some fluctuation of traits, in general there was more persistence than change.

PERSONALITY EVALUATION [8]

Following the development and use of objective methods of evaluating intelligence and educational achievement, objective techniques for evaluating personality have come into use. Such evaluations should provide a better basis for guiding the growth and development of adolescent boys and girls in their personal and social adjustments. The methods most commonly used are (1) personality questionnaires and inventories, (2) rating scales, (3) interviews, (4) anecdotal records, and (5) projective techniques. A brief description of each of these should suffice for our understanding of the personality characteristics of adolescents.

Personality questionnaire. One of the earliest personality inventories developed was the Thurstone Personality Schedule.[9] This scale gives a single score indicating the degree of the presence of neurotic tendencies. By means of this scale Thurstone was able to show that neurotic tendencies were largely independent of mental ability, but were definitely related to certain educational accomplishments. Since the advent of this schedule, many questionnaires and inventories have been developed. The California Test of Personality was developed as an instrument for identifying certain highly important factors in personality. The test is divided into two sections. The first section is designed to reveal how the individual feels about himself; the second section indicates how the individual functions as a social being.

Considerable research has been conducted with personality inventories. There is some evidence that such inventories when used judiciously may have value. However, Ellis points out that from 55 attempts to intercorrelate similar personality questionnaires, 9 gave

[7] Katherine E. Roberts and Virginia Fleming, "Persistence and Change in Personality Patterns," *Monographs of the Society for Research in Child Development,* Society for Research in Child Development, National Research Council, 1943, Vol. 8, No. 3.

[8] Some of these materials bearing on the evaluation of personality are adapted from materials presented in Chapter V of the writer's, *Psychology of Exceptional Children,* revised edition. Copyright 1950 by The Ronald Press Company.

[9] L. L. Thurstone and Thelma G. Thurstone, "Personality Schedule." Chicago: University of Chicago Press, 1929.

positive, 18 questionably positive, and 18 negative indications of validity. As a result of his studies, he states:

> It is concluded that group-administered paper-and-pencil personality questionnaires are of dubious value in distinguishing between groups of adjusted and maladjusted individuals, and that they are of much less value in the diagnosis of individual adjustment of personality.[10]

The personal interview. The personal interview furnishes a more personal basis for evaluating personality than does the paper-and-pencil test. One of the first essentials in the use of this method for evaluating personality is that the interviewer should be carefully chosen and trained for such a task. If trustworthy responses are to be secured, the interviewer must not resort to writing down everything that is said and done. A method that has been found useful is to have the subject complete an inquiry blank, such as that presented in Appendix B. The subject should be assured that any information presented on the blank will be confidential. In this connection, the subject may be told that he need not write his name on the blank, unless he prefers to do so.

The Woody Student Inquiry Blank may be administered in the form of a personal interview.[11] It purports to evaluate traits of social adaptability, dominance, inferiority, cooperativeness, and phantasy. From these scores evaluations of adjustments to home life, school life, social life, and vocational orientation are secured. A reliability coefficient of .92 has been reported by the author of this blank. This interview blank differs from that presented in Appendix B in that it is more quantitative in nature. This makes it especially useful in gathering data in which research materials are involved. However, in our quest for objectivity, we should not overlook the importance of qualitative data. It appears that by means of a well planned interview by a trained interviewer one should be able to get beyond the tangible manifestations of personality, and get into the individual's innermost feelings, urges, and aspirations. These aspects of the adolescent personality must be tapped if he is to be understood and dealt with on a rational basis.

The rating scale. In the application of the rating scales, these ratings

---

[10] A. Ellis, "The Validity of Personality Questionnaires," *Psychological Bulletin,* 1946, Vol. 43, p. 426. Reproduced by permission of the Psychological Bulletin and the American Psychological Association.

[11] C. Woody and R. Gatien, "The Sophomore and Freshman Testing Program," *Bureau of Educational Reference and Research Bulletin No. 155,* University of Michigan, 1943, pp. 111–197.

may be made by parents, teachers, or associates, or by the individual himself. Two tendencies have been noted in the use of the rating scale. The first is that of the "halo effect." This is a tendency for the rater to be influenced in the assignment of a rating to a specific trait by the ratings given other traits. For example, if an individual is rated high on one trait, the rater will be inclined to rate him high on other traits, whereas a low rating on the same trait would produce lower ratings on other traits than they would normally be given. A second tendency relates to self ratings. There is a pronounced tendency for individuals to rate themselves higher on those traits they consider desirable than on others they consider less desirable. The graphic scale has been developed to facilitate ratings. Also, the item being rated is clearly defined in an effort to secure more reliable and valid ratings. Ratings that are based upon actual descriptions of behavior are the most desirable, since they are probably more objective in nature. The following items, taken from a scale used by Newman,[12] illustrate one type of such a rating scale:

GROOMING ACTIVITY

1 2	3 4	5 6 7
Obviously spends a great deal of time in grooming self. Frequently arranges or combs hair, brushes off clothes, puts on make-up.	Offers evidence of some attention to clothes, hair, nails, shoes, etc.; but grooming not a major or very important activity.	Pays no attention to personal appearance. Can't be bothered about how appearance impresses others.

INTEREST IN OPPOSITE SEX (—*Social Participation*)

1 2	3 4	5 6 7
Continually initiates contacts with and takes every opportunity to attract attention of members of opposite sex. Preoccupied with preference for activities in which sexes are mixed.	Watches activities of mixed groups and occasionally participates. Drive for attention increased by presence of opposite sex.	Seems entirely unaware of presence of members of opposite sex, or exhibits a very casual, matter-of-fact attitude toward them.

Anecdotal records. The observation method is the oldest and perhaps most widely used method of all the research techniques. The procedures for using it have been standardized and refined so that the results thus obtained are oftentimes highly reliable. Within recent years the observational method in the form of anecdotes has been used in studying child and adolescent behavior in various situations. The anecdote is simply a descriptive account of some incident or occurrence

[12] F. B. Newman, "The Adolescent in Social Groups: Studies in the Observation of Personality," *Applied Psychology Monographs,* 1946, No. 9, Institute of Child Welfare, University of California.

in a child's life activities. This description is recorded in order to give it increased stability and permanence. The anecdotal method has been described by Randall as both objective and subjective:

1. It is objective to a degree approximating that of an X-ray photograph or motion picture in that it records, when they occur, the events that have attracted the attention of the observer, rather than the opinions of an observer as to the significance of what has happened. . . .
2. It is subjective to the degree that an artistically composed photograph is subjective. It sharply limits itself to a center of attention and subordinates inconsequential details. It has a satisfying completeness so far as the center of attention is concerned.[13]

In an attempt to evaluate adolescent personalities by means of anecdotes, a number of difficulties appear. In the first place, most teachers have not had the training that would provide them with the needed skill for recording and interpreting anecdotal behavior. Thus, when anecdotes are recorded by the teacher, his own subjective interpretation is too likely to be given to the anecdote. A second difficulty and perhaps a valid criticism of most anecdotes as well as so many observations of behavior is that only the extroverted individual is likely to be studied. Overt acts are more observable than are acts of daydreaming and various forms of withdrawal behavior. Anyone observing and attempting to evaluate personality should realize that withdrawal behavior and seclusiveness on the part of the individual adolescent are just as significant as fighting, bullying, and other forms of aggressive behavior. Another difficulty is that of *time*. Anecdotes should be collected over an extended period and should be looked upon and studied as longitudinal data on the individual. When gathered over a period of time and interpreted in the light of developmental changes occurring in the individual, they may be most useful in arriving at a better understanding of the personality characteristics of adolescents.

Projective techniques. No attempt is made here to describe the various projective techniques that have been developed for the measurement of personality. There is much evidence, both theoretical and practical, that the free associations and free responses of the individual to various situations have a definite place in the evaluation of personality characteristics. Within recent years considerable work has been done in this field. The various studies have been subjected to their

---

[13] J. A. Randall, "The Anecdotal Behavior Journal," *Progressive Education,* 1936, Vol. 13, p. 25.

share of criticism, much of which is based on their lack of objectivity and on their general reliance upon the subjective evaluation of the individual's responses. This criticism has merit; however, caution should be observed in attempting to arrive at a complete evaluation of an individual's personality characteristics on the basis of a quantitative score or a group of scores from paper-and-pencil tests or other quite objective procedures. The intangible and interdependent nature of personality makes it very difficult, almost impossible, to describe in purely quantitative terms.

The Rorschach ink-blot test is perhaps the best known projective technique.[14] This consists of a series of cards on each of which is an irregularly shaped ink-blot. The subject is shown the card and asked what the ink-blot might be. The subject then gives his interpretation of the ink blot. The subject's interpretation of the test will be based upon his experiences, conditioned ways of responding that have been established, his feelings and emotions, and other personal reactions and attitudes. The experimenter, who has been trained in administering and interpreting Rorschach tests, notes the nature of these responses and studies them from both a qualitative and a quantitative standpoint. The Thematic Apperception Test is another example of the application of the projective technique for evaluating personality. This test makes use of a series of pictures.[15] The subject taking the test is shown a picture and requested to indicate what is happening, and also to reconstruct what may have gone on before and what followed the action presented in the picture. In giving an interpretation to the picture and in imaginary reconstructions of events that preceded and followed what is shown in the picture, the subject is revealing his own inner self and the experiences that have affected him. Certainly such interpretations, like the interpretations given to the ink-blots, must come from the experiences of the subject and will be colored by his own personal and social needs.

SOME PERSONALITY CHARACTERISTICS
OF ADOLESCENTS

Contrasting phases of adolescent personality. Some elements characteristic of the personality of adolescents tend to make the individual

[14] For a more complete discussion of this technique see T. W. Richards, *Modern Clinical Psychology.* New York: McGraw-Hill Book Co., 1946.
[15] C. D. Morgan and H. A. Murry, "A Method for Investigating Fantasies," *Archives of Neurology and Psychiatry*, 1933, Vol. 34, pp. 289–306.

unstable in nature; these elements are here referred to as "contrasting phases." The importance of the emotional elements in the development of personality has already been considered. Furthermore, it might be pointed out here that emotional habits are the important factors upon which we judge the personality of those closest and best known to us; these elements stand out much clearer in some than in other individuals.

During the adolescent period some specific emotional characteristics are outstanding. Many drives of an instinctive or biological nature are held in restraint during adolescence because of various customs and other forces present in man's environment, but these become quite pronounced in other phases of the individual's life. G. Stanley Hall recognized the importance of emotion in adolescent life, and in one of his writings says: [16] "Youth loves intense states of mind and is passionately fond of excitement." Here we find a true and valid expression of the contrasting states of vitality and lassitude so characteristic of adolescents. The attitude of carefree individuals seeking joy and the company of others for the sake of excitement characterizes their play, social interests, and activities. The true gang and team loyalty has already been described as characteristic of this age.

Pleasure and pain are sometimes close together; tears and laughter may closely follow each other; elation and depression also are somewhat characteristic of this period of life. Egocentrism and sociability, ascendancy and submissiveness, selfishness and altruism, radicalism and conservatism, heightened ambitions and loss of interest—these tend to mark off this period of life as one of contrasts in moods, which are manifested by a single individual in slightly different situations. These contrasting moods probably make it more difficult to predict an individual's behavior during adolescence than at any other single period. Individual reactions are more transitory and less stable than they are at later stages of life; different traits will predominate under slightly different conditions; and their changes are likely to be very marked. As the individual has more and more social experiences, his manners of reaction change and his personality characteristics are increasingly modified and made more stable.

Anyone who studies the problems of young people becomes familiar with these common manifestations of behavior. Here is an individual in whom habit patterns have not fully developed. Because of his lack

[16] G. S. Hall, *Adolescence*. New York: Appleton-Century-Crofts, Inc., 1904, Vol. 2, Chap. X.

of maturity, he is sometimes characterized as "flighty." His work in school is not altogether steady; his activities on the plaground vary from time to time; his general attitude toward the school is often easily changed. Pride in dress is followed by extreme carelessness. While these particular sudden and extreme changes are the exception, the average adolescent has them to some degree. Bronner makes the following observation:

Today's enthusiasms may become matters of boredom before long. The desire one day may be to become a missionary, and e'er long this has been completely forgotten and the goal of life is to be a dancer. Many an adolescent has said, "I don't know what I want to be. One day I think I want to be one thing and the next day something else, only I want to be someone great." [17]

Analyzing the cause of this changefulness, one again turns to the newly developed interests and broadened outlook of these boys and girls as they reach maturity and come into contact with social reality. These changes in outlook take place more rapidly than habit systems change, develop, and become integrated into a unified personality. We therefore find individuals not only with often inconsistent attitudes, beliefs, outlooks, and emotions, but also strikingly contrasting moods and attitudes toward situations or topics not wholly different in nature. Not all of these inconsistencies and contrasting phases of life are finally eliminated, but many are substantially eliminated as the personality becomes more and more integrated into a general schema.

Adolescent instability. The adolescent is said to be impulsive and unstable in nature. Emotional expression, as we have seen, is largely a matter of habit, and from such habits develop behavior patterns characteristic of extroversion or introversion. As attested by the pointless giggling, impulsiveness, yelling, loud talking, and other symptoms of instability, extroversion usually appears to be more universal than introversion, which is manifested in relation to new situations and intensified by newly forming habits of a social nature. Habits of introversion are especially in evidence in individuals who are reaching maturity with poorly developed social and emotional habits. With the awakened social consciousness, the new physiologcal nature, and the wider social contacts there is naturally good reason for disturbances.

The instability of adolescence is especially marked by contrasting

---

[17] A. F. Bronner, "Emotional Problems of Adolescence," *The Child's Emotions.* Chicago: University of Chicago Press, 1930, p. 220.

personalities, heightened emotional behavior, religious enthusiasm, and juvenile behavior problems. Just how truly such conditions are a result of training is quite evident as we observe many adolescents with varying backgrounds who are socially well-adjusted, wholesome in attitude, courteous in manners, and stable in the exhibition of various habit systems. Far too many children, as they reach adolescence, are expected to assume the places of adults with only the training that would enable them to follow authority blindly. These individuals have not been given the opportunity for the development of habits of initiative and responsibility so essential in the ordinary pursuits of adult life; they are "too young" to do the things adults are doing and "too old" to act and play as children do. For many individuals this is, therefore, a period of bewilderment. If the individual desires to run and play the "kid-like" games, he is laughed at; if he offers his advice and counsel too freely to the adult group, he is reminded that he is still a child. Probably most persons soon pass through this transition and are able to establish themselves and their place in the social order. Naturally, a sort of training that will enable the individual to adjust his earlier habit patterns to those of the adult group will aid him to develop desirable social habits and attitudes. If the specific elements of the adolescent personality do not develop desirably, we should then search his past—or present—experience for the causes.

This question may be raised: To what extent are adolescent worries, doubts, and fears associated with moodiness? This problem was investigated by Fleege, when he asked 2,000 Catholic high school boys the question: "Do you ever get into moods when you can't seem to cheer up to save yourself?" [18] In answer to this query, 75.1 per cent of the boys replied "Yes." This furnishes a barometer concerning the amount of moodiness experienced among high school boys. On the basis of information presented in earlier chapters about the fears, worries, and anxieties of adolescent girls, one would expect the amount of moodiness experienced by high school girls to be in excess of that experienced by the high school boys.

The alleged causes for the moodiness experienced by the boys are listed in Table XLVII. An outstanding cause of moodiness during adolescence is *self-consciousness* about faults, weaknesses, failures, and the like. This was observed in Chapter III in connection with the physical and motor growth of boys; it is also reflected in Table XLVII.

[18] U. H. Fleege, *Self-Revelation of the Adolescent Boy.* Milwaukee: The Bruce Publishing Co., 1945, p. 321.

However, only one boy out of seven indicated that he was frequently depressed. In addition to the self-conscious feeling so characteristic of adolescents, we note that disappointment, deprivations, and feelings of guilt stand out as factors closely related to the onset of sadness and depressed states.

TABLE XLVII

Alleged Causes for Feelings of Sadness and Depression According to 2,000 High School Boys (*After Fleege*)

Cause	No. of Boys	Per Cent
1. Difficulties in studies and school, low marks, failure	298	14.9
2. Troubles in the home, arguments, debts, parental attitude	192	9.6
3. Disappointments, things go wrong	179	9.0
4. Sins, sex, self-abuse, guilty conscience, mistakes, wrong conduct	163	8.2
5. Deprivations, lack of social opportunities, curtailment of liberties	106	5.3
6. Sickness, death, mishap	103	5.2
7. Hurt feelings: because of a remark or because I have hurt those of others	102	5.1
8. Misunderstandings, quarrels with friends	102	5.1
9. Personality difficulties, inferiority complex, personal defects, lack of ability	80	4.0
10. Girl-friend troubles	65	3.2
11. Miscellaneous: nothing to do, worries, fears, lack of sleep, my future, etc.	117	5.8
12. No answer, or the statement, "I don't know"	451	22.6

The needs of the adolescent. Throughout this study the adolescent has been described as a dynamic individual growing and developing through an interaction with the conditions and forces in his environment. The dynamics of the human organism has been described in terms of certain basic needs, the significance of which has been presented by Prescott:

These needs are the basis of permanent adjustment problems which all of us face. They are more or less continuously with us. Our behavior is patterned in accordance with what experience has shown us to be the most satisfactory way of working them out, but, as conditions around us vary and change, we are continuously under the necessity of modifying our behavior. These needs become sources of unpleasant effect and even of serious personality maladjustments if they are not met adequately. Futhermore, our society is rich in circumstances which deny to individuals the fulfillment of one or several of these needs and quasi-needs for periods of varying length—this is what has happened to thousands of maladjusted children. There is a serious disharmony between the needs which they feel to be vital to themselves and the experiences in life as they meet them.[19]

Authorities are not in complete agreement relative to these basic needs, particularly those which may be termed personality needs and which must be met by the environment if the individual is to be happy and is to function effectively. The organic needs are better understood and more clearly defined. These include the need for food, rest, sleep, elimination, desirable temperature conditions, air, and water. The physiological basis for the sex drive is also well known. This was discussed in Chapter IV in connection with the physiological development of the adolescent.

There is evidence that these personality needs are not manifested in a similar manner among all groups of adolescents. However, to the extent that these are interrelated with the biological nature of the maturing dynamic individual they would appear in some form with all groups of adolescents. Concerning these Kurt Lewin has stated:

The needs of the individual are, to a very high degree, determined by social factors. The needs of the growing child are changed and new needs induced as a result of the many small and large social groups to which he belongs. His needs are much affected, also, by the ideology and conduct of those groups to which he would like to belong or from which he would like to be set apart. The effects of the advice of the mother, of the demand of a fellow child, or of what the psychoanalyst calls superego, all are closely interwoven with socially induced needs. We have seen that the level of aspiration is related to social facts. We may state more generally that the culture in which a child grows affects practically every need and all his behavior and that the problem of acculturation is one of the foremost in child psychology.

One can distinguish three types of cases where needs pertain to social relations: (1) the action of the individual may be performed for the benefit of someone else (in the manner of an altruistic act); (2) needs may be induced by the power field of another person or group (as a weaker person's obedience to a more powerful one); (3) needs may be created by belonging to a group and adhering to its goals. Actually, these three types are closely interwoven.[20]

[19] Daniel A. Prescott, *Emotion and the Educative Process*. Washington: American Council on Education, 1938, pp. 111–112.

[20] Reprinted by permission from Kurt Lewin, "Behavior and Development as a Function of the Total Situation," *Manual of Child Psychology* (D. L. Carmichael, Editor), published by John Wiley & Sons, Inc., 1947, p. 832.

An analysis of needs and conditions of adolescence shows that the individual, though physically reaching the stage of maturity, is compelled to delay the natural expression of certain drives now coming to play a large part in his everyday activities. Civilization has made it necessary that the training period of life be lengthened, but human biological development still proceeds at the rate of earlier times. At adolescence the individual is not established as a stable member of society. His habit systems, as was pointed out earlier, are in a formative stage; many of them are still unrelated and the process of generalization has not as yet carried over into broader social experience. His natural drives, which up to this period have found a greater freedom of outlet, are checked and modified by the great social organization in which he finds himself. Hence, the generally confusing and conflicting situations to which he must adjust himself often lead to certain forms of instability. However, we are not to despair of adolescents, since, out of this medley of circumstances and conditions, develop the age of youth and adulthood. Thompson says of this:

> The young person reared in a society which increasingly demands that he follow in an imitative manner its exemplar behaviors, expressed as tenderness, affection, and courteousness; cruelty, discontent, and hatred; emotional stability and a temper which is defensively rebellious; independence of and yet willingness to sacrifice himself for the group at large; a progressive interest in the opposite sex, regardless of the restraining taboos, maturing in marriage; and an insistence on individual financial success; these and a score of other similar behaviors make up the continuous barrage of traumatic experiences which assail the maturing individual. Out of this the individual resolves whatever problems afford him an accepted place in society and by so doing enters into adulthood with an integrated personality.[21]

SUMMARY

In all the various definitions of personality there appears, first, the notion of a *totality* of elements; then, in the second place, a general recognition of the *interrelation* of these various elements into a unified pattern. Furthermore, there is emphasis upon the *interaction* of these elements in the relationship between the individual subject and other persons. The totality may be made up of an abundance of some traits, other traits being lacking. Again, there may be a lack of harmonious interrelation of traits—conflicting values, or actually conflicting traits. Or, there may be a breakdown in the desirable interaction

[21] Charles E. Thompson, "The Personality of the Teacher as It Affects the Child," *The Educational Forum,* 1942, Vol. 6, p. 264.

of the individual's personality traits and the characteristics of others. The latter is sometimes referred to as "personality clashes."

Since the period of adolescence is one in which the personality traits are developing and finding expression in many directions, it becomes a period fraught with many problems and difficulties. It might be stated as a fundamental principle that *any period in life in which there is an undue physiological, social, or emotional stress for which the individual is not prepared, is a period at which mental abnormalities may and do appear, or at which those already in existence become more socially significant.*

Problems related to home, school, and peer relations were discussed in earlier chapters. Personal-social problems will be presented in several of the following chapters. These problems should not be conceived of, however, as detached and unrelated to each other or to the forces and conditions present in different aspects of the adolescent's life.

The various attempts to evaluate personality have served to emphasize the complex and interrelated nature of so-called personality traits. A word of caution should be given to those employing any of the devices described for evaluating personality. By recognizing the adolescent personality as dynamic rather than static, and by conceiving of the individual as a unified whole rather than as the sum total of so many discrete traits, one is in a better position to evaluate results obtained from tests, rating scales, inventories, and other devices for gathering data relative to personality.

THOUGHT PROBLEMS

1. Cite evidence from your own life for the persistence of certain personality traits.

2. Look up several definitions of personality, other than the ones presented in this chapter. Show how one's definition will affect his general treatment of this subject.

3. Why is it very difficult to measure personality traits? What experiences have you had with personality evaluations? What uses can be made of results from such evaluations? What cautions should be observed?

4. What "personality types" have you heard about other than those listed in this chapter?

5. Evaluate the personalities of several adolescents with whom you are acquainted, using the general outline given in Table XLVI. What difficulties are encountered in making these evaluations?

6. Can you cite evidence from your own life or someone with whom you are well acquainted for the persistence of basic personality characteristics. How do you account for any significant changes that might have occurred?

7. In what ways are the needs of adolescents different from those of the eight- and nine-year-old individuals? In what ways are they different from those of the mature adult of twenty-five or thirty years of age?

SELECTED REFERENCES

Arlitt, Ada H., *The Adolescent*. New York: McGraw-Hill Book Co., 1938, Chaps. XI and XII.

Blos, Peter, *The Adolescent Personality*. New York: Appleton-Century-Crofts, Inc., 1941.

Bogardus, E. S., and Lewis, R. H., *Social Life and Personality*. New York: Silver Burdett Company, 1938.

Havighurst, R. J., and Taba, Hilda, *Adolescent Character and Personality*. New York: John Wiley & Sons, 1949.

Hurlock, Elizabeth, *Adolescent Psychology*. New York: McGraw-Hill Book Co., 1949, Chap. XIV.

Meek, Lois H., and others, *Personal-Social Development of Boys and Girls*. New York: Progressive Education Association, 1940.

Milner, Esther, "Effects of Sex Role and Social Status on the Early Adolescent Personality," *Genetic Psychology Monographs*, 1949, Vol. 40, pp. 231–325.

Murphy, G., *Personality*. New York: Harper and Bros., 1947.

Sadler, W. S., *Adolescent Problems*. St. Louis: C. V. Mosby Co., 1949, Chap. VI.

Thom, D. A., *Normal Youth and Its Everyday Problems*. New York: Appleton-Century-Crofts, Inc., 1932.

Thorpe, Louis P., *Personality and Life*. New York: Longmans, Green and Co., 1941.

Zachry, Caroline B., "Customary Stresses and Strains of Adolescence," *The Annals of the American Academy of Political and Social Science,* November, 1944.

XV

Personal and Social Adjustments

It was pointed out in previous chapters that, as the adolescent's circle of friends widens, so widen also the areas of his problems, needs, and interests; as he grows toward maturity, he comes into contact with people who are unwilling to humor his childlike, egocentric tendencies. Then, as he becomes more mature, he frees himself from many of the close home ties and becomes more closely attached to various social groups and friends, loyalty, trustworthiness, sympathy, service to others, and other characteristics of man's social nature begin to develop into a fuller state. It is this change from the point of view of a self-centered individual to that of one who realizes the proper place of each individual in the life of the group that constitutes the essential social difference between the child and the adult. It is, then, at this time that personality change occurs, in harmony with the maturing of the social and physiological organism.

ADJUSTMENT PROBLEMS OF ADOLESCENTS

Every individual has adjustment problems, for it has been said that "Life is a continual process of adjustment, it is an incessant round of give and take." There are, of course, varying degrees in the intensity of conflicts that arise. Perhaps the best way to develop a wholesome personality in the adolescent is to maintain in that child feelings of *adequacy* and *value*. The adolescent meets many situations that tend to produce feelings of futility, inadequacy, and uselessness. If each time a child attempts to react he is criticized by his elders, teased by his associates, actually restrained by some obstacle stronger than he is, or made to suffer pain with each movement, he will be tempted to relax his efforts—to feel inadequate and useless. He must have enough successful experiences to offset the unfavorable ones if he is to develop a balanced personality. When the attention of the adolescent, or younger child, is called to his inadequacies, positive suggestions for overcoming them should be given. The recognition of previous ineffectiveness should be a stimulus to better adjustment.

Conflicts and adjustments. When the emotional conflicts of childhood are not solved in a satisfactory manner, they lead to symptoms of neuroses. Many of the difficulties experienced by adolescents and postadolescents are simply a continuation of these persistent unsolved problems of childhood—in many cases actually accentuated by changed social conditions and physiological maturation. Another group of problems experienced during adolescence may more correctly be labeled adolescent problems, since their origin is closely related to the development and ripening of the sex drive. The appearance of an increased sex drive, which is characteristic of the onset of pubescence, may seriously affect the harmony established between the socializing forces and the dynamic self. Thus, conflicts may appear between these forces and the ideals and concepts relative to the self that have developed during the years of growth. These conflicts appear in the form of feelings of guilt, depressed states, anxieties, and the like. The adolescent resorts to various forms of behavior in an attempt to resolve these conflicts. Open rebellion against parental restrictions may appear for the first time in the individual's life. When this is not feasible, more subtle procedures involving lying and deception may be resorted to in an effort to overcome some frustrating situation or condition. Withdrawal behavior, regression, and reversion to an early more secure pattern may follow an attempt at the solution of these conflicts. Some of these are discussed later in this chapter.

Adjustments of boys and girls. Interesting differences were found by Bell in the responses of boys and girls to certain social adjustment questions on the Adjustment Inventory.[1] High school boys gave significantly more maladjusted responses than did high school girls to the following social situations: enjoying social gatherings, meeting important people, introducing people at a party, enlivening a dull party, asking help of others, making plans for others, making contacts with opposite sex, speaking in public, leading at a social affair, and makings friends easily. On the other hand, high school girls gave more maladjusted responses than did high school boys to these social situations: afraid to speak out in class, fear of public speaking, timidity at important dinners, self-consciousness before people, avoiding calling

---

[1] H. M. Bell, *The Theory and Practice of Personal Counseling.* Stanford University, Calif.: The Stanford University Press, 1939, p. 113.

attention to oneself, hesitancy to ask favors, and conversational difficulty with strangers. A generalization from these findings indicates
that boys are more aggressive in social situations but have little interest
in them. Girls appear to have greater interest in social groups but
are more submissive by nature in social situations.

In so far as the Thurstone Personality Schedule measures maladjustments, high school girls—particularly sophomores and juniors—were
found in the study of Remmus, Whisler, and Duwald [2] to be more
maladjusted than boys. The fact that the greatest maladjustment is
present in the sophomore and junior years indicates that these are
years during which the girls are in the midst of the adjustment process.
At this period they are probably faced with more new and difficult
problem situations requiring social adjustments and direction of social
drives than at an earlier or later period. The differential motivating
influences exerted by different environmental stimuli form the most
plausible explanation for the greater maladjustments among the girls.
For example, the ideal of beauty and sex attraction looms larger in the
life of the girl. She has fewer vocational outlooks but a greater interest
in winning the approval of members of the opposite sex. The subterfuge outlets for drives, the greater socialization, and the more frequent
thwarting and blocking of the drives of the girl are influential factors
in affecting her maladjustments.

Intelligence and adjustments. The comparisons of behavior at different levels of intelligence indicate that children with average mentality
are better adjusted than children who are either brighter or duller.
This is to be expected in a world where the environment is adapted to
the average level of mentality. Though Terman reported only 11.9
per cent of unselected school children to have intelligence quotients
above 115, Anderson found 17.3 per cent of the children at a child-
guidance clinic at that level. Similarly, Terman found that 11.2 per
cent of unselected school children have intelligence quotients below 85;
in the child-guidance clinic 20.3 per cent were in that group. An
examination of Anderson's data indicates that among the deviations
from normal intelligence in either direction there was an increase
in the per cent of cases of problem behavior. (See Table XLVIII.) [3]
Some types of discipline problems somewhat common among the

[2] H. H. Remmus, L. Whisler, and V. F. Duwald, " 'Neurotic' Indicators at
the Adolescent Level," *Journal of Social Psychology,* 1938, Vol. 9, pp. 17–24.

[3] R. G. Anderson, "The Problem of the Dull-Normal Child," *Mental Hygiene,*
1927, Vol. 11, pp. 272–286.

gifted are: (1) disorderly discussion in classroom, (2) expression of disappointment at not being heard, (3) egotism, and (4) indolence. It is interesting to note that while the gifted often become problem children, they are not so often satisfactorily registered among the juvenile delinquents, indicating that some sort of adjustment is usually made.

TABLE XLVIII

A Comparison of the Intelligence of Children Referred to a Child-Guidance Clinic and of Unselected School Children

I.Q.	Terman's 1,000 Unselected School Children	320 Child-Guidance Clinic Problem Children	65 Juvenile Court Delinquents
125 and over	2.9	10.7	0
115–124	9.0	6.6	1.5
105–114	23.1	18.6	6.0
95–104	33.9	22.2	20.0
85–94	20.1	21.6	35.0
75–84	8.6	11.6	26.0
Below 75	2.6	8.7	11.0

Likewise, the subnormal group presents an abundance of maladjustments. The materials dealing with juvenile delinquency show that offenses are committed more frequently by those of inferior intelligence than by those of average or superior mental ability. This is to be expected, since in the first place those of inferior intelligence are not mentally equipped to assimilate and generalize ideals. Again, this group is usually made up of failures in the school program and thus finds opportunities for escape, adventure, and success in acts of mischief and crime. Probably the most serious result of school failure lies in its effects on the developing personality and character. Defeated and discouraged, failures resort to other activities, and develop out-of-school and all too often unwholesome attitudes toward society. They have learned thoroughly, through experience, the "opportunist doctrine." They take advantage wherever they can—of each other, the school, and the world outside. Their lot in life can only be that of the hewers of wood and drawers of water in the ranks of unskilled labor. Today, however, as the machine gradually displaces the unskilled worker, the individual developing as a failure in school life is likely to be less and less prepared to play a significant role in the world of work.

Symptoms of maladjustment. The symptoms of personal and social maladjustments will vary from individual to individual, and to some degree from one situation to another when the same individual is involved. However, some symptoms will appear more frequently than others. Studies show that it is the frequency of the appearance of certain symptoms that is especially important in evaluating personality maladjustments. Most children appear to show some of these symptoms at irregular intervals; but certain children display these symptoms in most situations and quite regularly. The persistence of such symptoms should be regarded as serious. Again, some symptoms are more readily observed than others. The inexperienced teacher, or the poorly trained teacher, or the teacher lacking in social sensitiveness may not recognize such symptoms when they appear; or, if she recognizes them, she may not be able to appreciate their meaning and import.[4] It was pointed out in the previous chapter that overt types of behavior are more discernible than withdrawing types. Maladjustments of the withdrawal type are likely to go unnoticed or be ignored by the busy teacher or the parent who fails to recognize this condition as a symptom of maladjustment. Some cases of maladjustments arise as a result of needs that the teacher and school situation can easily supply; while others will require the services of a psychiatric worker or a clinical psychologist.

Any form of behavior that is exaggerated beyond that found among individuals of the same age and cultural status should be given further study and consideration, since it is symptomatic, indicating that some of the individual's needs are not being satisfactorily met. The following behavior items, if exaggerated, were considered in the St. Paul experiment, referred to on page 317, as indicative of unfulfilled needs:[5]

Bashfulness	Crying	Drinking
Boastfulness	Daydreaming	Eating disturbances
Boisterousness	Deceit	Effeminate behavior (boys)
Bossiness	Defiance	Enuresis
Bullying	Dependence	Fabrication
Cheating	Destructiveness	Failure to perform assigned
Cruelty	Disobedience	tasks

[4] See C. E. Thompson, "The Attitudes of Various Groups Toward Behavior Problems of Children," *Journal of Abnormal and Social Psychology,* 1940, Vol. 35, pp. 120–125 for a worth-while study comparing the attitudes of teachers, parents, and mental hygiene workers toward various types of behavior usually classified as maladjusted or problem behavior.

[5] *Children in the Community: The St. Paul Experiment in Child Welfare.* Children's Bureau Publication 317, 1946, pp. 47–48.

Fighting	Quarreling	Temper displays
Finicalness	Roughness	Tics
Gambling	Selfishness	Timidity
Gate-crashing	Sex perversion	Thumbsucking
Hitching rides	Sex play	Truancy from home
Ill-mannered behavior	Sexual activity	Truancy from school
Impudence	Shifting activities	Uncleanliness
Inattentiveness	Show-off behavior	Uncouth personalities
Indolence	Silliness	Underactivity
Lack of orderliness	Sleep disturbances	Undesirable companions
Masturbation	Smoking	Undesirable recreation
Nailbiting	Speech disturbances	Unsportsmanship
Negativism	Stealing	Untidiness
Obscenity	Stubbornness	Violation of street-trades
Overactivity	Sullenness	regulations
Over-masculine behavior	Tardiness	Violation of traffic regu-
(girls)	Tattling	lations
Profanity	Teasing	

The problem child. In the study reported by Martens and Russ, 109 problem children brought before the behavior clinic of Berkeley, California were matched with 109 normal children exhibiting no problem behavior that warranted clinical attention.[6] These children were equated by age, sex, general level of intelligence, and grade in school. Furthermore, the equated problem and nonproblem children were chosen from the same school and teachers. Thus, sex, age, intelligence, and school environment were quite similar for the two groups.

The problem and nonproblem children were rated for a large number of traits that had been developed out of earlier studies. A comparison of the frequency of behavior problems reported for the problem and nonproblem children is presented in Table XLIX. The major differences between the two groups appear in the behavior activities frequently reported. Thus, it is the recurrent nature of the misdemeanor that should be regarded as serious. These data offer clear evidence that it is not the occasional occurrence of certain behavior activities that places the child, in the eyes of the teacher, in the problem group, but rather the persistent display of such activities.

The St. Paul experiment in child welfare was established by the Children's Bureau in order to acquire experiences in a city of medium size where the conditions would be typical of the average urban com-

[6] Elise H. Martens and Helen Russ, "Adjustment of Behavior Problems of School Children. A Description and Evaluation of the Clinical Program in Berkeley, California," *U. S. Office of Education Bulletin No. 18,* 1932.

TABLE XLIX

BEHAVIOR PROBLEM	Problem Children		Nonproblem Children	
	Frequently	Total	Frequently	Total
Inattention	77	106	14	78
Carelessness, slovenliness in work	62	96	10	48
Restlessness, talking, etc.	61	90	9	48
Bad posture, slumping in seat	51	90	14	61
Laziness	44	74	9	45
Forgetting notes or books	32	74	1	42
Doing work other than assigned	50	74	10	45
Daydreaming	35	69	20	57
Teasing	25	66	1	35
Dirty hands, face, etc.	32	65	3	34
Fighting	26	63	1	24
Exuberance (laughing, giggling)	36	62	6	35
Showing off	31	62	3	18
Lying	33	61	1	6
Tardiness	19	60	2	31
Temper outbreaks	20	59	0	25
Sulkiness	21	59	2	26
Eating candy, fruit, chewing gum	18	56	7	42
Dirty belongings, books	23	47	2	15
Excessive reticence (easily embarrassed)	21	46	20	57
Impertinence	17	44	2	14
Dirty clothes	20	41	3	13
Cheating in schoolwork	13	40	0	13
Deliberate refusal to obey	22	39	0	6
Cheating in play	14	38	0	7
Stealing	14	37	0	4
Injury to others (not smaller)	21	33	0	8
Bullying	19	33	0	4
Damage to school property	9	32	0	3
Resistance to punishment	13	32	1	5
Profanity	12	32	0	4
Weeping (cries easily)	10	31	2	21
Damage to personal property	9	26	0	6
Hurting small children	15	26	0	1
Vulgar speech	9	24	0	1
Writing notes	7	24	2	14
Truancy	8	23	1	3
Damage to neighborhood property	6	18	0	0
Masturbation (suspected)	2	15	0	2
Sexual pictures or stories	4	14	0	1
Hurting animals	4	8	0	1
Heterosexual activity	2	5	—	0
Masturbation (known)	2	4	0	0
Vermin	1	2	0	0

munity. Emphasis was placed on the study of social services to all children with personality and behavior problems, however mild, and special consideration was given to means of preventing the development of such problems. During the period of the project, 1,466 children were registered for services. Of these, 727 were from families

TABLE L

The General Nature of the Problem and the Number of Children Referred to the Community Service Center for Each Type of Problem

Academic difficulties	374
Conflict with authority	210
Undesirable personality traits	203
Stealing	178
Physical difficulties	177
Attendance irregularities	165
Conflict with other children	123
Habit problems	120
Failure to observe routines and regulations	114
Social withdrawal	109
Destruction of property	56
Running away	42
Dishonesties	34
Sex	26
Other	54

receiving services from the Community Service staff. The intelligence quotient of this group ranged from 42 to 149, with an average of 94. The particular difficulty that brought these children to the attention of these services varied considerably. The types of misbehavior and the number of children referred for each are presented in Table L.[7] The fact that some children were listed for more than one difficulty accounts for the fact that the total is more than the 727 children reported.

Personal maladjustments. It has been observed that personality maladjustments are closely related to difficulties and frustrations in life. Case studies of pupils having educational difficulties reveal that home conditions are important factors contributing to such difficulties. The anxieties built up in the home, at the club, or in some community activity becomes a source of frustration when such anxieties are beyond

[7] *Children in the Community: The St. Paul Experiment in Child Welfare.* Federal Security Agency, Social Security Administration, Children's Bureau Publication 317, 1946, p. 12.

the scope of likely or possible fulfillment. Often the child from the upper-lower class group has built up ambitions relative to education and to the future all of which are well-nigh impossible of attainment. Sometimes a boy or girl from the middle class group may have only average ability, but because of the expectations of the family and friends he may have developed an anxiety to reach a level equal to that attained by an older and more capable brother or sister. Any difficulty that interferes with the attainment of one's aspirations is a potential source of personal maladjustments. There is good evidence that failure in school may be a major catastrophe to many children and adolescents. In a study conducted by Preston[8] 100 normal children who were reading failures were compared with 76 controls. Each child was interviewed for an hour at school. Further interviews were conducted with the parents in the home. From these interviews it became apparent that insecurities at home and maladjustments at school are closely related to reading failure. Furthermore, a general insecurity was manifested toward the school program. Six cases were reported that showed a considerable improvement in a badly maladjusted state after being taught to read by methods suited to their individual learning capacities.

Too often the overly ambitious parent is the major cause of maladjustment. Cummings studied the incidence of emotional symptoms in elementary school children.[9] It was found that overprotected children revealed a preponderance of nervous difficulties, while neglected children were more aggressive in their behavior and were more given to lying, stealing, and cruelty. Some case studies indicate that maladjustments involve deep-seated emotional disturbances. These disturbances turn in different directions. Some of them tend toward *unsocialized aggressive tendencies,* observed in bullying, fighting, defiance of authority, and the like. Some have been termed *socialized delinquency behavior,* which is characterized by group activities contrary to established rules, such as group stealing, truancy from school, acts of mischief, and generally unwholesome gang activities. Others tend toward overinhibited tendencies revealed in shyness, seclusiveness, daydreaming, jealousy, and the like. The direction the malad-

[8] M. I. Preston, "Reading Failure and the Child's Security," *American Journal of Orthopsychiatry,* 1940, Vol. 10, pp. 239–253.

[9] Jean D. Cummings, "The Incidence of Emotional Symptoms in School Children," *British Journal of Educational Psychology,* 1944, Vol. 14, pp. 151–161.

justment takes will depend in a large measure upon social forces in one's environment—the general pattern being exemplified by those with whom the individual is in social contact.[10]

Unsocialized aggressive behavior. The unsocialized aggressive type of behavior is seldom found among girls, since they are under considerable pressure from their parents and peers to conform to well established patterns. Hewitt and Jenkins conclude from a study of these types:

> In each of the three behavior-situation pattern relationships there appears to be some evidence that not only is the behavior in question "provoked" by a peculiar type of frustration, but the general pattern of behavior itself is exemplified by other persons with whom the child is in close contact. Thus the resulting type of maladjustment would appear to be a "rational" reaction of the child to his distorted environment in a double fashion.
>
> Inasmuch as the behavior is rational in this sense, arises from a more or less identifiable type of circumstances, and represents a fundamental warping of the child's personality in a particular direction, the therapeutic implications must necessarily differ from those involved in other forms of maladjustment. That is, the three types of behavior distinguished in this analysis would seem to demand rather strikingly different methods of treatment. Both the pattern of *unsocialized aggressive* behavior and that of *overinhibited* behavior appear to involve deep seated emotional disturbances, but in different directions and for different reasons. Social delinquency, on the other hand, appears to involve the identification of loyalties and a positive response to the numerous deviation-pressure patterns displayed in the child's environment.[11]

Materials presented in Chapter XVI show that there is a close relation between emotional maladjustments and delinquency. In the studies reported by Bell,[12] high school boys gave significantly more maladjusted answers than delinquent boys to such social situations as the following: enjoying social gatherings, meeting people socially, introducing people, leading parties, engaging in conversation, asking for help, making plans for others, making contacts with girls, speaking in public, enjoying dancing, making friends, and being the center of attention at a party. Delinquent boys gave more maladjusted responses

[10] See L. E. Hewitt and R. L. Jenkins, *Fundamental Patterns of Maladjustment, The Dynamics of Their Origin,* State of Illinois, 1946. These three types are studied and described.

[11] *Ibid.,* p. 91.

[12] H. M. Bell, *The Theory and Practice of Personal Counseling.* Stanford University, Calif.: The Stanford University Press, 1939, p. 115.

than normal high school boys to these emotional situations: frightened when going to a doctor, discouraged easily, sorry for things done, feelings easily hurt, worry over misfortunes, disturbed by criticism, upset easily, and bothered by useless thoughts. Delinquent girls gave more maladjusted responses than normal girls on the question dealing with discouragement, the one on nervousness, and the one on being easily upset. Delinquent girls gave more maladjusted answers than delinquent boys with respect to the following home conditions: parents who expect too much of them, dominating mother, demand for complete obedience, fear of parents, quarrels at home, home conflicts, and divorce in the home.

Socialized delinquent behavior. While withdrawing, recessive personality traits may be serious from the viewpoint of mental health, it should be realized that the possession of certain aggressive types of conduct may also seriously handicap an adult in making adjustments. Ellis and Miller point this out when they state:

Present standards of society impose requirements for certain types of behavior and exact retribution from transgressors. Offenders who steal are in serious difficulty (if caught). The person who violates these standards of social conduct certainly is handicapped in his success in making adjustments to the social group. Such traits as impudence, impertinence, and temper outbursts are frowned on in adult society, and the person who habitually exhibits them is unpopular with his associates and finds difficulty in making happy adjustments in his contacts with society.[13]

With growth into adolescence occurs the first stage of the development of such habit systems as, when carried to an extreme, will bring the individual into direct conflict with the rules and regulations imposed by the social group. With the onset of such social conflicts we have a mental-hygiene case or a case of delinquency—a case of undesirable behavior, growing directly out of earlier failures in social adjustment. Earlier failures have many and varied causes depending upon the inherent qualities of the individual, the peculiarities in the situation, and the habit systems established earlier in life. Moreover, since their growth is gradual and continuous, habit patterns tend to become integrated into larger units, thus creating a specific type of disposition or attitude. It is therefore difficult to say at just what point in the life of the adolescent the wrong elements developed and became integrated into larger units.

[13] D. B. Ellis and L. W. Miller, "Teachers' Attitudes and Child Behavior Problems," *Journal of Educational Psychology,* 1936, Vol. 27, p. 508.

Failure in socialization. The study by Bonney[14] was designed to determine the type of individual who is generally well accepted socially as compared with the one who is socially unsuccessful. Two methods of gathering data were used in this study. These were: (1) trait ratings, on the part of both teachers and pupils, and (2) the pupil's choices of friends—a method referred to as a sociometric test. One fact emerging from this study was that the most popular children are more aggressive and overt in their responses. It was found that the highest social recognition does not go to children who are submissive and docile. It appears that to be well accepted as a child or adolescent, one needs to possess positive attributes that will make him count in the group. Popularity among children and adolescents is closely related to strong personalities, enthusiasm, friendliness, and marked abilities. Although there are changes with age, as suggested in an earlier chapter, there is good evidence that the individual at all age levels is popular because of desirable positive actions rather than because of inhibitions and restraints.

Deviated personalities begin to be observed to a large degree as the individual makes wider social contacts; the adolescent's physiological development, new contacts, heightened emotions, and enlarged mental life create a new self, and this new self seeks an expression that needs sympathetic guidance if it is to develop along desirable lines. Extreme introversion and daydreaming or antisocial tendencies are quite likely to arise when there is a failure in the socialization process.

The psychological pattern for seclusiveness, just as other forms of behavior, originates as a response to a motive or stimulus. In many instances the satisfying response or action is found in seclusiveness and timidity, forms of withdrawal behavior that will be discussed in the subsequent chapter. Shyness is an outstanding characteristic of a large number of pre-adolescents and adolescents.

Timidity is well illustrated in the case of a pre-adolescent girl described by Rivlin.[15] The method of handling the case through participation has been used by many teachers, and has been found to be very successful.

Stella, a ten-year-old pupil in the fourth grade, was referred to the school mental hygiene committee because she was "Very retiring, extremely quiet.

[14] M. E. Bonney, "Personality Traits of Socially Successful and Unsuccessful Children," *Journal of Educational Psychology,* 1943, Vol. 34, pp. 449–472.

[15] H. N. Rivlin, *Educating for Adjustment.* New York: Appleton-Century-Crofts, Inc., 1936, p. 375.

She does not volunteer answers in class and does not play with other children." The family background was good and her work and conduct marks were excellent. Her I.Q. was 111. The teacher appointed Stella leader of playground games and made a definite attempt to induce her to speak freely in class.

In the course of the lesson the teacher asked her a question, any answer to which was tenable. During a geography lesson she was asked, "Do you think you could be happy if you had to live in Africa?" The response called for is a simple one and need cause even the most timid child little embarrassment. Whatever answer is given can be commented on favorably by the teacher. "It would be difficult for us to learn to like Africa," or "We really can't tell how we would like it," would be acceptable. Stella responded by shrugging her shoulders. The children were then asked to tell why one didn't know how pleasant or unpleasant life in Africa could be. Probably for the first time in her school career the girl was treated to the sight of the entire class working on an answer she had given, a flattering situation. Stella was then asked which of these reasons she had in mind when she expressed her opinion. Since many possible answers had just been suggested by the other pupils and were still on the blackboard, it was not difficult for her to offer an appropriate answer, even though she may not have had any definite thought in mind when she first responded to the teacher's question.

This procedure was followed for several days until Stella grew accustomed to speaking in class and to having her answers taken seriously by teacher and pupils. About a week later, after an easy question had been asked, the teacher looked at the girl in an encouraging and expectant manner till the youngster sensed the teacher's belief that she had something worth offering to the others.

BEHAVIOR MECHANISMS

A useful classification of behavior mechanism has been developed by Thorpe and Katz.[16] This is presented in Table LI and is used as a basis for much of the discussion on adjustment mechanisms frequently used by adolescents. Thorpe and Katz grouped these according to the relative degree of social acceptance of the different mechanisms; although it should be pointed out that the actual toleration accorded to any of these mechanisms will depend upon a number of factors, some of which are largely independent of the individual concerned. It is important for the student of adolescent psychology to understand that these behavior mechanisms do have social implications, that some of these are acceptable in certain conditions, and that they are not acceptable under other conditions. However, there is

[16] Louis P. Thorpe and Barney Katz, *The Psychology of Abnormal Behavior,* p. 69. Copyright 1948 by The Ronald Press Company.

a tendency in our culture for the relative acceptability to follow the classification set forth in Table LI.

TABLE LI

CLASSIFICATION OF BEHAVIOR MECHANISMS

Socially approved adjustment mechanisms:
    Compensation
    Rationalization
    Substituted activities
Socially tolerated adjustment mechanisms:
    Identification
    Projection
    Egocentrism
Socially criticized adjustment mechanisms:
    Sympathism
    Regression
    Dissociation
Socially disapproved adjustmental mechanisms:
    Repression
    Negativism
    Fantasy (daydreaming)

Although twelve mechanisms are listed, these seldom occur in isolation. Furthermore, this is not necessarily the only possible classification, but rather a classification that furnishes a basis for studying the conflicts and behavior mechanisms of adolescents. Case studies of normal as well as pathological subjects show that certain mechanisms or combinations of mechanisms seem to be most frequently used by some individuals, while other mechanisms or combinations occur more frequently with others.

Socially approved mechanisms. Although it is not possible to set forth a specific list of mechanisms that is entirely approved and another list that is in the same way disapproved, certain behavior mechanisms are in general more acceptable than others. However, the ways these mechanisms are used and the degree to which they are exhibited will affect the attitudes of others toward them.

a. *The compensatory mechanism* is based on the inferiority-complex theory presented by Adler, in which it is pointed out that every individual sooner or later senses his inferiorities, real or imagined, and tends to compensate for them. Some people are able to compensate in desirable and others in socially undesirable ways. The degree of compensatory behavior depends upon the intensity of basic conflicts and upon the ability to find modes of expression that are socially acceptable.

The compensatory mechanism is seen most clearly in certain type of the neurotics. The question may be projected whether every person who makes an intense effort to gain an education, to develop a talent, or to accumulate excessive property is utilizing the compensatory mechanism. If this view is accepted as correct, every student who pursues an academic education despite economic and social difficulties must be compensating for some real or imagined inferiority.

b. *Substitute behavior* is in reality a form of compensation. In this form of behavior the individual attempts to achieve success (tension reduction) by substituting an activity that promises success and the attainment of the goal for the frustrating activity. The school can well afford to provide substitute activities for those whose aspirations for success in the traditional school subjects is beyond their capacities. During his childhood years many an individual has developed aspirations beyond his capacities. The teachers and counselors should attempt to help these youngsters find in some substitute activity a suitable outlet for their drives.

c. *Rationalization* is a characteristic form of human behavior, about as old as language itself. The young child blames the stone for sticking up out of the ground when he stubs his toe. The adolescent girl who is unwilling to exercise and diet in order to secure and maintain a desired figure gives as her reason that fat people are "good-natured," and that she wants to be jolly and happy. This is a process of self-justification. The individual explains to himself why certain things were done, and this explanation meets with his approval, whereas the true explanation probably would not. The use of the rationalization process is closely related to the superego, or what some would call a superiority complex. It frequently appears among adolescents, who are ever desirous of securing and maintaining peer approval of their motives and deeds.

Socially tolerated adjustment mechanisms. The mechanisms here discussed are usually tolerated although not wholly approved by others. Such activities may at times appear to be approved, since they are manifested in a manner acceptable to the group. However, the following activities are not generally regarded with approval by the group.

a. *Identification,* a more or less universal adjustment mechanism, is a strong force in the development of the child. The child early assumes the characteristics of his parents and takes on their mannerisms and ways of behaving. As the boy grows and develops he attempts to identify himself with the older and larger boys. This is the likely explanation of the ten- or twelve-year-old's desire to smoke a cigarette. The pre-adolescent attempts to emulate the adolescent and the grown-up. This identification may extend to characters seen in the movies or heard over the radio. The number of boys who have identified them-selves with Tarzan would run into the thousands. The number of young adolescent girls who have identified themselves with some movie star is too great even to attempt to estimate. The culture of adolescent groups is characterized by their efforts to identify themselves with older groups by engaging in the activities these older groups pursue, but at the same time to build up barriers for keeping out adults, es-pecially those who might be critical of their behavior.

b. *Projection* as a behavior mechanism is quite similar in nature to identificaton. This serves a useful purpose for the individual in that it rids him of knowledge of undesirable attributes or characteristics of things, and at the same time leaves him free to express himself about them. This has been referred to as "the use of scapegoats." Early man frequently ascribed evil thoughts and actions to wicked spirits. The adolescent boy or girl may project his or her trouble on the teacher, the parent, or a classmate. The type of projection revealed by an individual gives insight into his problems and nature. Certainly projection, when carried to an extreme, or when it tends to persist, is in itself a danger signal.

c. *Egocentrism* is an attempt to exaggerate and thus to enhance the importance of *the self*. It is usually a result of a lack of attention dur-ing the early years of life. The "pole-sitter," the "show-off" in class, and the "exhibitionist" are examples of the operation of egocentrism. Adolescents are extremely eager to have the approval of their peers and are anxious to get the attention of members of the opposite sex. When they are unable to secure the desired attention from their peers, they may resort to egocentrism, most of which is harmless in nature and is tolerated by their peers. Egocentrism is fostered through iso-lation and by autocratic controls. The unfortunate feature of it is that the results usually rebound—the individual finds disapproval rather than the approval he seeks.

Socially criticized adjustment mechanisms. Egocentrism and sympa-thism are sometimes very closely related. *Sympathism, regression,* and *dissociation* are subterfuge procedures for allowing an individual to carry out motives that are contrary to each other, but which operate successfully when one is held in abeyance through the operation of one of the behavior mechanisms. In sympathism the individual seeks comfort by placing himself in a position where he will be noticed. The case of a college student who persisted in fainting in the class where she had a "crush" on the instructor illustrates the operation of this mechanism. The girl would probably not resort to a more direct method of seeking the attention of the instructor. Perhaps exhibitionism is too direct in nature, but sympathism does not appear (to the self) to rely upon selfish or self-centered characteristics.

Likewise *regression* and *dissociation* serve as means of satisfying a motive in the face of contradictory motives. The regression aids in keeping in abeyance a motive that is contrary to the one being followed. These mechanisms, although frequently used by adolescents, are perhaps no more characteristic of this age than of any other age. The socially tolerated adjustment mechanisms are more frequently used by adolescents. Since this is an age in which peer status is sought and conformity is followed, the individual refrains from behavior that is likely to receive criticism or disapproval from his peers. It is likely that sympathism, regression, and dissociation are more frequently used at a later period of life when the individual's motives continue to be thwarted.

Socially disapproved adjustment mechanisms. These mechanisms include those oftentimes grouped under the title "escape–evasion of reality." The three forms that have been listed by Thorpe and Katz are *repression, negativism,* and *fantasy* (daydreaming). Of these, fantasy is perhaps the one most characteristic of adolescents. Many adolescents, unsuccessful in their attempts at social adjustments, find refuge in a "dream world"—a world of fantasy. Lewin has spoken of personalities as varying along a dimension of reality–irreality. The distinction between fact and fantasy is not clear to the small child; but with his development into adolescence the world of fantasy becomes similar in content to the world of reality. The difference lies primarily in the fact that the adolescent daydreams of dates, playing the piano, being a hero on the football field, and so on, rather than attempting to accomplish these feats in real life.

We are familiar with Mary. The teacher calls on her but she is staring into space, seeing nothing in particular but thinking of things she would like to have and things she would like to do. Mary doesn't understand what the teacher says, but she faintly recalls that her name was called and she becomes aware that the attention of the class is now focused on her. Mary might be said to be living in a world of fantasy. Perhaps, at times, all people may be said to live in a world of fantasy; however, the behavior just described for Mary is characteristic of her. When an individual persists in daydreaming, such as Mary seems to do, there is a definite indication that certain needs are not being fulfilled.

Another form of escape is that of complete repression and withdrawal. In the authoritarian schoolroom, the child may repress his feelings and desires and withdraw from the general classroom activities as much as possible. An entire class may display this form of behavior in response to a completely dominating, authoritarian teacher. Forgetting, sickness, and even sleep have been suggested as procedures used in the evasion process. An illustration of the operation of this is presented by Sherman:

> The boy had serious sex conflicts which increased his feeling of insecurity. These problems apparently were the cause of his flight into abnormality. His parents had never discussed questions of sex with him, had never given him any sex information, and disapproved of discussion of such topics. At the age of thirteen when his curiosity was intensely aroused, he had no one to ask for explanations or information. He was forced to solve his problems without help. Having a sense of guilt, he began to blame others for his deficiencies, to think boys did not like him and that the teachers discriminated against him. His paranoid ideas were not fully systematized because they did not sufficiently justify his supposed inferiorities. He then unconsciously attempted to solve his difficulties by withdrawing from normal social contacts into a world of phantasy in which he felt secure because he had no need to compare himself with others, to compete with others economically or socially, to wonder what others thought of him, his progress or his future. In his new phantasy world he felt completely at ease.[17]

SUMMARY AND CONCLUSIONS

There is ample evidence for the general conclusion that adequate and well adjusted personalities are largely the result of wholesome examples and training during the period of childhood and adolescence. Behavior patterns are established during these years that tend to persist, and to form the core of personality. The importance of the home, the school, the community, and one's peers in the formation of personality was emphasized in Chapters X, XI, XII, and XIII. Faulty habit formation during the growing years, undesirable reaction tendencies, and unfavorable attitudes toward the self and others are outcomes of earlier experiences.

[17] Adapted from Mandel Sherman, *Mental Hygiene and Education.* New York: Longmans, Green and Co., 1934, Chap. IV.

This chapter has emphasized the importance of successful social adjustment during the adolescent years. Various symptoms of maladjustment appear when there is a failure in the socialization process. The persistence of these symptoms is of greater importance than the mere appearance of the symptoms. Behavior mechanisms, resulting in many cases from trial-and-error experiences, are manifested by the individual and are definitely indicative of an effort to adjust to frustrations or conflicts. The mechanisms described in this chapter are common forms of behavior, and no doubt in many cases necessary forms of behavior. They are used by the stable, the unstable, and the neurotic. They do not suddenly appear in an individual's life; neither do they occur in some haphazard manner. Concerning these Warters has stated:

> They are learned habits acquired over a period of time through the process of social interaction. The individual learns them through imitation of others and through the guidance and instruction provided him directly or indirectly by others. He often acquires them, as he does many other habits, through the trial-and-error method. Once he discovers that a particular mechanism is a useful way to adjust to thwarting, he is likely to try it again when he again meets thwarting. Should at another time the mechanism not prove successful, he will first try altering it and then perhaps, if necessary, try changing it entirely.[18]

THOUGHT PROBLEMS

1. Compare the adjustment problems of boys and girls. How do you account for the differences?

2. Observe a child in school several times and note any symptoms indicative of maladjustments. Are these symptoms more or less universal among children? Elaborate upon the significance of any observations on this point.

3. What do you consider the most important idea suggested by the materials of Table XLIX comparing the frequency of behavior problems reported by problem and nonproblem children?

4. Consider one or more deliquents that you have observed during your life. Would you regard them as of the *socialized delinquent* type or the *unsocialized aggressive* type? What are the major differences?

5. List in order of seriousness what you believe to be the major causes of *failure in socialization*. Describe some individual in whom there has been a pronounced failure in socialization. Give any causes that you believe might have contributed to this failure.

---

[18] Jane Warters, *Achieving Maturity*. New York: McGraw-Hill Book Co., 1949, p. 209.

6. Give illustrations of the operation of the various mechanisms described on pages 323–327.

7. What uses can be made of the *Diagnostic Child Study Record,* presented in Appendix B?

SELECTED REFERENCES

Anderson, J. E., *The Psychology of Development and Personal Adjustment.* New York: Henry Holt and Co., 1949, Chaps. XVII and XVIII.

Baker, Harry J., *Introduction to Exceptional Children.* New York: The Macmillan Co., 1944, Chaps. XXII and XXIII.

Cole, Luella, *The Psychology of Adolescence* (Third Edition). New York: Rinehart and Co., 1948, Chap. VI.

Garrison, Karl C., *The Psychology of Exceptional Children* (Revised Edition). New York: The Ronald Press Co., 1950, Chap. XIX.

Landis, C., and Bolles, M. M., *Textbook of Abnormal Psychology.* New York: The Macmillan Co., 1946.

O'Kelly, L. I., *Introduction to Psychopathology.* New York: Prentice-Hall, Inc., 1949, Chap. IX.

Richards, T. W., *Modern Clinical Psychology.* New York: The Macmillan Co., 1946.

Sherman, M., *Basic Problems of Behavior.* New York: Longmans, Green and Co., 1941.

Stone, C. P., *Case Studies in Abnormal Psychology.* Stanford University, Calif.: Stanford University Press, 1943.

Thorpe, Louis P., *Child Psychology and Development.* New York: The Ronald Press Co., 1946, Chap. XV.

Travis, L. E., and Baruch, Dorothy W., *Personal Problems of Everyday Life.* New York: Appleton-Century-Crofts, Inc., 1941.

Warters, Jane, *Achieving Maturity.* New York: McGraw-Hill Book Co., 1949, Chaps. VII, VIII, IX, and X.

Some valuable suggestions for the development of a program to assist adolescents in making better emotional adjustments and in understanding behavior have been presented by R. H. Ojemann and Mildred I. Morgan, "Effects of a Program Designed to Assist Youth in an Understanding of Behavior," *Child Development,* 1942, Vol. 13, pp. 181–194.

XVI

Juvenile Delinquency

INTRODUCTION: DEFINITIONS AND EXPLANATIONS

Universality and normality of delinquency. Every year a quarter of a million boys and girls are arraigned before the juvenile courts of this country as delinquents. At least three times that number get into trouble serious enough to bring them to the attention of the police or school authorities. Many others are maladjusted but find an escape by dropping out of school, or by becoming a social menace. These individuals are basically *normal* individuals, but have become rebellious, irregular in attendance at school, careless of property rights, or inconsiderate of the rights and welfare of others. Their behavior is a result of their past experiences. The occurrence of such acts is an indication either that something was lacking in their early environment or that conditions of their early years were not conducive to the development of wholesome attitudes and desirable habit patterns.

The seriousness of introverted, withdrawing types of behavior has come to be appreciated only within the past fifty years. There is, therefore, a much better understanding of neurosis and infantile behavior today than formerly. However, there is still a strong tendency to regard delinquent acts in terms of the nature of the offense. Thus, the adolescent who is once arraigned before the courts is labeled a "delinquent" from then on. The delinquent act represents an effort toward adjustment as surely as does behavior regarded as normal.

Children in the courts. The juvenile courts provide a basis for determining the nature of delinquent acts committed by juveniles. The data presented in Figure 33 represent the juvenile delinquency cases disposed of from 1938 to 1947 by 76 courts serving areas with populations of 100,000 or more.[1] A complete inventory of police, juvenile

[1] *Preliminary Statement, Juvenile Court Statistics,* 1947, Children's Bureau, U. S. Department of Labor, Division of Statistical Research, 1945.

330

court, and other data on juvenile delinquency is not available for the
states or for local communities.   The available national statistics reveal,
however, some significant changes in the characteristics of the children
and adolescents coming to the attention of agencies concerned with
the problem of juvenile delinquency.   Some striking changes in the
ratio of boys to girls are indicated, although in general the curves have
followed a parallel course.   The proportion of girl cases to the total
number of cases has usually been around 1:6.   This proportion is consistently higher than among cases that come to the attention of the
police.   Thus, a girl's chance of being brought before the courts, once
she is called to the attention of the police, is greater than that for the
boy.   The median age of juvenile court cases under 18 years of age in
1945 was 15.3 years for boys and 15.5 years for girls.

Much publicity and various interpretations have been given to the
increase of juvenile delinquency with the entrance of the United
States into World War II.   This increase resulted from a multiplicity
of factors related to the war, including the fact that in many communities there was an increased effort to apprehend the juvenile delinquent and to refer him to law-enforcing agents and juvenile courts.
Statistics of juvenile delinquency cases disposed of by the courts do not
reveal the total amount of delinquent behavior.   Many cases of adolescents involved in delinquent behavior are not brought before the courts,
either because the delinquents are not apprehended, or because these
cases are handled by the police, schools, or community agencies especially concerned with this problem.   This fact must be considered in
the interpretation of the data presented on this problem.

Basic explanations of juvenile delinquency.   The development of
explanations of delinquent behavior seems to have gone through three
fairly distinct phases.   These have been referred to by Tappan[2] as:
"a prescientific mystical period, an early modern particularistic era,
and a contemporary quasi-scientific empirical period."   In each of
these, however, there has been considerable variation in the nature of
the explanation brought forth by its exponent, although there is a
special feature characteristic of each of these explanations that makes
such a classification possible.   Since our treatment of juvenile delinquency in this chapter will not be especially concerned with the first
two explanations, only a brief description of them will be presented.

The mystical concept ascribed delinquent behavior to some force

[2] Paul W. Tappan, *Juvenile Delinquency*.   New York: McGraw-Hill Book Co.,
1949, p. 74.

outside of and beyond the individual, such as evil spirits. Such spirits might, however, inhabit the body and thus affect one's actions. There were many varieties of this concept; but they were similar in the one respect just listed. The mystical concept held sway for many centuries. Its influence was extremely great until about a century ago. Beginning, however, around the middle of the nineteenth century, a

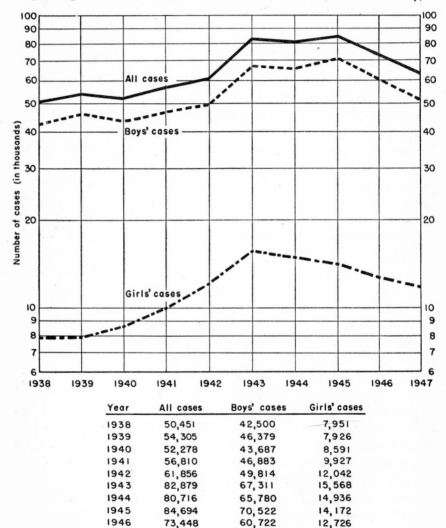

Year	All cases	Boys' cases	Girls' cases
1938	50,451	42,500	7,951
1939	54,305	46,379	7,926
1940	52,278	43,687	8,591
1941	56,810	46,883	9,927
1942	61,856	49,814	12,042
1943	82,879	67,311	15,568
1944	80,716	65,780	14,936
1945	84,694	70,522	14,172
1946	73,448	60,722	12,726
1947	62,911	51,067	11,844

Courtesy, Children's Bureau, U. S. Department of Labor.

FIG. 33. *Juvenile-delinquency Cases Disposed of from 1938 to 1947 by 76 Courts Serving Areas with Populations of 100,000 or more.* (Ratio Chart)

series of deterministic explanations appeared. These were also referred to as particularists, since they attempted to explain a specific behavior act on the basis of some particular factor. The exponents of the various deterministic viewpoints were aided in their explanations by objective testing techniques and statistical procedures. The particularists (deterministic concept) followed the pattern set forth by the mystics in oversimplifying the basis for delinquency. Adherents to the particularistic concepts appeared with many different causal interpretations, such as heredity, climate, endocrines, frustration, religion, ignorance, broken homes, slums, health, and the like. In fact some of the adherents would rely on less inclusive particulars, such as movies, comics, modern jazz, and other factors or conditions of a like nature.

With the greater use of scientific techniques combined with critical studies of the various theories that had been propounded by the particularists, it was soon recognized that these explanations were oversimplified. Thus, the theory of multiple causation has gradually come to be accepted by a large number of students of juvenile delinquency. A great deal of statistical work has been conducted in an effort to determine the relative importance of various factors in producing delinquency. The multicausal notion has revealed that crime results from many complicated factors, and that its prevention is not as simple as some of the earlier particularists had indicated. This theory, however, should not be viewed in a mechanistic manner. These various factors do not operate separately. The hyperthyroid adolescent lives in a home and neighborhood. His overactivity will be affected by the nature of his home, his size, his financial circumstances, his intelligence, and the neighborhood conditions. The girl is reared not simply in a poverty-stricken home, but in a neighborhood of a certain type. How she reacts to her home conditions will be affected by her emotional characteristics, her intelligence, her body build, the quality of her neighborhood, the attitudes of her parents, and many other factors too numerous to list. In the discussions of the influence of various factors on the development of delinquent behavior, one should realize at all times that these do not act separately upon the growing individual. There is a definite interrelationship existing at all times between all the forces or conditions that affect the behavior of the individual.

Motives underlying delinquent behavior. The motives underlying stealing or other delinquent acts are varied and are not always easy to determine. One cannot, therefore, resort to the more or less primitive tendency of reacting to stealing as a symptom of some specific character trait. Too frequently, stealing, like other forms of anti-social behavior, is looked upon by the observer as misconduct engaged in by someone who deserves punishment as a general retribution for the act. There are many factors or drives that may lead to the act of stealing. Howard W. Hightower has suggested three types of motivations underlying stealing. These are: "first, to have what others have; second, to gain attention; and, third, an emotional outlet for conflicts."[3] The case of Bob is here presented to show the conditions and motives found in one boy's life that contributed to delinquency.

Bob has been transferred from the eighth grade to the Frazier Home from which he has now twice run away and has been twice apprehended.

This action was taken as a result of a complaint brought against him by the Southern Pacific Company charging him with shooting rocks with a slingshot into train windows and injuring a passenger. Until this last offense, Bob has been reporting to the juvenile court concerning the list of delinquencies for which he has been held accountable.

Bob is not only physically healthy but also one who observes health rules. He is neat and clean, as well as attractive in appearance. Along with having a "wonderful set of teeth," as his dental record shows, he brushes them regularly. His muscular coordination, too, is excellent.

With an I.Q. of 97 Bob still seems unable to concentrate on school work, however, and his latest achievement test shows grade placement of sixth grade. Bob has attended several elementary schools in Portland, Oregon, in all of which he has had a truancy record until he reached the eighth grade. His attendance in this grade has been perfect. (His father promised him money and a watch if he would have perfect attendance.) Bob shows a lack of fundamental skills in reading, arithmetic, language, and social studies. Poor reading habits, little knowledge of the best methods of study, a slow rate of working, inability to organize material, and a certain satisfaction with doing superficial work—all these factors have furnished Bob with no persistence along academic lines.

Bob enjoys art, and even though it isn't of the creative type, he has produced several creditable pieces. He also seems to like physical education, in which he has shown progress in the development of skills. When requested to demonstrate these knowledges and skills to younger boys, he does so willingly and with enthusiasm. Bob is helpful and quite faithful, also, when he volunteers to work in the classroom. He serves well in keeping the library books in order and is most orderly about his desk and supplies. He often offers constructive suggestions to others in the class concerning improvement of methods of work. If Bob is serving on a committee on which someone is absent, he eagerly assumes the responsibility of inform-

[3] Howard W. Hightower, "School Problems of Pupils Who Steal," *Educational Administration and Supervision,* 1947, Vol. 33, p. 230.

ing the absent member, upon his appearance at the next committee meeting, concerning the progress of the committee. Bob has expressed a strong desire to go to a high school where agricultural courses are offered so he can learn about farming.

Bob's mother says that even though he tells her what happens, he has never been an affectionate child and that his four younger brothers and sisters are "easier to handle." Several years ago Bob's mother divorced his father, who was an habitual drinker and who gave her insufficient support for the children. The stepfather is disliked by Bob and is blamed by him for his tensions and unhappiness. According to Bob, his stepfather is so cruel to him that he does not want to live at home and will do anything "to get away from him." Apparently Bob has very little social life with his family because "the man who says he's my dad doesn't do anything like hunt or fish with us."

When listening to Bob tell of events which take place it would seem that he has no apparent knowledge of right and wrong. He doesn't hesitate to tell openly about stolen radiator caps and bicycle wheels, yet he always accompanies the story with, "Everybody steals." At this point, also, he insists that his stepfather is "mean, unreasonable, and crazy." He says that his brother steals too, but "I always get the blame just because I'm Bob." This feeling seems to be prevalent when Bob is accused by anyone in authority; he often insists, "The cops always grab me because I'm Bob, and let the rest of the gang go."

On the playground at school Bob generally aligns himself with the younger and smaller boys rather than with boys of his own age level. He is soon found fighting, and, in general, taking advantage of the situation. At this point he has a lack of approved goals or ideals, for regardless of all reasoning or bargaining, he continues to interfere with the play of the younger children. In spite of this he is not unpopular with his classmates and is frequently chosen by them to take part in both classroom and play activities.

In Bob's unwise use of leisure, in his poor choice of social activities, he has developed an irresponsibility in regard to most of his obligations to his family, to his school, and to his community. Bob's mother says that Bob, like his father, whom he admires, is proud and boastful each time he gets into serious trouble. She feels certain that Bob is of such nature that he will be tougher and merely brag to everyone about having been to Frazier Home.[4]

SEX AND JUVENILE CRIME

Despite the fact that there is almost no type of antisocial behavior committed by one sex that is not committed by the other, rather pronounced differences in the modal trends of the delinquencies of the two sexes exist. Here again it appears that such differences as exist are

---

[4] Quoted from Karl C. Garrison, *The Psychology of Exceptional Children,* revised edition. Copyright 1950 by The Ronald Press Company.

not inherent but only reflect the interaction of the various elements peculiar to the personalities of each sex.

An escape. That there are no sex differences with respect to mean age and mean IQ is what might have been expected on logical grounds. For the delinquent boy or girl the offense is most likely (at the time committed) to be an escape from an unsatisfactory system of behavior. It is apparent that girls have just about as great a need for escape as boys. The chief sex differences existing probably represent the types of escapes that are found most feasible. Some of the sex differences in offenses will be studied both quantitatively and qualitatively. Data bearing on this problem are not always available; this is especially true for girls, since some states make a tabulation of the offenses causing the commitment of boys but show no consistency in the tabulation and classification of crime data for girls.

Sex differences in offenses committed. According to reports from the Children's Bureau of Washington and data from other sources, five or six times as many boys as girls are arraigned before the juvenile courts, the ages for most of these boys being fourteen and fifteen. The two most common types of offenses by boys, as disposed of by the courts in 1945, are stealing or attempting to steal and acts of carelessness or mischief. With respect to the reasons for referring boys to the juvenile courts in 1938, the percentage has dropped slowly but consistently from 71 per cent to 61 per cent for these two activities combined. During this time there has been a gradual and continuous increase in traffic violations and in running away from home. The "hot rod" drivers have produced some real traffic hazards and traffic problems in some of our large cities. If adolescents are to be permitted to drive, it is necessary that they develop a greater sense of responsibility.

The data of Table LII show rather clearly that sex offenses bring many more girls than boys to the courts. It was suggested in Chapter X that changes have come about in the sex lives of growing boys and girls, affecting various types of behavior, including sex behavior itself. Thus, it is becoming increasingly difficult to determine just which sex acts may be regarded as delinquent forms of behavior. Some analytic investigations of crimes committed by girls have indicated, in fact, that sex is much more prominent in their commitments than records show. Many families would say "ungovernable" when the real delinquency is probably sex offenses. It appears, further, that in many cases of ungovernability or running away, the sex offense is probably prominent. Although a fairly large number of girls are affected, and immorality is admitted by a rather high percentage according to some studies, very few of the delinquent girls have fallen to the level of prostitution.

TABLE LII

JUVENILE-DELINQUENCY CASES, 1945: REASONS FOR REFERENCE TO COURT IN BOYS' AND GIRLS' CASES DISPOSED OF BY 374 COURTS [5]

REASON FOR REFERENCE TO COURT	JUVENILE-DELINQUENCY CASES					
	Number			Percent		
	Total	Boys	Girls	Total	Boys	Girls
Total cases	122,851	101,240	21,611	....	....	....
Reason for reference reported	111,939	92,671	19,268	100	100	100
Stealing	40,879	38,610	2,269	37	42	12
Act of carelessness or mischief	19,241	17,779	1,462	17	19	8
Traffic violation	9,852	9,659	193	9	10	1
Truancy	8,681	6,164	2,517	8	7	13
Running away	9,307	5,652	3,655	8	6	19
Being ungovernable	9,840	5,542	4,298	9	6	22
Sex offense	5,990	2,579	3,411	5	3	18
Injury to person	3,224	2,828	396	3	3	2
Other reason	4,925	3,858	1,067	4	4	5
Reason for reference not reported	10,912	8,569	2,343	....	....	....

[5] From: "Social Statistics," Supplement to Vol. 2 of *The Child,* November, 1946, Children's Bureau, U. S. Department of Labor.

This is quite an interesting commentary on the whole situation. Again, as was the case in connection with intelligence levels, the records of those actually sentenced will not give a true picture of the sex life of the entire number of girls appearing before the courts. The salvaging process is again at work, leaving for a final sentence the worst of the group appearing. The data on the 138 white women prisoners from North Carolina show that 51 per cent were committed for sex offenses, 18.1 per cent for delinquency, 5 per cent for larceny, and 3.6

per cent for robbery.[6] It is quite true, at any rate, that a very great number of the offenses of a non-sexual nature grow out of some sex situation. For various reasons boys of the adolescent age are seldom placed in institutions because of sex experiences, and especially is this true for heterosexual experiences. They, on the other hand, have been held more responsible for their own support, have probably been less protected in the home than the girls, and are faced with certain needs that they attempt to satisfy; hence they develop habits of stealing more than do the girls.

INTELLIGENCE AND CRIME

No criminal type. It is quite generally believed that most delinquents are feeble-minded or that delinquency and feeble-mindedness parallel each other. This belief is exceedingly unfortunate, because objectively obtained and carefully interpreted data do not substantiate it. It arose before modern intelligence tests had been developed or put into such actual, widespread use as would enable those using them to know the true meaning or import of data obtained. Lombroso's now thoroughly disproved idea that there is a definite criminal type did much to make people feel that delinquents and criminals were qualitatively different from those not so branded by the law. His discussion of the stigmata of the criminal type and his description of it as being possessed of "the characters of primitive men and of inferior animals" [7] went far toward making that part of the general public which is attentive really feel that the criminal and delinquent surely must be set apart as a separate type.

While Goring very conclusively demonstrated the falsity of Lombroso's concept of special physical stigmata, he himself is probably in part responsible for the previously mentioned current concept. In fact, though he denies it, he really took over Lombroso's qualitative position, simply substituting the term "defective intelligence" for Lombroso's "defective physique." Lombroso believed that the characteristics that he described were of an atavistic type, and thus inherited; and Goring, as previously mentioned, states that heredity and intelligence are the two main factors that differentiate the criminal from the non-criminal type. Since Goring's method of classifying prisoners by intelli-

[6] Lena B. Ladu and K. C. Garrison, "A Study of Emotional Instability and Intelligence of Women in the Penal Institutions of North Carolina," *Social Forces,* 1931, Vol. 10, pp. 209–216.

[7] Charles Goring, *The English Convict.* London: His Majesty's Stationery Office, p. 13. (Quoted from an address delivered by Lombroso in 1906 before the Congress of Criminal Anthropology at Turin.)

gence was wholly subjective, one cannot rely very much on its results.

Goddard's early work. In America, Goddard, more than anyone else, is responsible for the quite prevalent idea in some circles that the delinquent is defective. Contrary to his thought, the fact that any one element of personality is associated with crime is not proof in itself that such an element is the sole factor responsible for crime. Granted that mental deficiency is related to inferior social and environmental status, that a preponderance of crime exists in congested sections of inferior social and environmental status in our cities, and that therefore an abundance of crime is committed by those of defective mental ability— granted this, it does not follow from the mere association of the factors that mental defectiveness is itself a cause of the crime.

Goddard concludes from some rather early studies:

Every investigation of the mentality of criminals, misdemeanants, delinquents, and other antisocial groups has proved beyond the possibility of contradiction that nearly all persons in these classes, and in some cases all, are of low mentality. . . . The greatest single cause of delinquency and crime is low-grade mentality.[8]

Error in sampling. A number of investigators early pointed out that factors of intelligence and socio-economic status operate to select delinquents that are brought before the Juvenile Court. It is quite doubtful if we at any time have a truly representative group that can be labeled "juvenile delinquents." Since this is the case, we should exercise caution when we assume that the mean IQ of delinquents is between 80 and 90. A number of different investigators have presented data showing the average IQ of juvenile delinquents. The discrepancies found between the results from these studies may be explained on the basis of differences in sampling. To be sure, there is evidence from these studies that a larger percentage of boys and girls of low grade intelligence appear before the juvenile courts than would be expected on the basis of chance. However, Mann and Mann point out:

A closer approximation to a general rule is that delinquents having an IQ below 90—because of low intelligence, because of the area from which they come, or for some other reason or reasons—are more likely to be caught in their delinquencies than those whose IQ is higher.[9]

[8] H. H. Goddard, *Human Efficiency and Levels of Intelligence.* Princeton, N. J.: Princeton University Press, 1920, pp. 72–73.

[9] Cecil W. Mann and Helen Powner Mann, "Age and Intelligence of a Group of Juvenile Delinquents," *Journal of Abnormal and Social Psychology,* 1939, Vol. 34, pp. 351–360. A review of the average IQ's found by various investigators is presented in this study. Mann and Mann found the average IQ of 1,061 delinquent boys to be 84.88 and that of 670 delinquent girls to be 83.77.

If one bases his conclusion on children already committed to institutions, it is probably true that intelligence superiority among delinquents is rather rare. (Of course, it must be remembered that this group is not the entire body of delinquents in any state; the entire delinquent group, if *all* delinquents are considered, comes very close to being the entire population.) Among institutional cases the per cent of intelligence quotients in excess of 100 is small as compared with the per cent less than 100; however, for every intelligence quotient below 70 in the penal institution there can be found dozens or more, of comparably low-ability persons not in such an institution—and, from the standpoint of behavior activities, no more deserving of being there than the general average of the population. It is probably true, and in most cases proper, that many juvenile-court judges try to salvage from the human wreckage that is brought to their courts as many as possible who appear promising or capable of recognizing the nature and consequences of anti-social behavior—those who can profit from mistakes and thus give promise of making more adequate adjustments under some sort of supervision outside institutions. But these individuals are in most cases not retarded mentally and are therefore not counted among the institutional cases. Hence, counting methods decrease the average mental ability found in our institutions.

A fairly recent study reported by Maud A. Merrill[10] shows the average IQ of 500 cases referred to X County Juvenile Court to be 92.5; while the average IQ of 2,904 school children upon which the Stanford-Binet (Revised Form) was standardized was 101.8. A distribution of the IQ's for the two groups is presented in Figure 34, which shows that the IQ's of delinquents referred to the courts represent a distribution range rather than a type or types.

In evaluating the findings reported by various investigators of the intelligence of juvenile delinquents, an answer should first be sought to the questions: (1) What evaluation device was used for determining the level of intelligence of the delinquents? (2) Whom did the investigator test? Juvenile courts cases? Institutionalized cases? Cases referred to a psychological or guidance clinic?

The nature of the test used may affect considerably the intelligence quotients found. Since the delinquent is quite often retarded in his school work, he will be seriously handicapped on tests involving language activities and involving reading materials particularly. In spite

[10] Maud A. Merrill, *Problems of Child Delinquency.* Boston: Houghton Mifflin Co., 1947, p. 167.

of the fact that the various investigators have used different techniques for evaluating the intelligence of delinquents and have in many cases tested subjects in which different criteria have been used for classifying them as delinquents, their data agree in certain respects as follows:

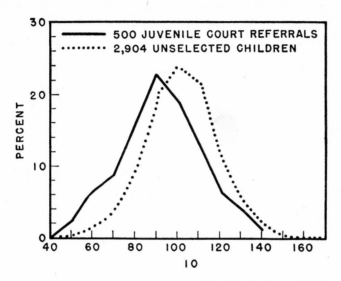

FIG. 34. *A Comparison of the IQ Distribution of 500 Consecutive Juvenile Court Referrals with that of 2,094 Unselected Children.* (After Merrill)

(1) There are more mental defectives among the delinquents tested than among unselected groups of children.

(2) The average intelligence level of children brought before the courts or institutionalized is less than the average for unselected school children of the same age level.

(3) The average educational retardation among the children regarded as delinquent is greater than that for public school children in general.

(4) There are delinquents with high levels of intelligence as well as delinquents with low intelligence levels.

(5) In line with item (4), we note that the distribution of IQ's among juvenile court cases tends to follow the normal probability curve (see Figure 34).

MALADJUSTMENTS AND DELINQUENCY [11]

It was emphasized earlier in this chapter that delinquent acts result from a multiplicity of factors operating in a unitary manner. Delinquent behavior may serve to express hostile retaliatory feelings against an institution, a society, or a person. The boy who breaks the window lights of the school building may be showing his resentment of the treatment received from the teacher or his classmates. Juvenile delinquent acts may be employed to resolve certain inner conflicts—aggressive acts release tension. Again, the boy who breaks out the window lights of the school building may find this an avenue for releasing tension resulting from a conflict between failure in his school activities and a desire for approval from his teachers, parents, or classmates. Juvenile delinquency may result from frustrations. The boy who is unable to attend the school dance may seek release from the tension thus developed by harming certain individuals attending or damaging property somewhat related to the dance. The girl who persists in indulging in sexual activities may be striving to satisfy a need for status or affection that is being denied her in the normal life pursuits.

Maladjustments of delinquents. In the study of Healy and Bronner, 105 delinquent children were matched with 105 nondelinquent siblings of as nearly the same age as possible. The two groups were then compared for degree of personal and home adjustments, for emotional difficulties, and the like. This comparison showed a preponderance of maladjustments among the delinquents. Over 90 per cent of them gave evidence of being or having been seriously maladjusted. The maladjustments discovered were classified as follows:

1. Feeling keenly either *rejected, deprived, insecure, not understood* in affectional relationships, unloved, or that love has been withdrawn.

2. Deep feeling of being *thwarted* other than affectionately; either (*a*) in normal impulses or desires for self-expression or other self-satifactions, (*b*) in unusual desires because earlier spoiled, or (*c*) in adolescent urges and desires—even when (as in five cases) desire for emancipation had been blocked only by the individual's counteractive pleasure in remaining childishly attached.

3. Feeling strongly either real or fancied *inadequacies or inferiorities* in home life, in school, or in relation to companionship or to sports.

[11] Karl F. Schuessler and Donald R. Cressey, "Personality Characteristics of Criminals," *The American Journal of Sociology,* 1950, Vol. 45, pp. 476–484. A critical analysis of the studies conducted on this problem is here presented. The results from 113 studies in which personality tests were employed are summarized. Of these 42 per cent differentiated between delinquents and nondelinquents.

4. Intense feelings of *discomfort about family disharmonies,* parental misconduct, the conditions of family life, or parental errors in management and discipline.

5. Bitter feelings of *jealousy* toward one or more siblings, or feelings of being markedly discriminated against because another in the family circle was favored.

6. Feelings of confused unhappiness due to some deep-seated, often repressed, *internal mental conflict*—expressed in various kinds of delinquent acts which often are seemingly unreasonable.

7. Conscious or unconscious *sense of guilt* about earlier delinquencies or about behavior which technically was not delinquency; the guilt sense directly or indirectly activating delinquency through the individual's feelings of the need of punishment (in nearly every instance this overlaps with the last category).[12]

Emotional immaturity and delinquency. Reference was made to the socialized delinquent in Chapter XV. The child who fails to develop this emotional control, who remains infantile in many of his reactions, and who is egocentric in his thoughts and concepts is referred to as emotionally immature. His reactions are likely to be determined by the immediate past and the very near future. Thus his conduct is geared to a large extent to means and activities for satisfying his immediate needs and desires, and toward the avoidance of unpleasant conditions in the immediate surroundings. Emotional maturity was described in Chapter V as closely related to emotional control. It is furthermore closely related to social maturity and may be evaluated in part by such an instrument as the Vineland Social Maturity Scale presented in Appendix D. The adolescent who is emotionally mature is motivated by social and altruistic values, not wholly by self-consideration. A fifteen-year-old boy, referred to by Hirschberg as B. S., illustrates the operation of emotional immaturity in producing delinquent behavior.

B. S. A small, thin 15-year-old boy of dull-normal intelligence. The fourth in a family of five children, two of whom have records of delinquency. Urban environment, high delinquency area. Associates with undesirable companions and spends most of his time on the streets. Goes to movies four times a week. Mother states he is interested in movies and athletics. School conduct "fair" and does passing work in spite of truancy. Subject disabilities in English and arithmetic. Goes to church regularly but has stolen while in church. In court for burglary (twice) and some petty stealing. Has run away from home, is unmanageable and disobedient. Stubborn attitude in court.

[12] William Healy and Augusta F. Bronner, *New Light on Delinquency and Its Treatment.* New Haven: Yale University Press, 1936.

When the psychiatrist had examined this boy he wrote: "From the intellectual standpoint one knows that B. is capable of better reasoning and a better sense of values than this incident demonstrates.   However, he does not formulate his attitudes and thinking on an intellectual basis.   So he explains his deviations by saying 'I do these things because I am just stupid,' yet he does not want us to think of him in terms of being stupid. Rather he is inclined to be quite insistent that his points and requests should be given serious consideration.   It is all part of the emotional immaturity of the boy, one who carries hostility within him, and when his emotional impulses and requests are not met, then his hostility shows forth. He is nearly sixteen years old, and he acts the part of a 13-year-old insofar as judgment, sense of values, etc. are concerned." [13]

In the study by Fertman, 180 delinquent girls from the Girls' Industrial School near Delaware, Ohio, between the ages of 14 years, 0 months, and 17 years, 11 months, were contrasted with an equivalent group of nondelinquents on the basis of responses to the Pressey Interest-Attitude Tests.[14]   "Age for age the delinquent girls demonstrated an emotional retardation of not less than 2.5 years, as measured by separate tests or by total scores of the Interest-Attitude Tests."   The greatest average retardation was made on Test I (things considered wrong).   Durea found an average emotional retardation of not less than 2.5 years for each life-age group of delinquent boys from 14 to 17 years old.[15]   However, a later study by Odoroff and Harris revealed that when delinquents and nondelinquents were matched for intelligence and socio-economic backgrounds, the nondelinquents were more retarded than the delinquents in all mental ages below fifteen.   The delinquents studied were 412 boys from the Minnesota Training School. The average IQ found from administering the Kuhlmann Tests of Mental Development of the boys was 83 (using sixteen as the maximum denominator).   The relationship of mental inferiority and emotional retardation was carefully studied.   They concluded: "The present study clearly indicates that delinquent boys are more like nondelinquents who come from similar social backgrounds than delinquents are like unselected boys in general." [16]

[13] Rudolph Hirschberg, "The Socialized Delinquent," *The Nervous Child,* 1947, Vol. 6, p. 464.

[14] M. H. Fertman, "Differentiating Personality Characteristics of Delinquent Girls," Master's Thesis, Ohio State University, 1939, p. 26.

[15] M. A. Durea, "The Emotional Maturity of Juvenile Delinquents," *Journal of Abnormal and Social Psychology,* 1937, Vol. 31, p. 4.

[16] M. E. Odoroff and D. A. Harris, "A Study of The Interest-Attitude Scores of Delinquent and Non-delinquent Boys," *Journal of Educational Psychology,* 1942, Vol. 33, pp. 13–23.

THE HOME AND JUVENILE DELINQUENCY

Interaction of the individual and social factors. Growth and development at any period of life must be evaluated in terms of the nature of the individual organism and the various forces that have operated in the organization and direction of growth. The study by Healy and Bronner shows how individual and social factors operate together in producing juvenile delinquents. Intensive studies were made of delinquents in Boston, New Haven, and Detroit. Those who had nondelinquent siblings were selected for special study, since a control group with social backgrounds similar to that of the offenders was available for study and comparison. The importance of the home and community background is evidenced by the fact that in only 22 cases of the 153 delinquents studied did they find the delinquent living in a favorable situation with reference to the following:

1. Reasonably good home conditions from the viewpoint of stability, normal recreational conditions, and normal physical needs and comforts.
2. Reasonably good attitudes of the parents from the viewpoint of freedom from family friction, normal attitude toward child care and treatment, and law abiding.
3. Normal neighborhood from the viewpoint of freedom from direct influences leading toward juvenile crime.

Parental attitudes. Entirely too many parents assume the attitude that a child is theirs, just as is any other personal property, and that they have the right to "cash in" on him at any time or in any way they please. As Van Waters says, "Some parents appear to think they have vested property right in their children and seek to coerce them when their rights are not yielding dividends." [17] This attitude is permissible insofar as discipline is concerned during the early years of training; but with growth into adolescence a rebellious child will result. A thwarting of the quest for independence (discussed in Chapter XI) is closely related to the early development of many juvenile crimes. It has been observed that at 10 or 11 years of age around 70 per cent of girls and 60 per cent of boys find greatest pleasure in the home and prefer to spend most of their leisure hours there. With the onset of puberty, the wider range of interests, and enlarged social contacts, adolescents begin to find more joys outside the home. It was suggested in Chapter XI that parents should not deplore this fact but, instead of thwarting

[17] Miriam Van Waters, *Youth in Conflict.* New York: Republic Publishing Co., 1925, p. 81.

adolescent desires, should aid the growing boy and girl in his or her emancipation from the dependency of childhood. The conflicts of adolescents with their parents result in most cases from a lack of adjustment in the emancipation process that should naturally take place as a part of the transition from childhood to adulthood.

A factor closely related to parental attitudes is the character of the parents. It has been pointed out that the child is imitative; especially does he imitate those whom he considers authorities. He comes to feel that their acts are an endorsement of such types of behavior. Imitation and suggestion in connection with drinking, immorality, or lawlessness aid in the establishment of delinquent tendencies in adolescent boys and girls. In a study by Lumpkin [18] the delinquent girls' parental background was found to be very unfavorable. Social defective tendencies such as crime, alcoholism, and sexual irregularity appeared 443 times in 189 families. In Schulman's study, "43 per cent of families of truants, 50 per cent of families of juvenile delinquents, 66 per cent of families of misdemeanants, and 83 per cent of families of felons had criminal records." [19] Although the objective information on this point is not wholly conclusive, it is the belief of this writer that criminality among parents is a very powerful conditioning factor for juvenile delinquency when it does appear. Goring [20] really classed such points as we are here considering under the head of heredity, a rather unfortunate classification because it is certainly true that the mere presence of certain characteristics in both parent and offspring does not prove they are inherited.

Family breakdown and delinquency. For many years the broken home has been pointed to as one of the main causes of juvenile delinquency. To substantiate this claim many studies of home backgrounds have appeared. These studies almost without exception show broken homes in the background of a large percentage of delinquent children. We should be extremely careful, however, in the interpretation of these studies. The broken home is, in most cases, the climax of a long series of events and simply indicates underlying adjustments that affect all members of the family. There is evidence that it is not the broken home so much as the factors often associated with this condition—

[18] K. D. Lumpkin, "Factors in the Commitment of Correctional School Girls in Wisconsin," *American Journal of Sociology,* 1931, Vol. 37, pp. 222–230.

[19] "Crime Prevention Through Education," *Research Bulletin of the National Education Association,* 1932, Vol. 10, p. 168.

[20] Charles Goring, *The English Convict.* London: His Majesty's Stationery Office, Chaps. V and VII.

especially among the lower economic groups. Furthermore, Campbell[21] presents evidence that it is the tension, neglect, and poverty accompanying broken home conditions that cause an increased percentage of delinquency in these groups. Based upon the records of 604 juvenile delinquents of both sexes, Hirsch[22] interprets the results as showing broken homes a consequence of constitutional abnormalities and temperamental instabilities of parents rather than a direct cause of delinquency. Many siblings of delinquents from broken homes are untouched by this factor.

It has been shown that economic inefficiency and educational deficiency, respectively, head the list of causes of domestic conflict. These same factors are closely related to juvenile delinquency. Sullenger[23] found, from studying the backgrounds of 500 delinquents brought before the courts in Omaha, that 25 per cent of the families were registered as having received some kind of aid from public and private agencies. The significant thing was that these registrations tended to occur near or just prior to the time when the behavior difficulties of the children brought them into court. This study is in agreement with other studies showing a large percentage of shiftless fathers and of mothers working outside the home. Poverty seems to bring with it or to include factors closely related to delinquent behavior. We usually find crime, disease, ignorance, and vice associated with poverty. These items are certainly conducive to the development of juvenile criminals.

Economic status. It is probable that a careful study of homes broken by the death of one parent would indicate that such a circumstance is apt to result in behavior difficulties in the children from families less favored financially more than in children at the other end of the economic scale. The death of the father in a laborer's family usually burdens the mother and the older children with the responsibility of furnishing a livelihood. Often no insurance or other form of security is available. The mother must be away from home during many hours of the day, so that close supervision is impossible. In such a situation it is not easy for the parent to establish a relationship of close confidence with the children, which is vitally necessary to harmonious home life.

[21] Marian W. Campbell, "The Effect of the Broken Home upon the Child in School," *Journal of Educational Sociology,* 1932, Vol. 5, pp. 274–281.

[22] N. D. M. Hirsch, *Dynamic Causes of Juvenile Crime.* Cambridge, Mass.: Sci-Art Publishers, 1937.

[23] T. E. Sullenger, "Economic Status as a Factor in Juvenile Delinquency," *Journal of Juvenile Research,* 1934, Vol. 18, pp. 233–245.

Even when both parents are alive, the poverty-stricken home presents a tremendous handicap to rearing the children as well-adjusted individuals. Often the father must work long hours without sufficient nourishment and recreation. His temper and training do not fit him for considerate handling of discipline situations and the result is constant tension. The children in such a family are denied the comforts and luxuries that some of their companions at school enjoy, with the result that many resort to dishonest means to attain these advantages. The death or desertion of the father in such a situation is the final stroke that brings on the more adverse conditions. Ettinger notes the following condition as closely associated with inferior economic status:

> In the slums and poverty-stricken areas of the cities are to be found ramshackle buildings, with well-defined types of submerged humanity. These determining areas are a world of foreign tongues, an area of cheap lodging-houses filled with economic failures—the broken family, the marooned family, and human derelicts.[24]

Overcrowding. Another home condition somewhat closely related to many of those already considered is overcrowding. This is especially likely to occur in circumstances of poverty and leads to stealing. Congested living conditions within the home or neighborhood may also throw children into undue contact with sexual stimulation and thereby result in increased immorality. Other causal home conditions exist that cannot be considered here; nor is there ample space to consider even the major studies that have been made of the subject. Rejection, the mother's being forced to work, lack of educational advantages, lack of recreational facilities, the broken home, and undesirable companions in relation to the home are all potent factors. However, there is considerable evidence from various studies that the most important home factor that influences the growing boys and girls is the relationship existing between parents and children, and between the children themselves. The influence of parental attitudes was discussed in Chapter XI. The importance of favorable and consistent attitudes of parents in relation to adolescent problems cannot be overemphasized.

NEIGHBORHOOD CONDITIONS

Congested neighborhoods. The detrimental effect of bad home conditions is usually supplemented by undesirable neighborhood influences. In the first place, congested home conditions are closely related

[24] C. J. Ettinger, *The Problem of Crime.* New York: R. Long and R. R. Smith, Inc., 1932, p. 151.

to congested neighborhood conditions.   It has been found from various studies that crime is relatively higher in populous territories.   A number of years ago it was shown that the four most populous counties of Tennessee contained about 27 per cent of the population (1920 census) but contributed over 50 per cent of the juvenile delinquents.[25]   Figures available from such states as Illinois, Indiana, Alabama, North Carolina, Missouri, Nebraska, Iowa, and other centers reveal the same tendency.

In Maller's[26] study of juvenile delinquency in New York City, it was observed that delinquency is largely concentrated in certain underprivileged areas.   Some conditions that characterize the neighborhoods with high delinquency are congested living conditions, poor housing, high adult crime rate, lack of recreational facilities, few Boy Scout troops, low rents, high infant mortality rate, and excessive retardation in school.   These conditions are similar in nature to the results obtained from surveys in other cities.   These studies indicate that delinquent areas fall into the following general types: (1) deteriorating residential areas in which business establishments are being organized, (2) manufacturing areas, and (3) districts characterized by an unstable population.

Gang activities.   Gang life forms the background of much delinquency and, as has already been pointed out, the gang is largely an adolescent phenomenon, originating mainly among boys.   It is formed in crowded territories and is characterized by the following types of behavior: meeting face to face, milling, movement through space as a unit, conflict, and planning.   The result of this collective behavior is the development of tradition, unreflective internal structure, *esprit de corps,* solidarity, morale, group awareness, and attachment to a local territory.[27]

Gangs, like most other social groups, originate under conditions that are typical for all groups of the same species—they develop in definite and predictable ways.   Their playgrounds are oftentimes areas where they find opportunities for amusement and adventure and too often escape from certain adversities of life.   However, it should be borne in mind that the gang is a protean manifestation: no two gangs are just alike; some are good; some are bad; each has to be considered on its

[25] *Biennial Report of the Department of Institutions of Tennessee,* 1926–1928.
[26] J. B. Maller, "Juvenile Delinquency in New York City, *Journal of Psychology,* 1936, Vol. 39, pp. 314–328.
[27] F. M. Thrasher, *The Gang.* Chicago: University of Chicago Press, 1927.

own merits. Strenuous efforts are being made at the present time to organize and direct these gang-like tendencies into more useful and desirable channels. It is well known that where this has been done juvenile crime has been considerably reduced.

Reinhardt and Harper[28] compared the club activities of 40 delinquent boys with 40 nondelinquent boys of the same age. The nondelinquents were members of all sorts of constructive clubs; only 5 did not belong to some club. In contrast, only 15 of the 40 delinquents belonged to some club; apparently the other 25 were left unsupervised and found recourse in gang activities. The Gluecks[29] found a similar condition, in that 75 per cent of 971 cases studied had never been associated with such organizations as the Y.M.C.A., Boy Scouts, and supervised playgrounds. On the other hand, there is considerable agreement among the various studies on the point that the delinquent usually has one or more companions in his delinquencies. The apparent exception in the case of Armstrong's study of 660 runaways is to be expected, since the nature of the offense itself would not call for companions. The data from the studies presented in Table LIII strongly suggest that companions are a significant factor in accounting for acts of delinquents.[30]

TABLE LIII

PERCENTAGE OF COMPANIONSHIP EXISTING AMONG ACTS OF DELINQUENTS

AUTHOR	DESCRIPTION	PER CENT
Healy and Bronner (1926)	3,000 cases	63.0
Illinois Crime Survey (1929)	6,000 cases of stealing	90.4
Shaw and McKay (1931)	3,517 offenses of 1,886 boys	81.0
Armstrong (1932)	660 runaways	13.5
	70 "unlawful entry"	84.3
Glueck and Glueck (1934)	823 cases	70.2
Fenton (1935)	282 boys	82.6

A number of years ago in rural America the problem of leisure time was not serious, since the adolescent was part of the economy. He did

[28] James M. Reinhardt and Fowler V. Harper, "Comparison of Environmental Factors of Delinquent and Non-delinquent Boys," *Journal of Juvenile Research,* 1931, Vol. 15, pp. 271–277.

[29] Sheldon Glueck and Eleanor T. Glueck, *One Thousand Juvenile Delinquents. Their Treatment by Court and Clinic.* Cambridge, Mass.: Harvard University Press, 1934.

[30] C. M. Louttit, *Clinical Psychology.* New York: Harper and Brothers, 1936, p. 377.

his share of the family work and thus shared in the prosperity or adversity of the family. The adolescent reared in the city is no longer part of the income-earning group, and he is finding it increasingly difficult to earn money through part-time employment. His needs for spending money, however, have increased, since he is to such a large degree dependent upon commercial types of amusements. It appears, then, that in gaining leisure time the adolescents have lost the opportunity for earning money. They are faced, however, with the need for having spending money. This presents a crucial problem to many adolescents. They seek to fill their leisure time with amusement and excitement. If they are unable to find this without money, they may resort to stealing in order to satisfy their desires for excitement or may find excitement in activities of an unwholesome nature. Many communities, recognizing this problem, are developing programs to meet the leisure-time needs of adolescents. These programs will be discussed further in Chapter XVIII.

THE SCHOOL AND DELINQUENCY

Its enlarged function. The school is becoming a potent force in the development and guidance of individual boys and girls into useful and worthy citizenship. It is sometimes thought of as one would think of a life insurance policy, except that in this case the state pays the premiums and is expecting returns in the form of better and more useful citizenship. One assumption here is that a citizen trained for earning a living will be a better citizen; the other is that a democratic state cannot afford to be controlled by the will of an ignorant demos.

Although the schools are playing an increasingly important role in the training of future citizens, they are also in many cases contributing to juvenile delinquency. Some of the major problems faced by adolescents have been listed in earlier chapters as school problems. It has been pointed out that many adolescents are almost doomed to failure because of an inadequate program, while another large group find themselves at odds with the teachers and school administration because they are not interested in, and in many cases actually dislike, the program in which they are required to participate at school. We note that the first step of many juvenile delinquents is truancy from school. Teachers and administrators must concern themselves with the causes of truancy, since truancy is so closely related to stealing and to sex offenses. The study by Williams shows the importance of truancy in relation to juvenile delinquency. His study is based on the results from

98 cases referred to a clinic during the school year 1944–45. Truancy, stealing, incorrigibility, and sex misdemeanors were the chief reasons for referring half the cases. There were few cases, however, where one factor alone was noted. Truancy was the chief complaint in 21 cases and there was a history of truancy present in 33 other cases. A further study of the 98 cases showed that certain factors seemed to favor truancy. Listed in order of frequency, these were: (1) poor parental control; (2) no goal; (3) gangs; (4) pushed against a low I.Q.; (5) low economic status, with desire to keep up with others as to style and dress; (6) inability to keep up with the progress of the class after a severe illness; (7) punishing parents; and (8) dislike of teacher.[31]

If the school program is well integrated with the life of the community, if the values of the teachers coincide with those of the community, and if there is a democratic and harmonious working relationship established between the pupils and teachers, the school will be a powerful agency for preventing juvenile delinquency and for developing desirable and wholesome personalities.

Effects of failure. More and more the problem of individual variation is receiving attention in an endeavor to interpret better the cause-and-effect relations in the development of behavior. The importance of this is indicated by the results of a study based upon surveys carried out in New York City Reformatory and the House of Refuge on Randall's Island, New York City. Peyser writes:

> School failure appears to be more highly correlated with the incidence of delinquency than is any other condition, including poverty, broken home, absence of religious association, physical defect, mental defectiveness, psychopathic condition, or truancy. Failure is written largely in the school histories of the great majority of the boys.[32]

His data indicated that 29 per cent of New York City elementary school children were retarded and that these children have contributed from 84.4 to 92.8 per cent of the delinquent groups that he investigated.

But school failures are largely conditioned by socio-economic circumstances and conditions. The relation between educational achieve-

---

[31] E. Y. Williams, "Truancy in Children Referred to a Clinic," *Mental Hygiene,* 1947, Vol. 31, p. 405.

[32] N. Peyser, "Character Building and Prevention of Crime," Unpublished Manuscript, 1933, p. 70.

ment and delinquency is shown again in a study by Moore.[33] The subjects in this study were 115 delinquent boys from the Tennessee State Training School and 122 orphans from the Tennessee State Industrial School. They were given the Otis Self-Administering Test of Intelligence and the Modern School Achievement Tests. The results show that the delinquent group was retarded mentally four years and the dependents were retarded one year, eight months. In educational achievement the delinquents were 33 points below the norm and 16 points below the control group.

Dorothy Kinzer Tyson, of the California Bureau of Juvenile Research, noted 33 commonly found behavior traits of delinquent boys. To find the occurrence of these traits in the cottage, the trade shop, and the classrooms, 246 boys of the Whittier State School were checked for a variety of traits. "The ten outstanding traits reported were laziness, disobedience, resentment toward discipline, inattentiveness, quarrelsomeness, lying, swearing, filthy language, instability of mood, and bullying."[34]

SUMMARY AND GENERALIZATION

It has already been pointed out that any attempt to explain delinquency on the basis of a single cause is fraught with difficulties. Throughout this volume there has been a continuous emphasis on the general concept that the development of behavior patterns is a result of forces and conditions both within and without the individual. The adolescent has been described as a dynamic individual in a state of transition from childhood to adulthood. His behavior at any particular time arises from a multiplicity of causes and conditions. Delinquent behavior, according to this viewpoint, is thus regarded as symptomatic of a great variety of conditions—among them physical conditions, emotional states, socio-economic status of the home, recreational needs and opportunities, educational attainments, relationships with peers, social and personal adjustments, and guidance. The conclusion of the personality structure of delinquent boys presented by Zakolski states: "The major difference between delinquents and non-delinquents seems initial psychological inadequacy plus the develop-

[33] J. E. Moore, "A Comparative Study of the Educational Achievement of Delinquent and Dependent Boys," *Peabody Journal of Education,* 1936, Vol. 14, pp. 1–6.

[34] "How Bad Boys Behave," *The Survey,* Vol. 15, 1931, p. 440.

ment of a new, socially unacceptable, adjustive reaction which society designates as delinquency." [35]   Ample evidence exists that boys commit crime from five to ten times as frequently as girls and that economic uncertainty at home—usually combined with such other factors as lack of education, inferior social status, and a broken home—is related to juvenile crime.   Neighborhoods in which there is extreme congestion, lack of wholesome recreational facilities, and a great deal of mobility tend to be centers in which juvenile crime is rather prominent. Truancy, failure in school, conflicts in school, and the display of psychoneurotic tendencies are closely related to the overt acts recognized as the beginning of a juvenile criminal career.

Sex alone does not cause delinquency; psychoneurotic tendencies alone do not cause delinquency; inferior intelligence alone does not cause delinquency.   It is not inherited; environment considered as an entirely isolated factor cannot give the whole story of delinquency. The delinquent personality is, in truth, as much a composite expression as the nondelinquent personality, and we do an injustice to any analysis when we consider only one element to the exclusion of all others.

THOUGHT PROBLEMS

1. List in order of importance the ten factors that you believe to be most closely associated with juvenile delinquency.

2. Describe some case of a juvenile deliquent that you are familiar with. Can you give the factors in his life that are probably responsible for his behavior?

3. What is wrong with some of the generalizations one might draw from some of the facts presented concerning elements of delinquency?

4. Elaborate on the thought that "badness" in behavior is symptomatic of a great variety of conditions affecting the individual.

5. Account for the increase in crime despite the development of public education.

6. Point out how three boys, each of whom had stolen a baseball glove, might have been motivated by different conditions or forces? Just what is the significance of this in relation to the treatment that should be accorded these three boys?

7. List several common misconceptions about the nature, characteristics, and problems of the juvenile delinquent.

[35] F. C. Zakolski, "Studies in Delinquency: I. Personality Structure of Delinquent Boys," *Journal of Genetic Psychology*, 1949, Vol. 74, pp. 109–117.

SELECTED REFERENCES

Bernard, William, *Jailbait! The Story of Juvenile Delinquency*. New York: Greenberg, 1949. Case studies are presented with the aid of the services of information supplied by various federal and state agencies and workers in the child welfare field.

Cabot, P. S. deQ., *Juvenile Delinquency: A Critical Annotated Bibliography*. New York: The H. W. Wilson Co., 1946. A critical and extensive bibliography of 972 titles is presented in this volume.

Cole, Luella, *Psychology of Adolescence* (Third Edition). New York: Farrar and Rinehart, 1948, Chap. X.

Fenton, N., "The Delinquent in the Classroom," *Juvenile Delinquency and the Schools. Forty-seventh Yearbook of the National Society for the Study of Education,* Part 1, 1948, Chap. III.

Hurlock, E. B., *Adolescent Psychology*. New York: McGraw-Hill Book Co., 1949, Chap. XI.

Landis, Paul H., *Adolescence and Youth*. New York: McGraw-Hill Book Co., 1945, Chap. II.

Porterfield, Austin, *Youth in Trouble: Studies in Delinquency and Despair*. Fort Worth: Leo Potishan Foundation, 1946. The stark realities presented by Porterfield relative to the causes of delinquency present a real social and educational challenge to everyone concerned with the education and guidance of adolescents.

Reckless, W. C., and Smith, M., *Juvenile Delinquency*. New York: McGraw-Hill Book Co., 1932.

"The School and Delinquency," *California Journal of Secondary Education,* 1943, Vol. 18, No. 8. (A symposium of articles from California leaders in the field of delinquency prevention and treatment. Both the general aspects of the problem and specific projects are presented in the discussions.)

Sellin, T. (Editor), *Juvenile Delinquency*. The Annals of the American Academy of Political and Social Science, 1949, Vol. 261.

Shaw, Clifford R., *Juvenile Delinquency and Urban Areas*. Chicago: University of Chicago Press, 1942.

Sheldon, William H., *Varieties of Delinquent Youth*. New York: Harper and Bros., 1949.

Sullenger, T. Earl, *Social Determinants in Juvenile Delinquency*. New York: John Wiley and Sons, 1936.

Tappan, P. W., *Juvenile Delinquency*. New York: McGraw-Hill Book Co., 1948. A valuable contribution to an understanding of the basic concepts of juvenile delinquency is presented in *The Nervous Child,* 1947, Vol. 6, No. 4.

PART IV

THE GUIDANCE OF ADOLESCENTS

XVII

The Hygiene of Adolescence: Healthful
Personal Living

INTRODUCTION: IMPORTANCE OF MENTAL HYGIENE

Advanced knowledge in the various sciences has provided an increased interest in man and an increased understanding of his behavior at his different periods of life. The accumulated knowledge of the forces that affect the growth of boys and girls has given renewed courage to those who would combat certain conditions that appear to affect adversely their physical and mental well-being. Today the term *hygiene* has taken on a very familiar meaning and is constantly found on the lips of the teacher, doctor, juvenile-court judge, social worker, and nurse; the term is also used by those who are attempting to minister to spiritual needs and values. More recently mental hygiene has come into prominence. Its nature and purpose have been presented by Isabel Parker as follows:

Mental hygiene offers a philosophy or objective in terms of optimum personality development. We may think of this objective from the standpoint of the individual in his striving to get along happily and effectively in his work or with his family and associates; or it may be considered from the standpoint of the study and prevention of the various forms of mental adjustment. The aim in either case—mental health—involves the adjustment of individuals to themselves and to the world at large with the maximum of effectiveness, satisfactions, and cheerfulness. It implies socially considerate behavior and the ability to face and accept reality.[1]

Although psychiatry in its earlier days emphasized responsibility and volitional control, and dealt almost exclusively with mental diseases, prevention of such diseases has gradually received more and more at-

[1] Isabel Parker, "Personality Problems and Reading Disability," *Nineteenth Yearbook of the National Elementary Principal,* 1940, Vol. 19, p. 603.

tention. With increased knowledge and understanding of the growth and development of the personality, especially in the area of the emotions, it was recognized that the best preventative measure against a personality disorder or maladjusted condition was a well developed personality.

The clinical approach. There is evidence from studies of the attitudes of teachers toward problem behavior that teachers have a better understanding of the needs of children and methods of dealing with problem behavior than they did two or three decades ago. Furthermore, teachers are more and more coming to regard certain forms of behavior, regarded as undesirable, as symptoms of maladjustments. Academic difficulties, stealing, social withdrawal, and other reasons for bringing the child to the attention of the teacher and school psychologist are only symptoms of maladjustment. A successful remedial program must first determine what factors or conditions are operating to produce the maladjustment. The clinician would seek the answer to such questions as: "Why is the child or adolescent failing in school? What factors are basic to his aggressiveness? Why does he withdraw from the play activities of the group?"

There will perhaps be as many answers to these questions as there are individuals concerned. The same forces and conditions have not affected all the children; neither have similar forces and conditions affected all the children in the same way. Adolescent boys and girls did not start off life with the same constitutional make-up. Thus, an environment that stimulates one individual to greater achievement may drive another individual into truancy or special acts of mischief and crime. The clinical approach attempts to analyze behavior patterns by means of any or all of the procedures that are available. All the materials that are available are brought to bear on the study of the child. The organismic view of the individual should never be forgotten amid the attention that may be given to data from special tests and examinations. The child's behavior at any stage in life must be considered and interpreted in the light of his past development.

Extent of mental health problems. There are some 600,000 hospital beds occupied by mental patients, and a quarter of a million new patients are admitted annually. This does not represent the total number of people suffering from mental and emotional difficulties; neither does it include the large number of mental cases resulting from World War II. Modern medicine reveals that many of the basic difficulties of those complaining of physical ills are to be found in mental and emotional conditions. Dr. Thomas Parran estimates that more than six per cent of the population suffers from some sort of mental illness. This would amount to approximately 8,000,000 Americans. The recent world war brought this to our attention in a still more convincing manner. Despite the fact that a nation draws upon its best manpower at the fittest age levels to do the fighting job, "30.4 per cent of the men who were tested for the fighting forces were found inadequate on one or another neuro-psychiatric basis." [2] Furthermore, despite the selective service screening, about 40 per cent of all men given medical discharges were declared unfit for service as a result of some mental or emotional difficulty. There is no way to calculate the cost of this in terms of human suffering and unhappiness.

The extent of maladjustments among adolescents enrolled in schools is much greater than the ordinary teacher would ever realize. The teacher is not expected to be a psychotherapist or a psychiatrist with responsibility for diagnosing and treating every problem that appears in her room. She can, however, use the clinical point of view in recognizing that a child's problem behavior is symptomatic of some problem condition or unsatisfied need and not an end in itself.[3] The moody child, the aggressive child, and the nonsocial child have adopted certain modes of behavior as the most satisfactory way thus far found for meeting certain life conditions. The understanding and sympathetic teacher may be able to determine some of the needs of the individual and thus help him in meeting his problems and directing his efforts into positive and useful channels. In nearly all school situations, however, there are cases that should receive the attention of a psychiatrist or a school psychologist, if such attention is available. The failure of the teacher to detect symptoms of maladjustments may lead to such serious consequences as the rape murder of a small child, moral degeneracy, or suicide.

Handicaps to the physical health of adolescents. There are several phases of this general problem dealing with handicaps to adolescent health, and in considering these we may tend to bring together some of the points already presented in connection with interests, volitions, religious growth, and personality development. Good health in its broadest aspect is essential to a well balanced personality, for on it

[2] Thomas R. C. Rennie, "Needed: 10,000 Psychiatrists," *Mental Hygiene,* 1945, Vol. 29, p. 644.
[3] See Mary C. Roland, "Help for Problem Children," *The Nation's Schools,* 1945, Vol. 36, p. 25.

depend to a large degree energy, volition, ideals, and happiness. People in poor health are often deficient in surplus energy, lacking in self-control, and pessimistically oriented toward life.

One of the most important handicaps to adolescent health is crowding. The theater, the school, the church, the dance, and the streetcar and the automobile all bring groups into close contact, within the range of spray ejected by sneezing and coughing. Many contacts are made also by handshaking and by transferring books and other articles. Such activities are very effective in spreading germs that develop into diseases.

Another handicap to adolescent health is suggested in relation to mental-hygiene principles. The beneficial effects of exercise have been emphasized, and rightly so, within recent years; but temperance in play activities is most desirable. There is grave danger that, with the onset of adolescence and the fuller development of the internal organs related to respiration, exercise will be insufficient; the more solitary and intellectual pursuits of high school years are likely to diminish this exercise unwarrantably. On the other hand, overexertion in team activities may exist; hence, boys and girls in good health who are especially interested in athletic performances should be cautioned against exertion to the point of physical injury. Students are appealed to through the force of rivalry and thus motivated to do their all for the sake of the school or organization to which they have firm loyalty. Exercise is often carried beyond the point of mere fatigue; some athletes resort to stimulants in order to overcome complete exhaustion. Of course, the adolescent, who is constantly building up additional energy, can stand a great deal of exertion; but it goes without saying that activity should not extend beyond one's physical powers. Physical maturity, in particular, should be taken into consideration in the control of recreational activities.

A third handicap of adolescent health is that of dissipation. Until the beginning of adolescence, the child has usually had some reasonable guidance and discipline; his hours of sleep and meals, along with his appetitive habits, have been observed and provided for. But with the onset of adolescence, and with group situations playing an increasingly prominent part, there is a tendency to live in conformity with group desires and activities, which quite commonly involve smoking, drinking, irregular hours of eating and sleeping, and exposure to colds and drafts. In this connection it is worthwhile to consider the minimum dietary requirements of adolescents. There is an enormous increase in the daily calorie needs of the adolescent as he progresses toward maturity.[4] Among boys, the total daily calorie need may run as high as 4,000; among girls it sometimes reaches a figure in excess of 3,000. This increase is reflected in the large appetites that characterize many adolescents. There is, of course, wide individual variation in the total daily calorie need; it is, however, of utmost importance that the adolescent have a well balanced and sufficient diet during this growing period. Studies of undernourished children in some of the European countries during the period of World War II reveal that stunted development, malformations, and susceptibility to disease result from deprivation. As late as 1930 the White House Conference estimated that 6,000,000 children in the United States 19 years of age were improperly nourished. Also, there is evidence that where a well balanced diet is supplied, there is less tendency on the child's part to choose peculiar combinations of sweets and sours, and to depend upon "soft drinks" and sweets between meals.

A fourth handicap of adolescent health has its source in accidents, which may cause body malformations during this stage of life. The daring spirit of adolescents will cause them to pursue activities that may and often do end in mishaps. The increased surplus energy, the greater emotional drive, the spirit of rivalry, and the new social contacts are all forces related to an accelerated and enlarged motor life. The awkwardness of the adolescent, his natural irresponsibility, and curiosity at this stage of his development tend to subject him to increased accident hazards. This fact has been observed in the play of boys and girls around home and school, and in their work in industry.

During adolescence the individual has a superior amount of energy and is able to withstand adverse environmental and physical conditions to a greater extent than at any other period of life. The death rate is very low during this stage despite all the handicaps we have noted. However, the effect of all these forces playing upon the adolescent is either indirect or delayed in results; the various situations and conditions with which the adolescent comes in contact will in all probability powerfully affect his further activities and tend to involve further changes of the personality. This development of further desirable or undesirable personality qualities should again be considered from the developmental viewpoint. It is here, as we have already seen, that a

---

[4] See Beatrice McLeod, *Teachers' Problems with Exceptional Children, VI. Children of Lowered Vitality.* U. S. Government Printing Office, Pamphlet No. 56, 1934.

well-grounded and firmly rooted religious attitude, nourished and empowered by a well-defined habit system of initiative and self-control, may further desirable deveopment.

Adolescence as a period of morbidity.[5] It might well be stated as a general principle that any period of life during which pronounced physiological changes are taking place is a period of morbidity to diseased conditions related to such changes. Using this as a basis in studying the physiological changes, it becomes possible to note the diseases that are likely to result. Because of the nature of the life activities of adolescents, theirs is a period of life susceptible to body malformations and various mental maladjustments. Round shoulders and spinal curvatures, for example, may develop at this time. Furthermore, owing to the frequency of exposure to somewhat dangerous environmental situations, deforming accidents are likely.

Headaches, eye troubles, indigestion, respiratory troubles, malformation of bones, and infections are especially prevalent among adolescents. These conditions result in a large measure from conditions and activities imposed upon them through our customs and institutions. Acne is perhaps the greatest hazard to a good complexion among adolescents. It is present in some degree during pubertal development in approximately three-fourths of them, and has been found to be closely related to the increased activity of the sebaceous glands during puberty. This thought was reviewed in connection with the skin glands in Chapter IV.

According to Kleinschmidt,[6] the two main causes for rejections of young men entering the armed forces in World War II were dental defects and eye defects. Table LIV shows that these, combined with cardiovascular diseases and hernia, account for more than half the total. It is well known that these conditions do not suddenly develop as a boy reaches seventeen or eighteen years of age. The conditions responsible for rejection by the armed forces were present among the school children two decades ago. Many educators, observing the deplorable physical condition of the youth of World War I, concerned themselves with a physical fitness program designed to develop stamina, strength, endurance, and agility; hence the increased interest in com-

[5] For a good abbreviated discussion of some mental-hygiene problems of adolescents see E. S. Conklin, "Mental Hygiene Problems of Adolescents," *Harvard Educational Review,* 1938, Vol. 8, pp. 343–352.

[6] Earl E. Kleinschmidt, "Meeting Today's Health Problems," *Phi Delta Kappan,* 1943, Vol. 26, p. 12. (Data for Table LIV are taken from materials presented in this article.)

petitive athletic programs. There is ample evidence, however, that such a program is far from adequate, and that it fails to take into account (a) individual differences, and (b) functional health needs.

TABLE LIV

CAUSES FOR REJECTION OF YOUNG MEN ENTERING THE ARMED FORCES
IN WORLD WAR II

CAUSES	PER CENT	CAUSES	PER CENT
Dental defects	20.9	Ear defects	4.6
Eye defects	13.7	Food defects	4.0
Cardiovascular diseases	10.6	Lung defects, including	
Hernia	7.1	tuberculosis	2.9
Mental and nervous		All others	24.5
diseases	6.3		
Venereal diseases	6.3		

CONDITIONS AFFECTING MENTAL HEALTH
OF ADOLESCENTS

Mental hygiene must begin during childhood. It must be remembered that the adolescent is but a product of earlier experiences and that his development is gradual and continuous. Thus, if we are to understand the mental hygiene of adolescents, it is necessary to study the influences that have thus far affected them. The problems of adolescent mental hygiene have their inception in most cases in the early fondling and feeding activities of the growing newborn baby. The mother who feeds her child every time he cries is certainly teaching the child to dominate situations by violence; she is also failing to develop in him desirable habits of digestion and habits of self-control. Such early habits will become a part of the child's growing personality, with the result that later he will probably shriek with rage if his desires are not satisfied. Such a child, unless his behavior is modified by some trying experiences with other children, will probably develop with very poor preparation for adjustment to a social world in which responsibility and self-control are essential.

It should be remembered, however, that only probabilities can serve as guides. And, although most of the difficulties encountered during adolescence can be traced, at least in part, to early childhood experiences, it would be a fallacy to conclude that all childhood disturbances lead to adolescent difficulties. The life development of the individual is not predictable by such a simple cause-and-effect formula. Many persons with unfortunate childhood experiences pass through adolescence without undue difficulties, whereas others, whose childhood was untroubled, encounter much turmoil during the transition from childhood into and through adolescence. Peter Blos has pointed out that:

The storms of this period are not the result of single causes; they arise, rather, from various pressures coinciding in time. For example, a boy of fourteen whose overdeveloped body is going through a phase of rapid growth may weather his adolescence without trouble; he is more likely to develop difficulties if, at the same time, he is experiencing the added strain of a family break-up. On the other hand, the girl whose physical development progresses very satisfactorily and smoothly is in a favorable position to work out the relationship problems which have been with her for many years.[7]

The school and mental hygiene. The function of the school in the development of the child has been emphasized throughout this study of growing boys and girls and is recognized as important by organizations concerned with problems of growth. Almost all mental-hygiene societies are using the schools as agencies for furthering their work; in fact, nearly all suggestions connected with mental-hygiene work include the use of the school in the program. Also, health clinics—designed rather to aid the child in adjusting the phases of his personality than to study behavior problems present in connection with environmental situations—are constantly held at schools to aid in the preservation of the health and sanity of youth.

No doubt many cases in need of mental-hygiene treatment will never come to the attention of a child-guidance clinic, a psychiatrist, or any other person or organization formally interested in these problems, but will have to be dealt with largely through a trial-and-error process carried on unconsciously in the home or school. The child spends the major portion of his time in the home and the school, where these problems are sure to be encountered in either a characteristic or a disguised form. Concerning this, Ryan states:

Mental health through education will be much farther advanced if the school becomes aware of its active function in community living and works systematically with other agencies, including churches and 'character-building agencies,' social workers, group education workers, the health forces, and other elements seeking a more wholesome life for human beings as

[7] Peter Blos, "Adolescence, Its Stimulations and Patterns," *Childhood Education,* 1941, Vol. 18, p. 83.

individuals and as members of the community. Instead of insisting upon their traditional separateness, the schools should welcome any movement to pool their resources with those of other developmental agencies in the community working in one way or another in behalf of mental health.[8]

Public schools could probably make no greater contribution to the welfare of the nation than to assume a reasonable amount of responsibility for the mental health of these maturing boys and girls. Often it is only through the agency of the school that enlightened influences can operate. The old Greek maxim, "A sound mind in a sound body," should become more the general aim of the present-day school. It appears quite likely that, following the Reformation, interest in education was overdeveloped intellectually and little thought was given to the physical and mental balance of students. It is only within recent years that the emphasis has begun to change, and that efforts have been made —and they are still being made—to develop the physical well-being of growing boys and girls. Special classes have been organized in a large number of cities to correct existing unhealthy conditions.

A systematized or unsystematized program of mental hygiene should be considered in every school. And any program should include the following elements:

1. Teachers trained in the principles of child and adolescent psychology and of mental health.

2. A psychophysical study of every beginning pupil.

3. A reorganization of primary grades in harmony with the interests and nature of children, along with an opportunity for more systematic and careful observation.

4. A consideration of the integrative nature of the various agencies dealing with the training and development of children.

5. The development of schools and classes to care for the handicapped and deficient.

6. The focusing of the attention upon the causes underlying maladjustments, rather than upon behavior disorders as such.

That education is most hygienic which provokes and promotes the child's innate abilities and disposes him to be a good citizen. Hence educators, by developing well balanced personalities among their pupils, may influence the ultimate mental vigor and health of the nation. Until the center of attention of the school is shifted from subject matter to pupils—to human beings—little progress will be made

---

[8] Carlson W. Ryan, *Mental Health Through Education*. New York: The Commonwealth Fund, 1938, p. 304.

in the better understanding and guidance of adolescents in the formation of personalities that will be able to adjust to a rapidly changing civilization. Teachers who are irritable, who have no appreciation of human nature, who are interested wholly in subject matter, who "don't have time" to study a problem case, who themselves are ill-adjusted, cannot apply principles of mental hygiene in their school work that will aid the maladjusted and prevent others from becoming maladjusted. Teachers who gain the confidence and good will of their pupils, who are eager to aid them in their problems, who are sympathetic with them in their troubles, and who manifest an interest in their interests will be able to exert a profound influence in the prevention and treatment of the growing problems of maladjustment.

The term *physically handicapped* may be used to include those children suffering from defects or deficiencies in their physical make-up that present a handicap in the normal processes of life. Such conditions would range from defects of the eyes and ears to deficiencies in vitality. Various estimates have been made of the number of children suffering from such conditions, and some of this material was presented at the beginning of this chapter. The estimates range from less than 50 per cent to as high as 72 per cent, the discrepancy thus revealed resulting from a number of factors, the most important of which is the lack of a standard by which to judge what constitutes a defect. There are also differences resulting from the use of different techniques for determining defects, as well as variations from place to place and from period to period. Climatic and other environmental conditions peculiar to one section of the country may cause an unusually large number of defects of certain types to occur in that region, which are almost unknown in other sections less favorable to their occurrence. Considering the findings from the various studies, it seems safe to assume that perhaps 60 per cent or more of American children have some kind of physical defect.

Mental hygiene of the teacher. A number of studies have been made of the personality adjustments of teachers. The influence of the teacher has been emphasized throughout the discussions in the previous chapters. The teacher may be endowed with an abundance of information and skills, have a thorough understanding of the best teaching techniques, and be well versed in child psychology but still be emotionally and socially maladjusted to such an extent that her classroom efforts are in the main ineffectual. The personality adjustments of the teacher are of utmost importance. It is through the teacher that the school's influences reach the child. Louttit[9] has presented an outline of attitude and behavior standards that should be helpful in evaluating the mental-hygiene relationships of the teacher in her classroom contacts with the pupils. These include:

1. Children must be accepted as they are—poor or rich, bright or dull, healthy or ill, clean or dirty. Whatever they may be, they are all growing human beings who must be trained, must be respected, and must be given every opportunity to find profit in the class.

2. Corollary to this is the principle of the equality of the children. Every child should be made to feel that he is an important member of the group. If the teacher does not accept all of the children, she will show favoritism toward one and neglect another.

3. Freedom from any feelings of being threatened by the children and by colleagues is necessary. . . . When she fears loss of dignity or status from the acts of children, or when she is jealous or suspicious of her teaching colleagues, the teacher has not the assurance of a well-integrated personality.

4. The teacher must have a sense of humor—not specifically wit, but that attitude toward one's self and one's work that puts them in proper relation to the world. To take one's self too seriously is another suggestive indication of insecurity. Unfortunately an attitude lacking in humor is apt to produce behavior in others, including children, that further blocks that sense of security.

5. Tolerance even toward persons whose ideas and behavior we disapprove of is also to be desired. Stubborn adherence to personal convictions does not make for classroom tolerance or for acceptance of the child who does not conform. . . .

6. The teacher's attitude toward the job will significantly affect her influence on the children. If teaching is a stepping stone to something else, or if it represents mere economic security, the children will suffer. If the subtle influence of teacher's attitude is to have the most favorable effects on children, that attitude must be one of vital interest in the task and enthusiasm in meeting the myriad adjustment problems a group of children presents.

7. Necessary to such vitality is a constant effort in the way of professional growth. The teacher, like the minister, the physician, or any other professional person, must constantly work to keep abreast of newer developments in her field. . . .

8. The teacher is not only a public employee and a mentor of children. She is a member of the civic community and as such has responsibilities the same as any other citizen. Recognition of and participation in activities concerning these responsibilities indicate a wholesome social maturity that makes its impress on children. The teacher must keep free, however, from

[9] C. M. Louttit, "The School as a Mental-Hygiene Factor," *Mental Hygiene,* 1947, Vol. 31, pp. 58–60.

any feeling of being compelled to attend every meeting or concert, because her time must be conserved as well as the physician's or the businessman's.

9. Personal appearance is another of the factors that have significance in an appraisal of the teacher's personality. Well-fitting, stylish clothes (not extreme), cleanliness, neat hair and hands, all suggest the person in tune with herself and the world. Clothing that is too conservative, especially if it is old-fashioned, or styles that are too extreme, suggest personality characteristics of narrowness and rigidity on the one hand and instability or insecurity on the other. . . .

10. Lastly, we must mention physical health. This is immediately related to personal appearance and has its significance in the establishment of teachers' attitudes. The actually ill teacher is certainly in no condition to meet the daily demands of a roomful of childen. The neurotically ill, the constant complainer, is obviously not the kind of personality who can give children the things that they need from the teacher.

HEALTHFUL PERSONAL LIVING

Developing a feeling of security. Individuals, especially children and adolescents, need to feel that they are secure. The small child needs to feel that he is wanted and will be cared for by his parents or someone. The adolescent needs to feel secure with his peers. There is considerable evidence that feelings of insecurity often have their roots in infancy. This need stems from the physiological drive connected with hunger and the sustenance of life itself. Whenever this need is disturbed, the infant will display various forms of disturbances bordering upon anxiety. He may develop such symptoms as eating or sleeping disturbances or general apathy. As he grows older he may become increasingly aggressive or stubborn. Security for the adolescent is enhanced through social participation and through successful achievement. It is closely related during this period to acceptability by one's peers—something very much desired and needed during adolescence. (Note the materials of Chapter X on this point.)

Very closely related to the feeling of security is the *feeling of belongingness*. The adolescent needs to feel that he is an integral part of the peer group as well as a member of a family group. His parents may not look and act as other parents do, but they are his parents and he belongs to the particular family group. He "belongs" not by virtue of what he does or does not do but by virtue of the particular person he happens to be. This feeling of belongingness furnishes him with a sense of security essential to good mental health. *Affection* goes along with this belonging. Every child needs this affection in order to develop a harmonious life.

Need for close friends. The need for affection, as the child begins to move away from the family circle, finds its satisfaction in close friendships found outside the home. The adolescent must achieve independence and ultimately break away from the close ties of the family circle. This break with the family was referred to in Chapter XI as achieving independence or emancipation. As he grows older, he will become more and more aware that he is a member of the peer group and will come to look more and more to the peer group for approval and affection. This does not mean that he should turn entirely away from the home, but rather that he should expand his horizon of interests and activities and gradually emancipate himself from the close family ties that were so binding and so important during the early years. It is not well in most cases for children of great difference in chronological or physiological age to form too close a friendship. This is true during the adolescent years more than at any other time of life, since differences stand out far more at this period than later. But contacts with others with similar interests, understandings, and problems are most important in the development of well adjusted and well balanced personalities.

Developing a sense of personal worth. Habits of initiative and responsibility tend to give an individual self-confidence and a feeling of personal worth. Such habits are developed gradually through practicing activities in which opportunities for the use of initiative and responsibility are present. The discovery and development of the latent abilities of the adolescent constitute a more important function than the recognition of abnormal behavior tendencies. It is on such a basis that a sense of personal worth is gained. The child who is constantly told that he is "good for nothing," will soon be just about that—good for nothing. Through satisfaction that comes from successful achievement, the child comes to feel that he is really good for something. As the child develops, he should come to recognize his abilities better, and to believe in himself and his abilities. This is what we sometimes call "self-confidence." He should be provided with situations in which he helps to plan for cooperative living—a living of sharing and participation by all the members. When his personal contributions are recognized and accepted, he eventually feels himself an accepted member of the group.

Self-confidence develops from the individual's learning through experience that he can do certain things satisfactorily. It is contingent upon his being able to do things well enough to satisfy his standards of achievement. These standards have developed largely as a result of what is expected of him. Thus, the individual should be given responsibilities in harmony with his abilities. The child who is encouraged to enter upon a curriculum not commensurate with his or her abilities is certainly not educationally adjusted. The child who is encouraged to direct his ambitions toward something out of harmony with his general aptitudes is not vocationally adjusted. Such conditions will contribute to the ill health of a child. Everyone should be given tasks that require effort and initiative, but the efforts required should yield returns in the form of success. Success in various tasks becomes a great motivating force for further effort in the same general direction. It is well known that the dull child who is not kept busy owing to his inability to understand work, and the bright child who is not kept busy owing to his ability to perform work quickly and with ease, are potential problem cases. From these sources arise many disciplinary problems. Sherman presents a splendid illustration of the prestige and better adjustment attained by a high school boy through his interest in collecting:

He was below the class average scholastically, and had failed in two subjects. He evidently was suffering from many conflicts of inferiority. He complained that the teachers paid little attention to him, that he was not popular in school, that his parents accused him of laziness because his grades were below standard. When asked if he had any trait which made him superior to others he brightened. He said that some of the boys were becoming interested in him because he had a number of antiques. For the past six months he had spent his allowance on antiques—miniatures, swords, coins, stamps. He said that he was studying their history and that he expected to become an expert in that type of work.

Through the possession of antiques this boy gained the prestige he was unable to attain in other ways. Attention from other students was a strong incentive to further interest in antiques. The attention from his fellow-students tended to decrease his feeling of inferiority in regard to his scholastic attainment.[10]

Maintaining optimum health. The close relation between physical well-being and mental well-being has been emphasized throughout this chapter, as well as indicated throughout our review of the growth and development of adolescent boys and girls. It is widely realized that we are not so well adjusted in our mental reactions when we are physically disturbed as we are when we have better physical balance. Further-

---

[10] Mandel Sherman, *Mental Hygiene and Education.* New York: Longmans, Green and Co., 1934, p. 173.

more, the alert physician is coming to recognize more and more the importance of a wholesome mental attitude in effecting better physical conditions.   The individual who is diseased is a more likely candidate for mental troubles than the healthy individual.   The human body is a totality, a completely integrated pattern of behavior, and the lack of balance in the activity of one part will in all likelihood have an ill effect upon the activity of all other parts.

Since the individual is to be regarded as a unit, and since the entire personality may be colored by some faulty element in this total pattern, one should not neglect health or physical examination in a careful personality diagnosis.   Behavior disorders, failure in schoolwork, or failure to adjust to fellow students are frequently to be understood only by reference to certain physical conditions—abnormalities and deficiency.   Here is a boy who does not mingle with others and takes no part in the various extracurricular activities of the school.   A careful physical examination reveals some internal disturbance that influences his whole personality.   Here is a pupil who is failing in her classwork; her teacher is unable to understand her slothfulness and lack of interest. An examination reveals chronic fatigue due to insufficient rest and sleep.   Pupils who are temperamental, slothful, untidy, or unkempt will probably be better understood once their physical condition is better known.

Understanding and accepting one's self.   The problem of understanding and accepting one's self is closely related to that of facing reality.   The adolescent should come to understand himself and accept himself and his role with his peers.   He should recognize his limitations and potentialities.   This implies a willingness to evaluate one's self free from prejudice, bias, or favorable or unfavorable attitudes. This objective attitude is characterized by an impartial, dispassionate regard for accurate, unbiased judgments; it assumes a scientific attitude toward the self.   Perhaps this is expecting too much of the adolescent; however, with sound and sane guidance the growing individual can come to weigh facts as they are rather than as he would like them to be. In the end this means being honest with the self.   It has been written: "To thine own self be true and it must follow as the night the day, thou cans't not then be false to any man." [11]

The child does not have this objective attitude by natural endowment, but has to develop it through his experiences.   It is extremely important that the child be guided into a fair and unprejudiced evaluation of

[11] Quoted from William Shakespeare's *Hamlet*.

his own worth. One of the first steps, perhaps, in training objectivity in a child is for him to develop many wholesome interests outside of himself. He will be much happier and will avoid needless emotional conflicts if he will accept himself at his own worth.

Understanding one's sex role. The sex life of the child begins at birth. Parents have the responsibility of giving the child instruction suited to his age and to his needs, as well as of giving him training in proper habits. The needed information should be given gradually, in proportion to the child's curiosity and capacity for understanding it. The information given the child of five or six will be different from that given a child of twelve, both in form and in certain features of content, but the one should be in harmony with the other. The child should feel perfectly free to ask his parents for information and he should feel confident that they will tell him the truth. His inquiries should be treated with candor because they are motivated by a natural unemotional curiosity. In this way the child will build up the right attitude toward sex and be prepared for puberty.

The sex role develops during the adolescent stage. It is during this time that a satisfactory relationship with the opposite sex should be attained. A healthy attitude will be maintained if the problems that arise at this time are treated positively and constructively.

Developing social consciousness. The child is born neither social nor antisocial. He is born into a society where certain cultural patterns are found in the home, at church, at school, on the streets, and elsewhere. At first he is more or less oblivious to most of the culture that surrounds him, although he reacts within his limits to certain aspects of it from the very beginning of life. His behavior is largely concerned with providing for his physical needs—eating, sleeping, exercising, and the like. Social consciousness is almost if not wholly lacking at this early period of life. The real beginnings of social consciousness are to be found in the activities of groups of children at play during the early school years. As they grow they come to realize what the group expects of them, and how they must behave with respect to the various members of the group.

Adolescence has often been described as a period of heightened social consciousness. This fact was brought out in the discussion of adolescent peer relations in Chapter X. Pre-adolescence is a period when heightened social consciousness is manifested in the formation of groups, gangs, and clubs. There is an extension and intensification of this during the adolescent period. If the adolescent is to secure and maintain a well adjusted personality, he must develop out of this early egocentric nature into a social being who recognizes and appreciates the personalities of others, and who is anxious to become a part of his peer group.

Achieving a consistent and unified philosophy of life. As the individual develops into adolescence he should have made a beginning in the development of some concepts of the nature of the world in which he lives and of some purposes of life. It is here that faiths and truths related to actual living may be able to function most effectively. The adolescent needs help in the development of some consistent attitudes that will give meaning to life. There are elements in life that make it difficult to reconcile various teaching in a way that will develop a unified viewpoint. The adolescent may know enough science to block his acceptance of a traditional religious concept but may not know enough to synthesize the two. In an effort to reconcile the two, he encounters a third. These various concepts create confusion and conflicts.

A philosophy may be regarded as a set of values and concepts. It provides a standard by means of which the individual is able to arrive at a clearer understanding and to make major decisions.

It provides a basis for passing judgments and for making evaluations. By having such standards, one's ideals and actions become stabilized and are made more consistent. The materials of Chapter IX indicated that adolescence is accompanied by an enlargement and by a unification of attitudes and beliefs. This is part of the development of a unified philosophy of life so essential for stability and growth of the adolescent boys and girls as healthy, well adjusted personalities.

SUMMARY

In our culture, growth into adolescence is accompanied by certain demands upon the individual. In the first place, he is required to accept more responsibility and, thus, to achieve increased independence. Secondly, he must affect a transition from interest in gang activities to interest in members of the opposite sex. Thirdly, he must adjust to his own capacities and limitations. Fourthly, he must learn to face reality.

Adolescence has been described as a period of danger for mental and physical health. Many health hazards appear at this age, and the adolescent needs guidance in meeting them. Childhood is regarded as the golden period for mental hygiene. Attitudes essential for desirable adjustments are in the formative stage at this time. Many of these are sufficiently formed to act as an aid or barrier to successful adjustments at a later stage.

The basis for the new impulses of adolescence is the development of the visceral organs of the body, and especially the sex and related glands. This development no doubt does much to give the adolescent a better and fuller understanding of his relation to others, and an admiration for those of the opposite sex. In fact, he experiences a new and heightened sensitiveness to all the phases of his personal and social environment. As was suggested in previous chapters, he now begins to heed the general approval of those about him; he begins to make a more careful inventory of his own personal qualities, and may easily develop a keener and more extensive display of achievement, in conflict with the fear of failure and thus of social disapproval or ridicule. This effort to adjust in harmony with the maturing self presents many vital problems to adolescents.

THOUGHT PROBLEMS

1. Look up further meanings of the term "mental hygiene." Why is it impossible to separate mental hygiene from physical hygiene?

2. Show how the stress and strain of modern life may affect the physical and mental health of adolescents.

3. What are some of the major problems of the hygiene of adolescence? Are any of these problems peculiar to this specific period of life?

4. In what way is the teacher's task much larger than mere teaching? Give concrete examples.

5. Consider several teachers of your acquaintance and evaluate their mental hygiene relationships in connection with their classroom contacts. What problems or difficulties most frequently appear among this group of teachers?

6. Show how the needs and principles involved in healthful personal living are interrelated.

7. What barriers does the adolescent often face in understanding and accepting himself?

8. What are some essentials for the development of a consistent and unified philosophy of life? What agencies or forces have been most helpful to you in the achievement of such a philosophy of life? What values have these had for you?

9. Just what is your interpretation of the phrase "the clinical point of view"? Why is it desirable for teachers to have this point of view?

10. Show how mental-hygiene problems follow the developmental idea presented throughout this text.

SELECTED REFERENCES

Association for Supervision and Curriculum Development, *Fostering Mental Health in Our Schools.* 1950 Yearbook of the Association, National Education Association, 1950.

Crow, L. D., and Crow, Alice, *Mental Hygiene in School and Home Life.* New York: McGraw-Hill Book Co., 1942. This is a presentation of the relation of mental hygiene to behavior problems under normal and critical conditions.

Douglass, Harl (Editor), *Education for Life Adjustment.* New York: The Ronald Press Co., 1950.

Garrison, Karl C., *The Psychology of Exceptional Children.* New York: The Ronald Press Co., 1950, Chap. XXII.

Klein, D. B., *Mental Hygiene: The Psychology of Personal Adjustment.* New York: Henry Holt & Co., 1944.

Myers, C. R., *Towards Mental Health in School.* Toronto: University of Toronto Press, 1939.

Steckle, L. C., *Problems of Human Adjustment.* New York: Harper and Bros., 1949.

Symonds, P. M., *Adolescent Fantasy.* New York: Columbia University Press, 1949.

Thorpe, Louis P., *Child Psychology and Development.* New York: The Ronald Press Co., 1946, Chap. XV.

Witty, Paul A., and Skinner, C. E. (Editors), *Mental Health in Modern Education.* New York: Farrar and Rinehart, Inc., 1939.

For a splendid review of studies bearing on mental health and adjustments, see: "Mental and Physical Health," *Review of Educational Research,* 1946, Vol. 16, No. 5; also, *Review of Educational Research,* 1949, Vol. 19, No. 5.

XVIII

Guidance: Moral Development and Character Formation

MEN AND NATIONS CAN ONLY BE REFORMED IN THEIR YOUTH;
THEY BECOME INCORRIGIBLE AS THEY GROW OLD. *Rousseau*

The previous chapter has emphasized the importance of healthy personal living in the lives of adolescents. The chapters of Part III have emphasized the importance of various institutions and good peer relations to the development of well-adjusted personalities. We have noted throughout this study of the growth and development of adolescents toward maturity that they are beset with many problems. The developmental concept of the individual suggests that moral concepts and ideals are learned during childhood and adolescence. Childish concepts are quite simple and usually concrete in nature. The young child accepts without question the concepts of his parents. With the onset of adolescence the individual comes to think for himself. This was referred to in Chapter XI as growth in independence.

Materials presented throughout Part III indicate that the adolescent stage is accompanied by many problems connected with the transition from childhood to adulthood. Chapter XVI presented data showing that there is a preponderance of crime during the period. There are many who despair of the frank self-expression of modern youth and its refusal to be blindly obedient to present-day customs and teachings. This attitude is by no means new, as was pointed out a number of years ago by V. K. Froula in his presidential address before the Washington Education Association.[1] He said:

Permit me to give you an example of a lamentation that is as old as the hills, but sounds like an excerpt from a fundamentalist's sermon: 'Our

[1] V. K. Froula, "Education and Public Morals," *Washington Educational Journal,* November, 1927.

earth is degenerate in these latter days. There are signs that the world is coming to an end. Children no longer obey their parents. The end of the world is manifestly drawing near.' The clay tablet upon which this inscription was made 6,000 years ago was found by archaeologists somewhere in the Mesopotamian Valley and now reposes in the British Museum with other relics of past times.

Need for guidance. The problems of freedom and authority have been given much consideration by those concerned with the guidance of boys and girls. Individual differences will be found here as elsewhere. One cannot set forth a rule or principle applicable to all cases, except in a very general way. Better adjusted boys and girls can be given greater freedom than those more poorly adjusted. However, there are many cases in which the poorly adjusted have had too much restraint and are in need of greater freedom. Since they have been given no opportunity to accept responsibility, there will be need of guidance in connection with these new liberties. There is good evidence from the results obtained in our modern educational programs that when pupils are given increased responsibility and freedom under guidance better social and personal adjustments result.

Various studies of maladjusted adolescents indicate that many of them have been dealt with in an autocratic manner. In some cases the adolescent has met this by means of a withdrawing mechanism; in other cases open rebellion has resulted. As the boys and girls grow out of childhood into adolescence, they are faced with many problems different from those met during the childhood years. Greater freedom and increased responsibility should come with growth and development. Thus, there is a constant need for guidance rather than unlimited freedom or an autocratic control. Social development is essential for adequate adjustment in our social order. Boys and girls must be taught that there are certain customs and conventions that must be followed if they expect to be accepted by the group. Social responsibility and social participation will aid in the development of these inner controls that are essential to group living in a democratic society.

JUVENILE CRIME PREVENTION

The home and delinquency. In viewing the home as it relates directly to the formation of delinquent habits, one should recognize in the beginning that the improvement that can be wrought here is seriously limited by many factors. In the first place, the sanctity of the home and marriage ties gives the home first claim to the development of the child's habit systems. This has been true all through the ages and is a fundamental factor in the problem of delinquency. Again, the secrecy and privacy of the home as a close-knit institution create the further problem of improving home attitudes and conditions. Finally, the child's earliest habits and attitudes are formed almost wholly in relation to or because of a lack of home contacts; the home is a primary group that operates face to face with the child over a great number of hours each day of the year. These factors, then, make the home a powerful, well-nigh impregnable force in the development of desirable or undesirable behavior patterns, barriers being set up against the intervention of society for the aid of the child.

Healy and Bronner[2] state as a result of their comparative study (previously mentioned) of 105 delinquents matched with the same number of nondelinquent siblings:

> As a logical outgrowth of our study which shows that parent-child relationships play such a part in the production of delinquent proclivities, we are inclined to believe that the single direct attack of greatest value may be through widespread parental education—to be sure not an easy task.

Treatment as shown from various follow-up studies is extremely uncertain. It must concern the family and neighborhood pattern situation, and this makes of it in so many cases a well-nigh hopeless task. However, the writer is not advocating that those concerned should adopt a fatalistic attitude toward the home and neighborhood situation; he would prefer to present this as a challenge to those who would adopt a positive, sympathetic, and understanding attitude.

Despite the development of the school as a secondary group in which character and personalities are often guided and molded along varied lines, the home is today the most potent factor in the building of character in growing youths. It is in the home that the child receives true lessons, indirect and unsystematic but meaningful, presented in the conduct of the father or mother. When the boy learns that the father considers it shrewd to keep extra money given to him by mistake in exchange for goods, to evade taxes, to misrepresent values in a "deal," his ideals are usually being established. The school might furnish ideas, information, and skills to aid the child in life's struggle; but the standard of conduct and ideals is really set forth by the examples of parents.

The school and delinquency. The failure of the junior and senior

[2] William Healy and A. F. Bronner, *New Light on Delinquency and Its Treatment*. New Haven: Yale University Press, 1936, p. 217.

high schools to adjust their programs in harmony with the interests and abilities of the increasing number of pupils entering high school is probably the greatest accusation that might be heaped upon them. This failure is probably a result of a false application of our democratic ideal. It should be the aim of the school to give the child the opportunity to develop those abilities he possesses, rather than to set up a great educational ladder to fit the abilities of all.

The case of an adolescent boy who was pushed by well meaning and fairly intelligent parents beyond his ability, described by Slattery, illustrates a common failure of both parents and schools. It is in this connection that report cards in the hands of many parents become a source of confusion and disturbance.

Rodger had an IQ of about 80. His father, a high school principal, wanted his son to be a white collar worker. He was blind to the fact that the boy could not make high school. Owing to his father's position, the high school covered up Rodger's failure by giving him passing grades. The further Rodger went in high school, the more at sea he became.

Rodger did possess a fair degree of mechanical ability and great interest along mechanical lines. Bewildered by academic subjects and frustrated in his efforts to express his natural tendencies, Rodger expressed his interest in an underhanded fashion.

He began by stealing animal traps and concealing the identity of the other boy's traps by taking them apart and assembling parts of various traps to make a new one. Success along these lines encouraged him to more ambitious thefts. Gossip did not reach the father until things got to the point at which the father's position was threatened. The father was beside himself with rage. . . . It took a great deal of persuasion to induce this father to send his son to a trade school, but when he did the behavior difficulties of the boy abated.[3]

Chapter XII presents some findings showing that dissatisfaction with school is the most important reason for pupils' leaving high school. Dissatisfaction with school work usually grows out of a combination of several things, among which are:

(1) lack of ability to do the work,
(2) lack of interest in the work,
(3) influence of the home or some companion in connection with the school program, and
(4) a general personality conflict with those in charge of the school work that must be met in a face-to-face manner.

[3] R. J. Slattery, "Spotting the Maladjusted Pupil," *The Nation's Schools,* 1942, Vol. 30, pp. 45–46.

Truancy from school on the part of one individual and then others is quite often the beginning of mischief that leads to the juvenile courts. Hence the school, by directing a part of its attention to problems related to truancy, can aid considerably in stopping crime at its very beginning. Truancy and delinquency constitute a problem directly related to the program of educational and vocational guidance. Investigations bear testimony to the effects of failure in school work, emotional disturbances that may be aggravated by school situations, and the harmful effects that companionships established at school have on the development of acts of mischief and ultimately delinquent behavior patterns. The school, therefore, in all its drama of social duties and privileges and with all its desire for high standards of achievement, has a greater significance than that of a mere dispenser of a traditional academic education.

The neighborhood. The social environment of adolescents is not restricted to the home and the school; another primary determinant, the neighborhood, also exerts a powerful influence. It was pointed out in an earlier chapter that the school grade and neighborhood were the great determining factors in the choice of chums. Yet the neighborhood is not only an important factor in the choice of playmates; its ideals in connection with the community are forces that determine to a large degree the behavior activities of growing boys and girls.

The various studies reported throughout this chapter have brought forth the close relation between overt delinquent behavior and specific personal and environmental factors. Within recent years attempts have been made to take boys who are "criminals in the making" and train them into good future citizens. Illustrative cases show that efforts have not been in vain; however, too often conclusions relative to a program are based upon faith or wishful thinking fortified by one or more cases of boys whose lives were directed into more useful channels.

An interesting and noteworthy experiment in crime prevention was launched by Dr. R. C. Cabot in the fall of 1937 in the Cambridge, Massachusetts area.[4] The project was a study of delinquency-prevention-selected boys under 12 years of age who had not yet become delinquent. Names of boys from Cambridge and Sommerville who were

---

[4] See Edwin Powers, "An Experiment in Prevention of Delinquency," *The Annals of the American Academy of Political and Social Science,* January, 1949, Vol. 261, pp. 77–88. The January number of *The Annals* contains valuable information relative to the juvenile delinquent, the juvenile court, juvenile detention, and juvenile crime control and prevention.

believed destined to become delinquent were secured during a two-year search, and an equal number were included who were thought to be nondelinquent. This plan called for 650 boys to be divided equally into an experimental group and a control group. To do this, it was necessary to set forth a careful selection and screening process. The matching process was used in securing the two groups. The 325 T boys (experimental group) were then assigned to the counselors. Treatment consisted in the establishment of a friendship relation with the counselor and the application by him of whatever skills he was capable of applying. Thus a wide variety of activities were pursued. Differences in the methods of counselors as well as in their effectiveness were no doubt apparent here. The C boys (control group) received no help of this nature.

A comparison of the official records of the T-C groups made within a few years after the termination of the treatment program revealed that the treatment program was no more effective than the usual forces in the community in preventing boys from committing delinquent acts. The records of the Crime Prevention Bureau show that C boys are more frequently brought in for repeated offenses; more C boys have been sent to more than one institution; more C boys have committed serious offenses known to authorities.

It should be pointed out in connection with this study as well as in relation to juvenile crime in general that many community forces operate to prevent juvenile delinquency. These deterrent forces were operating on the C group as well as on the T group. Since a juvenile criminal is an outgrowth of many forces and conditions, prevention must take into consideration these various conditions. Thus, the counselor who is able to establish rapport with a boy, whether in the scout group, in the classroom, or at the pool hall, becomes one more force influencing his life.

Recreation or delinquency? Recreation and delinquency have much in common. In the early stage, delinquency is usually an acceptable form of recreation or play. The boy who steals a peach from an orchard or who swipes an apple from the fruit stand may consider this a minor act of mischief of a play type. The parent may even be amused over some daring escapade of his adventurous son, recalling some of his own boyhood days of adventure.

However, the daring nature of some forms of play tends to become more pronounced. The juvenile delinquent develops out of these early acts of mischief. Both activities essentially involve groups and adventure.   Concerning the effects of groups and community forces, McKay states:

Both delinquency and recreation are essentially group activities.   Each can be participated in alone, but in the more prevalent and meaningful forms, two or more persons usually are involved.   Each type of activity has a tradition.   Children's groups are the recipients and bearers of tradition governing rules, regulations, and mode of play of a great variety of games and means of entertainment, ranging from the rhymes which are sung while skipping rope to the techniques for playing third base.   Similarly, in those neighborhoods where there are delinquent groups, the members are the recipients and bearers of a tradition on such subjects as how to break into a car, shoplift from a store, or avoid a policeman.   The latter groups may be the recipients, also, of the conventional traditions.[5]

Despite the fact that delinquency groups and their activities are quite similar in nature to those of recreational groups, from the standpoint of the larger community and the development of future citizens there are some pronounced differences.   Delinquent boys motivated by adventure might be very destructive to an old, vacant building; the group of boys guided by the desire to establish a recreational center may be motivated to repair and remodel it for their general use.   In this connection, it should be pointed out that the the likelihood that the random behavior of a group of boys might become a force for constructiveness rather than destructiveness will vary with the extent to which the community has provided for the needs of these boys along conventional lines.   We see in this case that both recreational activities and delinquent pursuits are carried on by groups.   Both are motivated by certain needs.   Both comprise leaders as well as followers.   It might be said that in the small town, at least, both would represent a cross section of the boys of the town.

There is no inherent reason that one group of boys turns to delinquency while the other group engages in wholesome recreational pursuits.   Experiences on the part of various organizations that have concerned themselves with the welfare of these boys and their development into useful citizens show that, when they are given the proper guidance and when their needs are given adequate consideration, potential delinquent-producing areas become builders of good citizens.

Recreational facilities are being developed more and more, and it has

[5] H. D. McKay, "The Neighborhood and Child Conduct," *The Annals of the American Academy of Political and Social Science,* January, 1949, Vol. 261, pp. 32–41.

been well demonstrated that directed activities through recreational programs will do much toward thwarting the adolescent pranks and mischief that may not appear bad in themselves but are often quite costly and, still worse, lead too often to dire consequences. The ways in which some communities direct the energies of boys and girls is well illustrated in connection with Halloween activities. The mischief associated with the mystic orange-and-black traditions has in many cases been very expensive as well as annoying. Many cities organize costume parades, various types of contests, and directed games. The picture on this page shows a crowd of children from the schools of Kansas City enjoying themselves during a picturesque Goblin parade.

A wholesome Halloween occasion.

THE PROBLEM OF RESTRAINT

Discipline and character. Closely related to the general problem of remedial treatment is discipline. Discipline in connection with antisocial behavior in school, in the home, and on the playgrounds is usually thought of as related to the milder forms of antisocial behavior. Thus, the breaking of some rule at school, the infringement upon the good will of some other member of the home or school, many acts of mischief, and other forms of behavior many of which are not necessarily antisocial behavior manifestations, are considered by someone in authority as undesirable and thus the subject concerned is disciplined by some means. The problem of discipline as it relates to the development of conduct in harmony with the mores of the group has been recognized in all emotional processes. Needless to say, the method of punishment has varied considerably from period to period. Not quite a century ago a rather detailed plan of discipline was established in our secondary schools. The following is a partial list of punishments that were in effect in an academy in Stokes County, North Carolina, in 1848:[6]

1. Boys and girls playing together	4 lashes
2. Quarrelling	4 "
7. Playing at cards at school	10 "
9. Telling lies	7 "
14. Swearing at school	8 "
16. For misbehaving to girls	10 "
19. For drinking liquors at school	8 "
22. For wearing long finger nails	2 "
31. For blotting your copy book	2 "
33. For wrestling at school	4 "
35. For not making a bow when going out to go home	2 "
43. For not saying "Yes sir" or "No sir," etc.	2 "
45. For not washing at playtime when going to books	4 "
46. For going and playing about the mill or creek	6 "

Modern conceptions of child training lay stress on the fact that morality is not developed by rules, creeds, dogmas, or the establishment of specific amounts of punishment for various acts of mischief. If the disciplinary act strikes deep into the innermost life and feelings of the individual and leads him to recognize that the antisocial behavior act will not be tolerated, probably some good effects will result. But

[6] C. L. Coon, *North Carolina Schools and Academies: A Documentary History,* p. 763. (State document, 1915.)

too often discipline is looked upon by the adolescent boy as a punishment for getting caught or as a means set forth by the teacher or parent for paying for some behavior act—a form of vengeance.

Bad habits are not usually formed overnight; neither are they likely to be broken in so short a period. Like other forms of behavior patterns, changes in conduct follow the general laws of learning and occur gradually. Parents often express amazement at the apparent onset of some maladaptive form of behavior on the part of the growing boy or girl, but usually this maladaptive form of behavior has not been so sudden as it appears. Here is, in most cases, an illustration of the failure of the parent to understand the other habits that have been established prior to the appearance of unadaptive habit. Discipline, if it is to be of value, must (1) be administered in terms of the past life of the child, (2) be based upon understanding rather than emotions, (3) be understood by the subject concerned, (4) relate to the behavior act from which it resulted rather than to the one administering the act, and (5) follow immediately after the act. Discipline is related to conduct in that, through purposive activity, habits of a desirable nature are established and maintained. Discipline is therefore directly related to self-control, and in this all discipline should have both its beginning and its ending.

Guidance in relation to other groups. If the companions and the play life of the individual are so important in the development of desirable or undesirable behavior traits, it is well to attempt to control, at least in part, these factors. Probably the greatest value to be attained from the adolescent's attendance at Sunday school is the fact that he is likely to be grouped with children who possess desirable behavior patterns. Yet a single child in a community can and often will interfere with the development of the proper habits in the other children; hence supervised play and the general supervision of the activities of adolescents should have as their main purpose the organization of a group into wholesome and desirable activities. Any member tending to interfere with the development of desirable habits in the group should be so supervised that his activities will not help to develop undesirable attitudes in the lives of the other members of the group.

Today we are studying more carefully the problem of juvenile delinquency, and it is fairly well recognized that if society is to do any constructive work for this group of young offenders, it must segregate them from those with more firmly established habits of an undesirable nature. The truant and the juvenile delinquent have not as a rule developed such habits as are firmly established and beyond modification. Still it must not be overlooked that the influence of members of the group upon each other is also likely to be detrimental. Probation, with reward for good conduct, should be offered in our dealings with adolescents who have established undesirable traits. We should "make sure of the nature of the urge back of an undesirable act and then . . . furnish the child with a more desirable outlet at the same time that the undesirable one is blocked." [7] The delinquent child can be trained, for he is a plastic individual in whom undesirable habit patterns have been developed. Certain motivating forces are already operating in bringing about specifically desired ends; once we decide upon what ends ought to be desired, through substitutions desirable activities can well take the place of undesirable ones.

In the treatment of the juvenile offender two problems are encountered: (1) the welfare of the group and (2) the restraint of the offender. The first of these is primarily a social problem; and since our social structure is so definitely related, this should receive major interest and effort. But the habit system of the juvenile offender is plastic, and through proper guidance and training he may well be made an individual who will take his place in society as a desirable citizen. The restraining phase is to be considered as an individual rather than a social problem.

Satisfying activities. The value of the positive phase of conduct has already been emphasized, especially for the endeavor to establish desirable forms of behavior. Various types of rewards, either direct or indirect, are constantly being introduced in the effort to relate the element of satisfaction to the performance of the desired act. When undesirable behavior is allowed to bring about satisfaction or reward, this will naturally be the form of behavior established. Situations should be so set up that there is a natural reward for doing the desirable thing. This reward, as it concerns adolescents, may be of an abstract nature involving ideals and attitudes of a worth-while and wholesome type. It has been found that beliefs are directly related to desires; thus one can well say that ideals are directly related to desires. Desires are established in part through a conditioning and directing of the natural impulses of the individual along lines in harmony with the ideals set forth by the group. Desires can and should be guided; but this guid-

[7] J. J. B. Morgan, *The Psychology of the Unadjusted School Child.* New York: The Macmillan Company, 1924, p. 289.

ance cannot best secure its end unless the desires are established in relation to situations for which there is an ultimate reward or form of satisfaction.

Adolescents should gradually learn through experience that antisocial conduct leads to their own misery and unpleasant experiences. It appears likely that one of the chief difficulties met by pre-adolescent and adolescent boys is the lack of men to idealize, who understand boys and their problems and are able to win their confidence and admiration. Men teachers, physical education and recreational directors, 4-H Club leaders, the vocational agriculture teacher, the boys' counselor, and others who work closely with boys have opportunities to influence their ideals and attitudes significantly. They meet these boys at a period when masculine contacts are desperately needed.

Self-realization. The juvenile delinquent should not be led to believe that he is suffering from a condition beyond cure. He should realize that he is indeed very similar to those who are not classed as delinquents. What he should know is his point of weakness. The concept of self-analysis brought out in an earlier chapter should here again be introduced and practical results obtained from its administration. The child should be led to realize that happiness, reward, and ultimate success are to be gained through desirable traits. He should come to realize that those dealing with him are neither spies nor policemen but individuals interested in the welfare of the group.

Again, the responsibility for the reward and satisfaction to be gained from group participation should be placed in the hands of the individual concerned. Those in charge should lead him to realize that he is somewhat on probation, and that he is expected to try to do the right thing in order that he may be happier and the group be generally better off. The important thing is to realize the point of weakness, to recognize that habits are built up through practice, and to be motivated toward the strengthening of good habits.

The adolescent is likely to resent authoritative control. The self-conscious attitude so clearly displayed at this stage of life tends to mark him as an individual on the alert, watching for someone to consider him as a child and thus boss him around. He is idealistic in nature and expects the teacher to play fair with him in his activities; he may question many of the procedures of the teacher for this reason. His personal manner of regarding everything as directed toward the self is a factor that should be watched. The adolescent is impulsive, oversensitive, and impressionable to mistreatment or unfair dealings. He will readily respond to group treatment and approval because of his developed social consciousness.

Need of follow-up work. In the Oaks School, juvenile delinquents of borderline intelligence are trained, supervised, and studied. The aim is to develop in the boys a sense of responsibility to one another and a desire to become socially acceptable, as well as to develop their abilities in order that they may be more nearly self-supporting. The Gluecks made a follow-up study, after an interval of from 10 to 15 years following the expiration of their sentences, of as many of 500 original juvenile delinquents as could be located (439 of the 500 were available for study).[8] The environmental circumstances, family relations, work history, use of leisure time, and antisocial (criminal) conduct were among the factors studied in this investigation. The traits of those who survived the period without being brought before the courts again were meticulously compared with the traits of those who did not. It was noted that 44 per cent of the group studied had had institutional experience during this period, and that over half had been arrested for various offenses. The study revealed a lack of adjustment on the part of those who appeared before the courts during this interim; about two-thirds were found to possess "abnormal mental conditions."

One might well conclude from these results that the adoption of a philosophy of punishment based upon the severity of the crime committed fails to transform the juvenile delinquent into a well adjusted personality. There is a need for law enforcement officers who will give more consideration to the offender than to the offense. Indeed, there is a pronounced trend among juvenile-court judges to consider the individual subject, and to fit the punishment to the individual, with the ultimate goal of rehabilitating the delinquent as well as of protecting the welfare of the group.

The change in attitudes toward juvenile crime is having its effect upon the treatment of juvenile offenders. Social welfare workers have found a close relationship between home and neighborhood conditions and juvenile conduct. The beginning point for most corrective treatment should be the home, and some juvenile-court judges, recognizing this fact, have placed the responsibility on the parents in various ways. The ignorance, indifference, carelessness, and distorted sense of values

---

[8] Sheldon and Eleanor Glueck, *Criminal Careers in Retrospect.* New York: Commonwealth Fund, 1943.

of parents account for a large share of juvenile crime. This circumstance presents a challenge to society to conduct an adult educational program dealing with child training. The schools must accept part of the responsibility—since these parents were in school just a decade or more ago—and the school program provided very little training in parenthood.

Again, educational adjustments are essential, if the adolescent boy or girl delinquent is to remain in school and out of mischief. This fact is revealed by the data presented in Table LV.[9] Physical well-being

TABLE LV

Suggested Treatment Procedures for Juvenile Delinquents

	Index Total
1. Adjustment of home situation	3.38
a. Social or educational work in home	1.48
b. Advice regarding methods of training	1.32
c. Consideration of placement	.40
d. Suggestions regarding sibling relationships	.17
e. Interests	.01
2. Educational adjustment	2.64
a. Modification of curriculum and instruction	1.37
b. Classroom management	.67
c. Placement and progress	.39
d. Special individual guidance	.21
3. Concerning physical well-being	1.46
a. Specific treatments	.71
b. Supplementary examinations	.53
c. Operative therapy	.22
4. Social adjustment in home, school, and community	1.02
a. Opportunities for adequate social relationships	.52
b. Development of recreational and other special interests	.27
c. Opportunity for employment	.12
d. Special summer program	.09
e. Enlistment of community aid	.02
5. Miscellaneous	.21

and adjustment to the community are other conditions related to the prevention and treatment of juvenile crime. These things must be taken into account in the training program for delinquents. The value of the training program will depend in a very large measure upon the follow-up work. If boys and girls are sent back into the community with a stigma attached to them and if the people of the

[9] Norman Fenton, "Treatment Procedures Suggested by Bureau Workers," *Journal of Juvenile Research*, 1937, Vol. 21.

community look upon them with disfavor, fear, or even curiousity, they will surely not be given the aid and encouragement they will need in order to follow a desirable course of behavior. Again, if they are sent back into a home where the same unfavorable attitudes prevail that were such prominent factors in affecting the development of delinquent trends, one might expect an early return of such trends. It may be stated that those conditions sufficiently undesirable to produce behavior patterns will in most cases be of sufficient quality and quantity to effect the reappearance of such a pattern.

Whatever method society might endeavor to formulate in dealing with the delinquent adolescent should furthermore embody the following general concepts:

1. Reward for the desirable act and a form of disapproval and probably punishment for the undesirable act.

2. A self-realization on the part of the individual of his possibilities for good behavior.

3. A segregation of those with undesirable traits in order to eliminate their influence upon the individual with desirable traits.

4. A separate method of control for the feeble-minded.

5. Worthwhile and probably gainful activities in harmony with certain needs and interests of the subjects concerned.

6. Punishment should not be administered in a spirit of anger or vengeance.

7. Confidence and fairness of those dealing with the subjects as to the outcome of efforts directed in the right manner along the correct path.

THE MORAL SELF

Attitudes and moral behavior. The younger the child, the more difficult it is to label certain behavior activities as moral and others as acts of behavior performed, without any special intent or purpose, because of habit patterns built up through practice. This fact may be illustrated by an example of a certain act of behavior performed by two boys. The one child may be unwilling to play with a toy of another because he has built habits, as a result of training by his parents, to play only with his own toys. The other child may be unwilling to play with the toy, too, not, however, because of a sort of blind habit-pattern formed, but because he has been taught that it would make another unhappy if he took his toy from him and played with it. In this connection it should be pointed out that moral and religious behavior grow gradually, thus resembling true friendship, social understanding, and social attitudes. In fact, the latter are vital elements in moral and religious development, and are important in the development of a child's religious concepts. Support for this view is found in one of the great religious precepts, "Love thy neighbor as thyself," for complete understanding and practice of this command can be attained only through ripened and guided experiences.

An ideal that has become more or less a part of the individual will be reflected in his attitudes and behavior. The ideal, when interrelated with attitudes and behavior, acts as an inner drive in controlling human behavior. An established attitude is a habit system built out of lesser habit systems into a hierarchy, in which generalizations are present. The boy who is taught to be courteous to his mother and has established this as an attitude rather than a stereotyped habit does not need to generalize so much when he faces his teacher. The child who has developed the attitude of truthfulness in his home will meet other social situations with an attitude of truthfulness.

Several problems of importance are encountered as we study the moral life of the adolescent: (1) What are the desirable attitudes that the home, school, church, and other agencies should strive to establish? (2) What specific habit patterns, when integrated, tend to produce such ideals and attitudes? (3) How can these specific habits best be acquired and integrated into a general attitude?

Moral development. The most important place in a list of environmental factors influencing moral behavior for most children is that of the home. From the period of their first perceptions they look first of all to their parents for guidance by precept and example. A child gets his first impressions in the home, and these impressions are made during that period when the foundational habits and moral attitudes are being formed. Undoubtedly, a good home is the greatest asset, and a bad one the greatest liability. Moral and religious values will be found in concrete social relations of daily living in a growing and expanding life rather than in meaningless creeds or stories or emotional exhortations. "Children who have immoral surroundings, whose struggle to exist involves corrupt practices, whose whole horizon is dark with foreboding shadows cannot have healthy social attitudes." [10]

There is much evidence that Sunday school and classroom instruction, which have relied largely upon verbal teachings, have been ineffective in meeting the moral demands of modern life.[11] Moral de-

[10] E. J. Chave, *Personality Development in Children.* Chicago: University of Chicago Press, 1937, p. 270.
[11] Hugh Hartshorne, *Character in Human Relations.* New York: Charles Scribner's Sons, 1932.

velopment, like the development of social habits and attitudes, will be most effective when it takes place in connection with situations arising naturally in the classroom or on the playground. The Sunday school can teach appreciation of one another and respect for the rights and feelings of others; but if this is done in a vacuum, and children see no relation between such teachings and the problems they meet on the street, at school, and in the park, the teaching will be so much babbling.[12]

Another essential in moral teachings, if they are to be effective, is the harmonious correlation of all agencies affecting the moral life of boys and girls. The concepts presented in the home, on the playground, in school, and at church are usually too unrelated to have any great functional significance. The program of the church is in so many cases too far divorced from the other interests of the child, and the materials presented are too archaic to have any meaning for him in connection with present-day living. There is evidence, however, of a change in the materials and methods of the church. Adolescents do not want the church to lose its significance as a place of worship and reverence, but they do want the church to aid them in solving some of their major problems. Some of these thoughts will be presented at a later point in this chapter. A survey of materials and methods used in Protestant religious education a number of years ago indicated that children were taught too much through the formal question-and-answer method, so widely used in our day schools, and that there was very little discussion of the bases for current concepts of God.[13] The child's ideas of God are somewhat realistic and concrete in nature; children are interested in His physical being and place of abode. At this stage He may be thought of as a powerful man, very like a king, who sits on a throne and wields great power. The child's ideas and questions become more involved as he grows intellectually; this is to be expected, and is an indication of religious growth. The concept having the most mean-

[12] H. Hartshorne and M. May, *Studies in Deceit*. New York: The Macmillan Company, 1928, p. 411; H. Hartshorne and M. May, *Studies in Service and Self-Control*. New York: The Macmillan Company, 1929, p. 268. These studies are based upon results from 11,000 children 8 to 16 years of age. They show that there are no significant relationships existing between Sunday school attendance of these children tested and their willingness to cheat, give services to others, and cooperate with others. One should be careful in generalizing from this as to the basis for this lack of correlation; however, the association is interesting and should present a challenge.

[13] A. H. MacLean, *The Idea of God in Protestant Religious Education*. New York: Columbia University Press, 1930.

ing and giving the greatest satisfaction at any particular stage of life is the one most acceptable to the individual at that stage.

During the earlier period of life the individual is neither moral nor immoral; he is to a large degree unmoral. Whatever his conduct may be, it is largely the result of simple forces that have played upon and thus conditioned his behavior during the earlier years of life. This is not so true of later adolescence, for now the period of habitual morality has closed. Whatever his actions may be, we are certain that the adolescent is thinking and reacting to various situations in terms of ideals that are being established as a unified part of his personality. Then it is not so strange that he turns part of his newly acquired abilities and interests toward problems more far-reaching, involving common ideals, and pertaining to conduct.

This was clearly shown in a survey of the opinions of 5,500 high school students (the majority of whom were seniors) from the state of Washington.[14] The problems relating to the morals, religion, ideals, and the future most commonly checked by these high school youths are presented in Table LVI. These results indicate that moral concepts and problems came into prominence at this age that had not been felt and recognized earlier. Students appeared to desire to achieve a better understanding of themselves, their behavior, and the actions of others.

Religious education. Religion, if properly taught, certainly would help young people to grasp the meaning and values of life. Too frequently the truth that our religion is evolutionary, that religion is still in the making, is not made clear to youth. The real assault upon religious opinions is not made by scholars but by the daily life and experience of the common people. Contact with any life situation tends to develop new interpretations of so-called spiritual matters. New standards of living mean the visualization of new meanings in religion. In this connection, Kuhlen and Arnold have set forth two implications from their study that should be of interest and value to those concerned with a religious education program. These are:

First, those issues represented by statements which are increasingly 'wondered about' as age increases may give clues as to appropriate topics for consideration in the teen years in both Sunday School classes and young people's groups. Second, beliefs discarded by children as they grow older may well be studied for their implications for teaching at earlier ages.

[14] L. J. Elias, *High School Youth Look at Their Problems.* Pullman, Washington: The State College of Washington, 1949.

Children's concepts regarding religion are more concrete and specific than are those of adults, the latter tending to be abstract and general. This change represents the normal growth of concepts. It would seem desirable that the specific and concrete beliefs taught to children be beliefs compatible with the more abstract adult views, and not beliefs later to be discarded because of incompatibility.[15]

TABLE LVI

The Frequency of Responses of High School Youth to Certain Problems Relating to Morals, Ideals, Religion, and the Future (*After Elias*)

PROBLEM	PER CENT		
	Boys	Girls	TOTAL
Making something of myself	27.2	33.0	30.4
Worrying about mistakes I've made	14.5	23.3	19.3
How to do my best	11.9	18.3	15.4
Concerned about life and death	9.7	18.9	14.7
What's happening in the world	11.4	16.6	14.2
I swear too much	20.8	8.3	14.0
Puzzled about religion	13.5	14.4	14.0
Worried about some bad habits	16.3	11.3	13.5
Understanding things people do	6.9	14.9	11.3
Learning how to enjoy life	11.5	10.7	11.0
Students cheating in school	8.3	12.6	10.6
Having high ideals	8.0	11.6	10.0
The morals of my crowd	6.0	11.0	8.7
Losing faith in religion	9.6	7.4	8.4
Kind of life kids lead	6.2	10.3	8.4
Not facing problems squarely	7.8	8.2	8.0
Worried about my reputation	7.5	7.3	7.4
Prejudice and intolerance	4.8	9.1	7.2
About going to church	6.7	6.8	6.8
Embarrassed by friends' action	4.7	7.6	6.3
Worried about my morals	6.4	4.3	5.2
Always alibiing	4.2	3.4	3.8
Religion and school conflict	2.7	4.8	3.8
Interested in dirty stories	5.4	2.0	3.6
People are cruel and selfish	2.9	2.2	2.5
Getting into trouble a lot	3.4	.9	2.0

The effectiveness of the child's religious training is contingent upon the application of the important facts and principles of learning in all religious teaching in home, school, or church. To say there is no need

[15] R. G. Kuhlen and A. Arnold, "Age Differences in Religious Beliefs and Problems during Adolescence," *Pedagogical Seminar and Journal of Genetic Psychology*, 1944, Vol. 65, p. 297.

for religion in an age of science is a distortion of the aspirations, desires, and values of man. Children need to develop a concept of the purpose of life and faith in worthy and desirable human relations. There is a need today, because of our dependency on each other and increased group activities, for a religion that will transform empty words and dogma into a fuller realization of spiritual values through adventurous living and experiences. Tuttle pointed this out when he stated: "To the degree that religious training emphasizes the cultivation and application of values, rather than the acceptance of dogma, the full potentialities of religion as a motivating force will be realized." [16]

The following set of standards by Hartshorne and Lotz [17] is an excellent formulation of criteria for evaluating religious instruction in weekday schools and Sunday schools:

1. The pupils show increasing respect for one another and for those with whom their activities bring them into real or imaginative contact.

2. The pupils are in real situations and are responding to the situations rather than to the teacher, for it is the function of the teacher to bring the pupils into vital relationship with these situations.

3. The situation, while continuous with out-of-school situations, is simplified so as to make possible the maximum freedom of the child without confusion or disaster.

4. The pupils view the situation objectively rather than through their prejudices and emotions.

5. Those phases of experience which are primarily acts of appreciation are so handled as to permit the children to make their own evaluations and to compare their judgments with those of others.

6. In facing new situations, the process of thought is such as to lead to valid conclusions. That is, the scientific method is used.

7. In facing new situations, the pupils make use of relevant past experience, so far as they can gain access to it.

8. Problem solving includes foresight of consequences of various possible procedures and a choice of one or the other in terms of their believed harmony with the general direction of the life unit or phase of which it is a part. When issues are critical such evaluation takes the form of worship, and is in terms of the value of persons.

9. The conclusion of a project is the occasion of measurement of progress in skill and appraisal of results in terms of objectives. This latter may involve worship when the results are of sufficient importance.

10. The pupils' responsibility includes the experiencing of the results of their experiments as well as the planning of them.

[16] H. S. Tuttle, "Religion as Motivation," *Journal of Social Psychology*, 1942, Vol. 15, pp. 255–264.

[17] *Case Studies of Present-Day Religious Teaching*. New Haven: Yale University Press, pp. 8–9.

Building spiritual values. It has already been pointed out that as the individual emerges from childhood into adolescence new outlooks appear and new moral concepts must be established. This does not mean that all early teachings are to be laid aside, but rather that with the growth of independence the individual must accept responsibilities and meet problems formerly met with the aid of others, as well as new problems. He must now adjust to making decisions where he is largely the judge of whether the behavior act is right or wrong. He must furthermore adjust himself to behaving in accordance with group standards of conduct as well as his own.

The development of moral concepts and spiritual values is not carried on in a vacuum. Neither are these developments unrelated to the mental, social, emotional, and physical development of the adolescent. The modern school emphasizes the growth of the total child, and is concerned with his total development—including spiritual development. Through associating with his peers and teachers under ideals set forth for living together in the school community, the child learns to live with others, to have consideration for their feelings, rights, and happiness, to gain satisfaction from achievement by the self or the group, to understand the orderliness of nature, and to recognize and accept ideals for guiding his daily activities. He learns how he can be helpful and how others can be helpful to him. He builds a framework of values from which to judge himself and his peers.

Values are learned in the same manner as other learnings, through experience. However, the nature of the values a child acquires must not be left to chance. The individual pupil has many needs in common with lower forms of life. In addition, he has insights, aspirations, and possibilities for learning and development that are distinctly human. Because of this he is capable, through experiences, of acquiring habits of initiative and responsibility relative to his own behavior. This gives him a measure of self-control not to be found among lower forms of life. Ideals of honesty, fair play, and consideration for the feelings of others are acquired through experiences with others in situations in which such ideals are guiding forces. These ideals become a part of the growing child and are controlling forces in his conduct. The importance of parents, teachers, and peers in the formation of ideals and the development of spiritual values has been shown through countless studies of the growth and development of boys and girls from childhood into adolescence and finally through adolescence into maturity.

SUMMARY OF PRINCIPLES

It may be stated that delinquency and maladjustments are factors resulting from a multiplicity of elements in the hereditary make-up and life history of each individual. It was stated in connection with "The Hygiene of Adolescence" that each case represents a distinct problem. The same statement should be made with reference to delinquents. It was pointed out in the preceding chapter that among the causes of delinquency are: low mentality, poverty, undesirable home conditions, bad parental attitudes, lack of facilities for wholesome recreational activities, bad companions, and emotional instability. We might list more specifically the strain placed upon adolescents by trying to make adequate adjustments to the existing social order. Any program of prevention and control must take these factors into consideration if it is going to be effective.

As the world of the adolescent expands and new drives appear, he is faced with many problems that he has not yet established adequate means for meeting. He will therefore need to seek aid and advice from those with a greater breadth of knowledge and experience, and it is in this capacity that the counselor should be able to work most effectively. In doing this he should keep in mind the following points: (1) He should recognize the adolescent as an individual and give serious consideration to the fact that the problem is one that belongs to him and may be very important in his life, even though it is of negligible importance to the adult counselor. (2) In beginning a conference with the adolescent, it is well to state some of his positive virtues first, and if possible make use of these in aiding him to establish confidence in himself and his possibilities for good behavior. (3) A specific plan of attack is most desirable. Just as the doctor makes his attack directly upon the disease or organ affected, so the counselor in many cases should attack the disorder itself. It is not enough to say to the boy or girl, "You should speak out in class"; on the contrary, the pupil must be given an opportunity to do this and should come to recognize that it is expected of him. (4) There should be frequent rechecking by the adviser. The adolescent needs a continued renewal of encouragement and guidance. (5) The degree to which cooperation can be secured and the extent to which responsibility can be placed upon the adolescent depends in part upon his physical, mental, and emotional maturity. Methods, therefore, should vary according to the maturity of the individual concerned.

The adolescent must be taught to have respect for himself and his abilities. He must come to realize that he has a definite contribution to make to society. If there are physical defects, he must learn to overcome these—not to overcompensate or use some defense mechanism in an effort to cover them up. The mental-hygiene principles set forth in the previous chapter should be followed if he is to develop the ability to function harmoniously in his social relations. Guidance and control should have as their function the bringing out of those qualities and assets of the individual that will be of greatest service to the self and society.

THOUGHT PROBLEMS

1. Just how is the control of the adolescent's behavior activities related to mental hygiene? To moral and religious growth? Illustrate.

2. Why is the cooperation of all agencies essential if juvenile crime is to be more closely controlled? What are some of the agencies that would be involved?

3. Give an illustration, from your own observation, of how juvenile mischief has been directed into more wholesome channels through the development of recreational activities for adolescents.

4. Show how self-realization is important in the development of a well-adjusted individual. What are some needs of the adolescent, if he is to develop a wholesome attitude toward himself and others?

5. Study the treatment procedures suggested for juvenile delinquents that are presented in Table LV. Which of these are being neglected in the general procedures of today? How would you account for this?

6. What do you understand the term *spiritual values* to mean? Can you see any relation between spiritual values and attaining a consistent and unified philosophy of life, referred to in the previous chapter? Explain.

SELECTED REFERENCES

Averill, L. A., *Adolescence*. Boston: Houghton Mifflin Company, 1936, Chap. XIII.

Baker, Harold J., "Spiritual Values Give Life Its Highest Meaning," *Nineteenth Yearbook of the Elementary Principal,* 1947, Vol. 27.

Brill, J. G., and Payne, E. G., *The Adolescent Court and Crime Prevention.* New York: Pitman Publishing Corporation, 1938.

Carr, Lowell J., *Delinquency Control.* New York: Harper and Bros., 1941. A handbook presenting data relative to the factors involved in juvenile crime. It points a way to the solution of juvenile crime by combining (1) scientific research, (2) adjustive and preventive techniques, (3) the art of social action, and (4) the art of social organization.

Chave, Ernest J., *A Functional Approach to Religious Education.* Chicago: University of Chicago Press, 1947. This book is written from an educational point of view. The purpose of religion in relation to everyday problems is clearly revealed in the presentation.

Fedder, Ruth, *Guiding Club and Homeroom Activities*. New York: McGraw-Hill Book Co., 1948.

Garinger, E. H., *The Administration of Discipline in the High School*. Bureau of Publication, Teachers College, Columbia University, 1936.

Hatfield, M., *Children in Court*. New York: Paebar Co., 1938. Condemning the "old court system" for falling short in its purpose of correcting youthful offenders, a juvenile-court judge offers, nontechnically and non-statistically, his personal observations of children in court in the form of anecdotes and brief commentaries.

Healy, William, and Bronner, A. F., *New Light on Delinquency and Its Treatment*. New Haven: Yale University Press, 1936.

Healy, William, and Bronner, A. F., *Treatment and What Happened Afterwards*. New York: Judge Baker Guidance Center, 1939.

Hurlock, E. B., *Adolescent Psychology*. New York: McGraw-Hill Book Co., 1949, Chap. XI.

Rogers, Carl R., *Counseling and Psychotherapy*. Boston: Houghton Mifflin Co., 1942. Through the nondirective counseling procedures, Rogers emphasizes the importance of the acceptance of the individual as a person different from the counselor, and the further acceptance of his right to be different.

Williamson, E. G., *Counseling Adolescents*. New York: McGraw-Hill Book Co., 1950. This is a splendid source for studying counseling techniques. The author has related the methodology and techniques of counseling to the philosophy and goals of a democratic society.

XIX

Educational Needs of the Adolescent

WHAT GOES INTO THE TRAINING OF YOUTH EMERGES IN THE LIFE OF THE NATION. *(Message of President S. C. Garrison to the graduating class at George Peabody College for Teachers, August 22, 1939.)*

Education and individual development. The cardinal principles of education state that in the school we are striving to inculcate in the child sound health habits, to give him command of fundamental processes, to prepare him for worthy home membership, to educate him to be a good citizen, to help him to choose and prepare himself for a vocation, to teach him to use his leisure time wisely, and to build his ethical character. This is indeed a large order and one that the school cannot and does not hope to do alone. However, with the increasing changes in home life a larger share of the burden has been thrown on the schools, and it is a share they must accept. We know that a large per cent of our adolescent population is now registered in our schools. If the schools are to be able to do their part in fulfilling the objectives set forth, those entrusted with educational responsibilities must have a thorough understanding of the nature and characteristics of children and adolescents. The curriculum of the school should be based upon the nature and needs of growing boys and girls. It should evolve around (a) their personal development; (b) their personal-social development; (c) their development toward social-civic efficiency; and (d) their growth in civic-vocational efficiency.

The understanding of the human material with which he must work and the development of adequate materials and effective techniques for shaping it are essential to the educator in the performance of his task. However, before he can do an effective job in developing these human resources, he must have a clear-cut picture of the goal toward which he is working. It is in this respect that the problem of providing a training program for tomorrow is more difficult than at any previous time in our history. Although our history is filled with events that betray fast-moving changes, the revolution that is now underway in our ways of living is more vast and far-reaching than any that took place before. It should be unnecessary to say in this connection that such changes are rendering the simple conditioning techniques of the past inadequate. And, although it is true that no culture remains completely static, there are vast differences in the rate of change. A realistic educator will not attempt to oversimplify the problems that have arisen as a result of technological developments. Neither will he be able to offer a complete solution to the problems that are appearing on the horizon. He will, however, recognize that there are many situations today for which no ready-made answers are available. The cultural patterns that are transmitted from one generation to another will not be sufficient for the solutions of the problems that will arise during the next decade. It becomes the task of the educator to study these problems intelligently and to be better prepared to meet them as they arise. He should strive further to train boys and girls to recognize the interrelations present in various world-wide problems and attempt to meet them with intelligence and insight. Whether or not intelligence can be developed is still a controversial question, but the possibility of training individuals to use their innate intelligence in the solution of problems is well known.

There is a growing recognition of individual differences in abilities, interests, and needs, and this recognition is affecting the activities of the schools to the extent that educational programs are being organized to suit the needs and conditions confronting the pupils; this, rather than compelling the students to conform to some rigid system designed and controlled by standards. Pupils with mediocre ability in studying literature in school should not be deprived of the opportunity to develop a richer life of appreciation because of some standard of conformity imposed upon them in the name of high standards. High standards of scholarship are desirable, but they should not be used to deprive students of opportunities for desirable educational growth in harmony with their individual abilities and needs. One of the major educational problems that has arisen as a result of developments in genetic psychology and measurements is that of providing materials of varying degrees of difficulty, and letting the pupil develop unhampered by standards impossible for him to attain.

Why pupils leave high school. The great increase in high school enrollment was pointed out in Part I. This wide distribution is in contrast to the early history of education in this country, when only a selected few were fortunate enough to secure more than an abbreviated elementary education. As late as 1880, after public high schools had been established very widely, only about 3 per cent of the youth of high school age, or about 110,000 persons, continued their education beyond the elementary school period.[1]

The American theory of education maintains that the public school system, extending through the high school, is designed to serve all; yet at least half of those who reach the fifth grade drop out before completing high school. Too often these drop-outs are looked upon merely as statistics, rather than as growing boys and girls who will be assuming increased responsibilities tomorrow. A study of the reasons these young people left school will reveal many unrecognized conditions that influence the motives of growing boys and girls. It was with the idea of studying these motives that the United States Department of Labor, in the spring of 1947, interviewed a sample of young people in Louisville, Kentucky—524 boys and girls, 440 of whom had not yet completed high school.[2] The questions asked of these young people dealt with their educational background, reasons for leaving school, work experiences, and plans for the future.

The survey showed that dissatisfaction with school was given as the major factor by 47.7 per cent of the young people, while economic need and lure of a job ranked second and third in importance. These factors accounted for 31.1 per cent of the entire group. A further analysis of the secondary reason for leaving school showed that over two-thirds of the group interviewed left school wholly or partly because of some dissatisfaction with some phase of school life. The main reasons were failure, dissatisfaction with courses, and dislike of the social situation. The number giving these and other elements in the school situation as the reason for leaving school is given in Table LVII.

Reticence of youth to include poverty as the major reason for leaving school may have been responsible for an underemphasis on lack of money as the major cause, in comparison with the number listing dissatisfaction with school. Owing in a large measure to our emphasis upon individual freedom and individual responsibility, economic hard-

[1] Grayson N. Kefauver, Victor H. Noll, and Drake C. Elwood, "The Secondary-School Population," *National Survey of Secondary Education Monograph No. 4,* United States Office of Education Bulletin No. 17, 1932.

[2] Elizabeth S. Johnson and Caroline Legg, *Why Young People Leave School.* National Association of Secondary School Principals, 1948.

ships are often taken for granted as a necessary and continuing element in the school-leaving situation. There are some who would rationalize the situation by merely stating that "If a boy or girl has it in him,

TABLE LVII

OUTSTANDING ELEMENT IN REASONS FOR LEAVING SCHOOL
AS GIVEN BY NONGRADUATES

Nature of dissatisfaction *	Young people who gave dissatisfaction with school as—		
	Principal Reason for Leaving	Contributory Reason for Leaving †	Either Principal or Contributory †
TOTAL	209	84	293
Failing grades—discouraged	38	22	60
Dissatisfied with courses	29	25	54
Disliked teachers or teaching methods	25	40	65
Disliked social relations, or the non-coed system	13	23	36
Unable to adjust after transfer	8	2	10
Thought disclipline too severe	5	4	9
Other miscellaneous reasons	17	16	33
Disliked school generally—no specific reason given	74	33	107

* Excludes dissatisfaction specifically due to lack of personal funds, which is included with economic reasons.

† In this column one individual may appear one or several times, according to the number of ways in which dissatisfied; hence the figures add to more than the total here shown.

he will be able to rise above such hardships and obstacles." The importance of financial circumstances in relation to a boy's educational opportunities is illustrated in the cases of Harold and Tracy.

Harold, a better-than-average student, was in the tenth grade at the age of fifteen, when his father deserted the family. Harold's mother was working but did not earn enough to support herself and the boy. So Harold left school and obtained a full-time job in a chain grocery at $20 a week. Clinging to his ambition to become a lawyer, he enrolled in the academic course at night school. This double load was a serious tax on his strength, and when interviewed in the early spring, it seemed doubtful whether his health would hold.

When Tracy's father became ill and was unable to work, the family rented rooms to pay for the rent of their house, and Tracy's mother did washing and ironing to earn money but could not earn enough to pay for their food. So fifteen-year-old Tracy left school at the end of the eighth grade and got a job helping a vegetable peddler. He turned his earnings over to his mother, who supplied him with money for cigarettes and picture shows.[3]

Courtesy, Baltimore Bulletin of Education.

Adolescents attain satisfaction through the development of special abilities.

Education and class status. With the constantly increasing number of adolescents remaining in school, teachers and school officials are confronted with problems that did not exist several decades ago. An important problem, often not recognized by the teachers and others concerned with the school program, is that of reconciling its middle-class point of view and values with the lower-class culture of so many of its pupils. This failure on the part of the schools creates conditions that drive the lower-class student from high school. The writer observed a case in which the small-town school had expanded and a number of boys and girls from the rural area were transported to school by school bus. Several of these students had not had the social and cultural opportunities of the average child. They were at first unable to enter into many of the social aspects of the school life. The teachers tended to ignore them and the students through their cliques built up

[3] *Ibid.,* p. 19.

barriers that excluded the farm children from their activities. Most of the underprivileged rural group dropped out of school. At a later date the school district was further expanded to include a much larger group of rural boys and girls. The program was expanded to provide vocational agriculture, industrial arts, home economics, and commercial courses. Many of these students were able to find courses and activities that were of interest to them. Perhaps a changed point of view was adopted by the teachers, without their realizing it, and after a few years the valedictorian and vice-president of the senior class came from the group of rural students.

A study of school retention among Elmtown[4] youth, referred to and described more completely in Chapter XII, showed that the dream of equality of educational opportunity is to a large extent a myth. This is shown in the analysis of the per cent of adolescents from each of the five social-economic cultural classes found in Elmtown presented in Table LVIII. There are two rather widespread misconceptions about

TABLE LVIII

School Enrollment of Elmtown Youth According to
Class Status (*After Hollingshead*)

CLASS	IN SCHOOL		OUT OF SCHOOL	
	Number	Per Cent	Number	Per Cent
I	4	100.0	0	00.0
II	31	100.0	0	00.0
III	146	92.4	12	7.6
IV	183	58.7	120	41.3
V	26	11.3	204	88.7

the problem of school leavers. First, there are many upper-class city dwellers who pass up the problem by pointing out that these leavers are from the rural areas. Surveys conducted at Elmtown and other points show that this is not the case. The great preponderance of school leavers appear among the underprivileged groups in our rural areas, towns, and cities. The second misconception relates to the time of school leaving. The Elmtown study shows that despite compulsory

[4] Table LVIII is reprinted by permission from *Elmtown's Youth* by A. B. Hollingshead, published by Wiley & Sons, Inc., 1949. Elmtown is a fictitious name given to a midwestern town of around 10,000 population.

school laws requiring boys and girls to remain in school until they are 16 years of age, 74 per cent of the 345 young people out of school in the spring of 1942 had dropped out before they had reached their sixteenth birthday.   Neither the Elmtowners nor the school authorities were aware of the large number of young people who had dropped from school.   Such a condition may become even more exaggerated in a larger urban area.   The number of children from the different age groups not enrolled in school for October 1947 has been listed as follows:[5]

> 7 to 9 years, inclusive.........................1.6 per cent
> 10 to 13 years, inclusive....................1.4 per cent
> 14 and 15 years .... 8.4 per cent
> 16 and 17 years ....32.4 per cent
> 18 and 19 years ....75.7 per cent

Withdrawal from school is a complex process; any attempt to explain it must take into account many intangible forces that begin playing upon the child from the date of his first entrance into school. These forces come into focus as he reaches the adolescent age and the upper elementary grades in school.   Although money, clothes, grades, and the like may be listed as the major contributing factors, it appears that it is the attitudes of his family, his teachers, and his peers that finally determine in the main whether or not he will remain in school. Thus, the concepts of education handed down through the family pattern, the attitudes and actions of the teachers and school officials, and the attitudes of his peers affect the individual's attitude toward remaining in school.   Needless to say, the lower the class status, the less likely that the teachers, community leaders, and peers will help him to remain in school.   Thus, many youngsters from the most underprivileged class begin remaining away from school by the time they are 12 years of age.   The irregularity of school attendance becomes greater for them with an advancing age level.

Educational choices of adolescents.   In a study conducted by Donald C. Doane,[6] a group of high school boys and girls was confronted with

---

[5] Data taken from Current Population Reports, Population Characteristics Series P-20, No. 19, July 30, 1948.   *School Enrollment of the Civilian Population: October, 1947.*

[6] Donald C. Doane, "The Needs of Youth: An Evaluation for Curriculum Purposes," Teachers College, Columbia University, *Contributions to Education,* No. 848, 1942.

an inventory of nineteen courses of action embodying the needs and problems of youth as found from previous studies, from which certain selections were to be made. These boys and girls were instructed to select the five areas (described as *courses* that might be pursued in school) that they would want most and the five that they would want least. The *course* "Vocational Choice and Placement" was selected by a larger percentage of both boys and girls than any other course. Furthermore, there was little difference between the choices of those found to be in the high IQ group and the choices of those in the low IQ group (see Table LIX). The results of a survey of 5,500 high school students in the state of Washington support the general findings of the study by Doane. The percentage of boys and girls checking the different things they thought the high school could have done to make them better prepared are listed in Table LX.[7] The need for more courses, especially vocational courses, was checked by half of these high school students. Also, a significant number checked items dealing with voca-

TABLE LIX

Comparison of the Choices of Courses of High
and Low IQ Groups (*After Doane*)

Course	High IQ (N 164)	Low IQ (N 164)
Vocational Choice and Placement	65%	68%
Getting Along with People	52	48
Health	37	49
Sex	22	30
Relationships with Opposite Sex	29	30
Finances	22	25
Plans for Marriage and Family	25	24
Philosophy of Life: Mental Hygiene	18	20
Relationships with Family	16	19
Leisure Time and Recreation	10	19
Morals	16	13
Religion	11	5
History	27	23
Government	29	27
Current Problems	10	18
Music, Art, Dramatics	42	30
Sciences	25	24
Languages	26	16
Literature	13	9

[7] L. J. Elias, *High School Youth Look at Their Problems.* Pullman, Washington: The State College of Washington, 1949.

tional guidance, vocational experience, better understanding of world problems, and follow-up work after graduation. The results of these studies are in harmony with those obtained from the Purdue University Opinion Panel for Young People, referred to in Chapter X. The poll showed that one-third of the pupils desired courses and opportunities for work experiences. Many expressed a desire for help in vocational selection, understanding themselves, choosing a college, and securing a job.

TABLE LX

THINGS HIGH SCHOOL YOUTH THINK THE SCHOOL COULD
DO TO MAKE THEM BETTER PREPARED (*After Elias*)

	Boys	Girls	Total
Offer more courses	50.2%	46.8%	48.3%
Give more vocational courses	45.6	52.4	43.8
Teachers could be more friendly	16.8	18.6	17.8
Provide more guidance and counseling	29.2	29.9	29.6
Show interest in what they do after graduation	13.3	7.1	9.9
Tell them what vocation to follow	5.2	3.7	4.4
Give them practical vocational experience	28.8	26.9	27.8
More help with personal problems	11.2	9.9	10.5
Give them understanding of world problems	11.6	12.8	12.2
School and teachers are too strict	4.6	6.3	5.3

From an analysis of the replies of over 2,100 high school students (about equally divided between sophomores and seniors) from 10 high schools of western Pennsylvania to a questionnaire submitted relative to their participation in varied aspects of the social and recreational program of their school, some interesting conclusions may be drawn and some constructive generalizations made.[8] The results of this study again bring out the need for the school to plan definite programs to satisfy the recreational and social needs of adolescents. Such a need is certainly to be found more often in the larger high schools than in the smaller ones. Those students questioned suggested, in addition, a need for get-acquainted parties during the early part of the year, in order that students from the different elementary schools who now made up the high school group might come to know one another better. It is pointed out that informal dances should have a part in

[8] Dan R. Kovar, "Student Attitudes Toward the Program of Social Recreation in Certain High Schools of Western Pennsylvania," Doctor's Thesis, University of Pittsburgh, 1939.

the social program, but that these should not make up the whole program. There seems to be a need for activities that will include all the pupils. In addition to these needs for social and recreational activities, there is felt to be a need for the teachers to offer help through clubs, the homeroom, or regular classes in social usage. Along with all this, there should be a democratic spirit existing within the school, and the faculty should display an interest in and appreciation of the problems and aspirations of the adolescent boys and girls.

In the organization of the well adjusted personality, there must be a consideration of all the factors that go to make up personality. Each individual represents a relational pattern of traits. These traits are more or less peculiar to the particular individual concerned, and their organization into a pattern that will produce a well integrated personality should be the major goal of education. Club activities, homeroom projects, socialized classroom procedures, and other methods and conditions characteristic of the modern school are designed to aid in the development of the individual pupil as a whole. We have come to realize that human health, happiness, and success depend upon more than intellectual development. A student may be able to solve intricate problems in physics and chemistry, or be able to write an essay rated as very superior, and still be so badly adjusted emotionally that he is incapable of solving the simplest problem of his own everyday living. The home and the school, as well as other educational and social agencies, must come to recognize the problems of social, educational, vocational, and health adjustments in relation to each other and to a well-adjusted personality. In the study of Doane,

Help in development of social abilities—making friends, popularity, manners and etiquette, etc.—was indicated as desired by about one half to three fourths of the girls, depending upon the particular topics, and by about one fourth to one half of the boys. *How to make friends* was checked by 74 per cent of the girls and 55 per cent of the boys, and *manners and etiquette,* etc., by 65 per cent of the girls and 43 per cent of the boys.[9]

NEEDS AND GOALS

Education and technology. Since the school is often so completely separated from the world of work, the adolescent may emerge with a store of abstract knowledge, but with very little notion about life itself. It is also likely that those who immersed themselves most completely in their studies are the ones most unfit for the ordinary day-to-day

[9] *Op. cit.,* p. 116.

activities. Furthermore, there is such a long time lag between the learning of certain facts and their application that the student must be motivated by various extrinsic and artificial incentives. The cry of unreality is raised against the school system by many who would integrate the school more closely with life. Many reforms have been proposed as the real solution to the problem. The real trouble seems to have developed as a by-product of technology. Technology has brought about an economic condition wherein the labor of children is no longer needed for the production and distribution of goods.

This has created important readjustments in the school program to meet the needs of boys and girls in a technological age. This need is especially evidenced when it is realized that most high schools are still geared to the goal of preparing students for college. Most high school students take the college-preparatory course; but only 20 per cent of them ordinarily go on to college. Another 20 per cent are prepared for a trade or vocation. But what about the other 60 per cent? Should they not be provided with a *real* education suitable to their present and future needs? The Life Adjustment Education Commission appointed by U. S. Education Commissioner John W. Studebaker undertook to plan a program that would prepare boys and girls for college and also for a life of usefulness and service. Such a program would have functional value for the 60 per cent who are neither going to college nor at the present time receiving training of a vocational nature. A summary of the type of training that would be included in such a program is here presented:

1. *Family-life training.* Such learning emphasizes facts about marriage, the family, and child care and training as well as materials involving English literature and mathematics.

2. *Consumer education.* Learning how to get your money's worth at the store counter is recognized as just as essential as learning skills in order to make money.

3. *Good habits of work.* These involve learning how to study, how to budget one's time, and how to get along with one's fellows in the performance of cooperative tasks.

4. *Creative use of leisure time.* Recreation involves more than rest, or just going places. Skills are acquired, interests are developed, and varied activities are pursued as means of developing worthwhile interests and creative abilities.

5. *Citizenship training.* An individual's obligations to his neighbors, to the community, and to society at large are learned. Such learning prepares for citizenship in a cooperative democratic society, which is dependent upon individual initiative, responsibility, and cooperation with others in many enterprises.

In addition to this learning for maturity, the immediate needs of the individual are given special consideration. In many respects it is recognized that the best preparation for adult living is that of preparing to live successfully and harmoniously today. The individual who learns to solve the problems encountered during the teen years is not likely to meet with insurmountable problems at maturity. The individual who learns to work harmoniously with his peers in high school should not find it difficult to work cooperatively and successfully with his fellow workers in the factory or at the office. The modern high school is giving increased attention to vocational guidance, job placement, social relations, family relations, and, perhaps above all, to helping the boys and girls to understand themselves.

Need for family-life training. The schools are inclined to point an accusing finger at the parents when a problem child comes to their attention. Materials presented in earlier chapters do suggest that the home is the most important institution affecting the attitudes and behavior of growing boys and girls. The statement, "We don't have problem children but rather problem parents," has considerable truth in it. However, the school cannot place the future blame on the parents. It has already been pointed out that *dissatisfaction in school* is one of the prime reasons for adolescents' dropping from school. Thus, the influence of the school becomes limited to that group remaining in school.

Also, the schools have not in general accepted the responsibility for preparing high school boys and girls for family living. It has been recognized since the early part of this century that this must be an important objective of every high school; but studies of school practices will reveal that a tremendous lag exists in the high schools in the attainment of this objective.

Harold Punke[10] conducted a study of the attitudes and activities of high school youth, in which several thousand youths participated. The schools were distributed over nine states and the data were secured primarily from schools with an enrollment of 200 to 600 students. This study was concerned with the ages that youth considered best for marriage; with whether the youth themselves expected to marry and if so, at what age. The group tested were freshmen and seniors. It was found that, in general, high school freshmen placed a younger age as the best age for marriage than did the seniors. Also, the freshmen were less uniform in their judgments on the subject of marriage than

[10] H. H. Punke, "Attitudes and Ideas of High School Youth in Regard to Marriage," *School and Society,* September 12, 1942, pp. 221–224.

were the seniors.   Some of the reasons why freshmen placed the best age for marriage lower than that indicated by the seniors may be:

1. As pupils grow older the ages 22–25 seem increasingly nearer.
2. Seniors think more of family life from the standpoint of responsibility than do the freshmen.
3. Freshmen may think more of sex, since the sex drive has appeared rather rapidly along with physical maturity.
4. Freshmen have less understanding of the biological and socio-economic aspects of married life.
5. Seniors are more concerned with further training or with entering a suitable vocation.

Need for a functional health program.   The schools could probably make no greater contribution to the welfare of the nation than to assume a reasonable amount of responsibility for the mental health and for the personality adjustments of growing boys and girls.   It is through the agency of the school that enlightened influences can best operate for the development of wholesome personalities and of well adjusted individuals.   The schools should more closely adhere to the old Greek maxim "a sound mind in a sound body."   It appears quite likely that, following the Renaissance and the Reformation, interest in education stressed intellectuality and gave little consideration to the physical and mental health of students.   It has only been within recent years that the emphasis has begun to change and that efforts have been made to develop the physical well-being of growing boys and girls.

An educational program should have as its first concern the development of the individual, a process sometimes referred to as personal development, and one which demands that the schools provide a functional health program.   There is evidence on hand from many sources that very often health education either has been neglected in our schools, or has not been organized and presented in a way to become effective in the lives of the pupils.   The health program should include more than a periodic examination, a course in hygiene, some formal physical-education activities, and an athletic program that touches the lives of only a small percentage of high school pupils.   Rather, it should function throughout all the school activities, and should have as its objectives the development of healthy individuals, desirable health attitudes and habits, and a recognition on the part of the pupils of the nature and importance of sanitation and community health problems.   The general aims and responsibilities of the school health program may be summarized as follows:

1. To provide a healthy environment for pupils and teachers.   This involves elements related to the school program and the emotional tone of the school as well as to good sanitary conditions.

2. To have a planned program and facilities for taking care of accident victims at school and for cases of sudden illness.

3. To teach pupils facts relative to the causes of diseases, the ways diseases are spread, and the known methods of preventing diseases.

4. To develop habits and attitudes conducive to the maintenance of good health, from the point of view of both the individual and the community.

5. To provide periodic health examinations of pupils and teachers, and to keep a cumulative record of the findings and recommendations.

6. To give special attention to those in need of medical or dental care. Where the pupils are not financially able to provide for their needs, the community should use its resources for this purpose.

7. To provide special educational programs adapted to the needs of the handicapped.

8. To cooperate with the community in community health programs and in the control of contagious diseases.

9. To provide for in-service growth of the teachers to the end that all teachers may realize that the development of a healthy individual is one of the fundamental aims of the school, and that every teacher is to some extent responsible for the health program.   However, the specific responsibility for certain courses and for the coordination of all school health activities and for relating these to community health programs should be that of some special teacher.[11]

The importance of citizenship training.   The complexity of our social order, the growing interconnections of our various social and economic units, and the increase in governmental activities have tended to increase the necessity for citizenship training.   Any educational movement that is to function in developing worth-while civic attitudes and habits should begin in an elementary way in the earlier years of life.   These are formative years for the development of attitudes and character, and therefore are important in relation to future developments.

Respect for authority, a feeling of part-ownership of public property, and a pride in civic cleanliness and beauty should be a part of this early training.   In the upper elementary grades and in junior high school, the pupils should be given a more specific and detailed account of civic

[11] For a good presentation of the need for a functional approach to this problem, see Archer Willis Hurd, "Post War Health Education," *Education,* 1945, Vol. 65, pp. 445–448.

problems. Courses designed for this purpose should not be purely theoretical or drawn from a textbook that takes a model community for its standard. These courses should be somewhat of a laboratory nature, and the community studies should be made at first hand by the maturing boys and girls.

A reasonable amount of the training that leads to definite conformity with customs is desirable. This is especially true where others are involved, and a fixed time- or place-schedule is necessary. However, there is a strong tendency for society to fix definite ways of doing things, and to enforce habit formation by interference and taboos, instead of allowing the child to learn by doing the task unhindered. Only when the pupil is allowed to choose his own plans of work and his own tasks, under sympathetic guidance, can he understand the true social significance of his actions. It is in this connection that certain aspects of student government, homeroom organizations, and the like should become most effective in civic development for democratic living.

Probably one reason why we have made no greater progress in social development lies in the fact that, in most situations in life, the child is responsible to some autocrat—the teacher, gang leader, or parent—instead of to a high notion of social responsibility. If social education is to mean anything, inner controls must be substituted for such outer ones. Herein lies the opportunity of the school in social education. Only by allowing the child freedom in the selection and performance of tasks is it possible to transfer authority from other persons to personal standards and ideals. In teaching the more formal subjects of the curriculum, teachers rarely provide opportunities for the development of personal standards. The subjects are so well organized and the school is so definitely standardized that any tendency toward originality is frowned upon. Fortunately, extracurricular activities are becoming increasingly important and are obtaining more than a minimum of teacher guidance.When aims, methods, and materials are organized with citizenship as a goal, better results will be obtained.[12]

Social versus individual development. The modern trend in education has been toward socialization. It has been recognized that individuals must be more socially conscious and develop a higher degree of social intelligence than was necessary when each family lived almost entirely unto itself and each community lived almost wholly within its

[12] See the procedure given by Martha P. McMillan, "A Project in School Citizenship for the Eighth Grade," *The Instructor,* November 1940, p. 28.

own limits. We have passed beyond the period of social isolation, which has been characterized as one of "rugged individualism," into a state of social organization and reorganization. There is today less emphasis placed upon tribal pride and more given to civic duty and responsibility. *Organization, cooperation,* and the like are terms that express our national outlook, and that are rapidly affecting the materials and procedures that make up our school program. However, this increased emphasis upon the socialization process may overleap its proper bounds. The school must not assume that the individual is entirely a social product. In earlier chapters it was pointed out that individual differences in mental and physical development are very great. Likewise, individual differences in the possibilities for the development of social behavior are exceedingly great. The school must not, in its efforts to develop socialized behavior that will prepare the pupil for the social world of today and tomorrow, attempt to produce standardized social products from one stereotype. It is not desirable to stamp out individuality and substitute for it some common form of sociability.

EDUCATIONAL GUIDANCE

What shall be the basis for promotion? A study of the theories and practices of student promotion reveals the existence of three fairly well defined points of view and practices. The first is based upon "grade standards." This is the basis for report cards and is in theory adhered to by a large percentage of teachers of the traditional school of thought. According to this theory, the child is supposed to master a prescribed amount of subject matter before being passed on to the next grade. As a consequence of his failure to meet the grade standard he must repeat the grade the next year. The second theory may be labeled "no failure." According to this practice each child is promoted to the next grade after spending one year in each grade. This procedure has in many places superseded the "grade standard" theory.

A third theory, now being introduced by many educators, is based upon the organismic concept of growth and upon a recognition of individual differences in mental, emotional, educational, and physical development. The child, according to this theory, will be placed in that group which will provide for him the best opportunities for his total growth and development. This theory has much to offer; however, it is much more difficult to administer than either of the two systems previously described. Such a theory, when followed with both the retarded and the gifted child, will do much to eliminate many educational problems related to his growth and development. La Baron has listed nine principles that should be followed in connection with this theory.[13]   These are as follows:

1. "Grade standards" in terms of achievement in skill subjects will be abandoned.

2. The curriculum will be developed continuously, and each child will progress through it at his optimum rate.

3. Adjustments from group to group and from grade to grade will be a continuous process and will not be solely a time period consideration.

4. Acceleration or double promotion will seldom be used.

5. Generally speaking, adolescents will not be kept in the elementary grades (kindergarten through grade six) but will be placed in junior high schools.

6. Special classes will be provided for the mentally deficient (those with intelligence quotient of 56 to 70 or 75).   Special classes will be provided for the physically handicapped whenever possible.

7. Promotion to junior high school will be made not in terms of readiness for a set curriculum, but rather in terms of the developmental needs of the individual.

8. The length of time each person spends in the elementary school will be determined by a careful estimate of his needs in light of his chronological age, mental age, achievement, physical development, and social and emotional maturity.

9. Methods of reporting to parents and techniques for testing parental understanding and cooperation with the program will be carefully revised and developed in light of these emphases in pupil progress.

Guidance in relation to the educational process.   The objectives of adolescent guidance must find their counterparts in the objectives set forth for the education of growing boys and girls.   These objectives might be thought of as large divisions of life activities, each related to the other.   Concerning guidance as it relates to these life activities, Kefauver and Hand say:

Guidance, then, will have contributions to make to education for social, health, recreational, and vocational activities.   One might appropriately refer to guidance in relation to each of these objectives as social, health, recreational, and vocational guidance.   A combination of these four objectives constitutes the objectives of the total educational program.   Similarly, a combination of the four types of guidance would give us the total guidance program of the school.   This total concept might be characterized as educational guidance—that which serves all parts of the educational program.[14]

[13] W. L. La Baron, "Developing a Program of Continuous Progress," *Elementary School Journal,* 1945, Vol. 46, pp. 89–96.

[14] G. N. Kefauver and H. C. Hand, "Objectives of Guidance in Secondary Schools," *Teachers College Record,* 1933, Vol. 34, p. 381.

The problem of individual variations in our schools has become more acute as a result of (1) the increased enrollment, (2) the lengthening of the period spent in schools, and (3) the enlarged program of the schools.   A democratic system of education should provide opportunities for each child to develop his abilities and potentialities. The dull child, the neuropathic child, the defective child, and the gifted child alike should receive consideration in our school program.   Complete recognition of individual differences means a recognition of these deviations in intelligence, in aptitudes, in temperament, as well as in goals and purposes in individual cases.   Freeman points out concerning this:

> In short, guidance in education is fundamentally a matter of understanding and utilizing our knowledge of human variability in mentality, in traits of personality and temperament; in understanding the causes of failure in school, as well as of success; in appreciating the fact that the human organism is not merely a mosaic of a variety of intellectual, temperamental, and physical traits, but is rather an integrated unit, the effectiveness of whose intellect may be increased or vitiated by other aspects of his individuality.[15]

Counseling the individual pupil.   Counseling is as old as formal education, if not older.   It has recently been described as "a personal and dynamic relationship between two people who approach a mutually defined problem with mutual consideration for each other to the end that the younger, or less mature, or more troubled, of the two is aided to a self-determined resolution of the problem. . . ."[16]   Counseling is primarily an individual matter and is more apt to be successful when conducted on that basis.   In connection with the school environment it implies greater maturity and understanding on the part of the teacher or adult.   In a study of 1,500 fifteen-year-old boys in Detroit,[17] 82 per cent wanted adult companionship and counsel.   Those who had least companionship and poor family adjustment were most eager for this adult counseling.   Counseling is not synonymous with interviewing, since the latter is a technique for some specific purpose.   Referring to Wrenn, we find the following diagrams with explanations:

[15] F. N. Freeman, "Contribution to Education of Scientific Knowledge About Individual Differences," *Thirty-seventh Yearbook of the National Society for the Study of Education,* Part 11.   Bloomington, Ill.: Public School Publishing Company, 1938, pp. 418–419.

[16] C. G. Wrenn, "Counseling with Students," *Thirty-seventh Yearbook of the National Society for the Study of Education,* Part 1.   Bloomington, Ill.: Public School Publishing Company, 1938, p. 121.

[17] K. Layton Warren, "Guidance Needs of Detroit's 15-Year Old Pupils," *Occupations, The Vocational Guidance Magazine,* 1936, Vol. 15, pp. 215–220.

A too common type of interview is information and advice given, thus:

Counselor $\longrightarrow$ Student

A less common and sometimes quite justifiable interview is the information-getting situation, thus:

Counselor $\longleftarrow\longrightarrow$ Student

The interview as it should be used in counseling must be represented:

Counselor $\longleftarrow\longrightarrow$ Student

It was pointed out in Chapter II that educational and vocational problems loom large in the lives of adolescents. The materials of Table LXI show that these are the types of problems encountered by counselors, whereas visiting teachers more often deal with problems involving social and personal factors. Too often social and personal problems are not given the attention and consideration they deserve. It is only when these reach the teacher or administrator that they are

TABLE LXI

PERCENTAGES OF TYPES OF PROBLEMS HANDLED BY COUNSELORS AND VISITING TEACHERS [18]

TYPE OF PROBLEM	PERCENTAGES	
	Counselors	Teachers
Educational and Vocational Guidance		
Educational and vocational plans	42.2	2.6
Poor scholarship	23.1	7.9
Vocational information	7.8	0.0
Program adjustment	3.5	0.0
Placement	2.7	0.2
Work permits	0.3	0.2
Total	79.6	10.9
Social and Personality Problems		
Behavior	5.6	27.3
Home conditions	.8	20.4
Attendance	1.4	16.8
Truancy	0.3	3.5
Relief	2.4	8.3
Health	1.3	8.6
Home placement	0.2	0.6
Total	12.0	85.5
Other Problems		

[18] "Activities of Counselors and Visiting Teachers in Minneapolis," *School Review*, 1931, Vol. 39, p. 488.

recognized. There, they have usually been treated as behavior activities, without much consideration of the drives back of such activities.

The need for "belongingness" or being accepted is very important during the adolescent years. Teachers must be aware of this need, and must recognize the problems of adolescence as real, even though they may appear trivial to the adult. This ability to understand the nature of the problems of others is a prerequisite for counseling. When the problems of others are considerably removed from one's own life, the ability to understand such problems becomes more difficult of attainment. Adolescents' problems are not synonymous with the problems of a teacher on the job. This in itself presents a challenge to the teacher, if his work in the guidance and direction of adolescents is to be effective:

Each boy and girl in the classroom wants to be accepted as a unique individual. He wants to be accepted *as a person* and not for what he can accomplish. Students like to think of a teacher as someone whom they can trust. To violate a confidence will freeze the channels of human relationships immediately. What might seem insignificant and unimportant to the teacher may be of vital importance to the child.[19]

Importance of records. A basic principle of guidance is that we must secure definite information about the individual before effective plans can be formulated to meet his needs. The kinds of information that should be secured in order that the counselor or teacher may understand the student and the student understand himself have been listed as follows:[20]

1. The record of his previous school experiences.
2. His aptitudes and abilities.
3. His home background and community environment.
4. His goals and purposes.
5. His likes and dislikes.
6. His social development and adjustment.
7. His emotional status.
8. His health record and present health status.
9. His economic and financial status.

Previous school experience is not by any means an absolute, safe basis for the prediction of future development. However, the pupil's

[19] Edwin C. Morgenroth, "Relationships between Teachers and Students in Secondary Schools," *Progressive Education,* April, 1939, pp. 248–249.

[20] See Eurich, A. C., and Wrenn, C. G., "Appraisal of Student Characteristics and Needs," *Thirty-seventh Yearbook of the National Society for the Study of Education,* Part 1. Bloomington, Ill.: Public School Publishing Co., 1938.

response to certain types of courses, his areas of high and low achievement, and his participation in extracurricular activities all show definite trends that are a source of help to the counselor. Records should be kept on mental and educational growth from time to time. Educational achievement tests furnish records of growth in the school subjects. Personality rating scales provide the teacher with a means for making reports on attitudes and personality development.[21]

Techniques for use in securing information on aptitudes and abilities are manifold. Such tests, however, may be dangerous in the hands of an untrained person. Some occupations require only a special type of intelligence while others demand only skills of some particular type. Facts about past activities, interests, health, and educational growth should serve as valuable information on a cumulative record. Through interviews, questionnaires, tests, rating devices, and observations, information is obtained. When it is organized in an understandable and usable manner, such information should serve as a basis for more accurate guidance and counseling of adolescent boys and girls.

The chief advantage of the cumulative record is the possibility of combining the separate items it contains into an integrated picture of the whole individual. The meaning of various patterns of interest records, as well as of patterns of abilities and activities, and ways of combining these most effectively is a matter demanding further study. Concerning the importance of having such information available, in order to assist the students in their educational problems, Ruth Strang has stated:

> Assistance to students must be thorough but not superfluous. One cannot have too much information about a student, but one can give him too much advice. His present level of maturity must be ascertained; his values, goals, aims and purposes recognized and respected. Counseling, in part, is instruction in self-direction. It is a process, not a conclusion.[22]

GENERALIZED SUMMARY

The lack of opportunities for employment in business and industry keep in school many young people who otherwise would go to work. Some of the increased school enrollment during the past two decades can be attributed to financial aid and encouragement, through grants

---

[21] See the form of scales in Paul W. Chapman's *Guidance Programs for Rural High Schools,* U. S. Office of Education, August, 1939. Misc. 2196.

[22] Ruth Strang, *Counseling Technics in College and Secondary School.* New York: Harper and Brothers, 1937, p. 130.

of money from the government for students.   However, the most important influence of all is the determination of most parents to give their children all the educational advantages within their means. There is an almost universal educational consciousness, and a general recognition that an elementary education is not sufficient.   The great increase in high school enrollment, like the widespread possession of radios and the increased use of electrical appliances, is a part of the total increase in the complexity of our civilization and the development of better standards of living.

In the next place, new curriculum materials are in the process of making.   The curriculum of the past decade represented units designed for a select group of boys and girls, and such curricula were not planned for the needs of our present school population.   Manual activities were looked upon with distaste; most of the work in the fine arts was thought of as suitable only for a few; education for social understanding was almost unheard of; and the materials that acquaint us with our present socialized civilization were considered as inferior in quality and value.   The schools have tended to emphasize short-time learning involving the acquisition of factual materials and skills.   The more significant long-time learnings involving attitudes and ideals have been regarded as incidental learnings.   The developmental concepts emphasized throughout this volume would emphasize the long-time learnings, and would evaluate growth in terms of long-time rather than immediate outcomes.   This does not mean that periodic evaluations should never be made, but rather that such evaluations should be interpreted in the light of the individual's development over a long period of time.   True, the high school of the future has its roots implanted in the present.   These roots are held sturdy by the spirit of freedom and research that characterizes the American educator today.   The nature of the high school of the future has been well summarized by Galen Jones:

We see a school where all normal youth remain in senior high school with individual profit, where the needs of all youth are appropriately met and served.   We see a school educating for unity by means of a program of general education to be found most frequently as a common language core, and continuously vitalized and dynamic by reason of the cooperative undertaking of its teachers to that end.   We see a school in search of uniqueness in every youth, strengthening desirable diversity through broad curricular provisions, particularly in the upper years of the secondary school, and thus continuously promoting freedom.   We see a school which knows its youth as individuals, and in which by means of competent guidance service all are aided to formulate meaningful goals toward which each works with purpose. And finally, we see a school which takes its pride in the quality of the citizens it has nurtured.[23]

THOUGHT PROBLEMS

1. What items are common to the aims of education listed by different students of education? What trends may be noted?

2. What educational aims have probably been most neglected in our high school program? How would you account for this?

3. Elaborate on the meaning and significance of the following statement: "In the future, schools will be more closely integrated with the life of the community than they were in the past."

4. It has been stated that, "Education comes out of all kinds of experiences and affects all phases of one's life." What effect should the acceptance of such a viewpoint have on the work of the schools?

5. How is counseling related to educational needs? Study the types of problems encountered by counselors. (See Table LXI.) What would you conclude concerning the nature of these problems?

6. List several problems suggested to you from an analysis of the data of Table LVIII, showing the school enrollment by socio-economic classes of Elmtown youth.

7. Review the article by Galen Jones from which the materials on page 423 are quoted. What five or six important characteristics of the high school of the future are suggested by this article?

8. Study the materials of Table LX showing the results of the survey of high school youth in the state of Washington. What are the implications of these results to the changing high school program?

SELECTED REFERENCES

Alberty, Harold, *Reorganizing the High School Curriculum*. New York: The Macmillan Co., 1947.

Briggs, Thomas H.; Leonard, J. Paul; and Justman, Joseph, *Secondary Education*. New York: The Macmillan Co., 1950.

Cole, Luella, *Psychology of Adolescence* (Third Edition). New York: Farrar and Rinehart, Inc., 1948, Chap. XVII.

Crow, L. D., and Crow, Alice, *Our Teen-Age Boys and Girls*. New York: McGraw-Hill Book Co., 1945.

Douglas, Harl R., *Secondary Education for Youth in Modern America*. Washington: American Youth Commission, 1937.

Douglas, Harl R. (Editor), *Education for Life Adjustment*. New York: The Ronald Press Co., 1950.

Educational Policies Commission: *Education for All American Youth*. Washington: National Education Association, 1944.

[23] Galen Jones, "The High School of the Future," *Teachers College Record*, 1949, Vol. 50, p. 456.

Landis, Paul H., *Adolescence and Youth*.   New York: McGraw-Hill Book Co., 1945, Chaps. XVII, XVIII, and XIX.

MacLean, M. S., "Adolescent Needs and Building the Curriculum," in E. G. Williamson (Editor), *Trends in Student Personnel Work*.   Minneapolis: University of Minnesota Press, 1949, pp. 27–39.

Olsen, Edward G., *School and Community Programs*. New York: Prentice-Hall, Inc., 1949.

Spears, Harold, *The High School for Today*.   New York: American Book Co., 1950.

XX

Vocational Choice and Adjustment

Every sign points to an increased use of scientific procedure in the ordinary life-activities of man. The World War of the recent past seems destined to strengthen this trend. The machine is itself one of the consequential products of the scientific age. Concerning it, Faires says:

The effects of the machine on society have been aggravated by two factors: (1) a more comprehensive use of mechanical and electrical power, and (2) a more general utilization of the principle of so-called 'scientific' management. We all know how these particular factors have resulted in a vast increase in the output of factories per man-hour of labor.[1]

This increase has brought with it changed modes of living, changed occupational conditions, more widespread leisure, increased luxuries, and a change in our social-economic structure that is now our undoing. All these have had an important influence upon family relations, work opportunities, and the economic needs and conditions of adolescents and youths.

The materials presented throughout the previous chapter indicated the educational needs of the adolescent. A review of the materials of this chapter shows that

(1) many young people lack the information needed for making sound decisions about occupations;

(2) services for counseling young people and giving them adequate placement assistance are in many instances woefully inadequate;

[1] V. M. Faires, "The Role of the Engineer of Tomorrow," *The Journal of Engineering Education,* 1938, Vol. 28, p. 365.

(3) young people are at an extreme disadvantage in the labor market, especially during a period of general unemployment;

(4) although there has been an enormous decline in child labor, many school age boys and girls are still employed under substandard working conditions; and

(5) there is a great need for the opportunity for young people to secure work experiences.

These problems are of prime importance in the development and guidance of adolescent boys and girls. This chapter is devoted to a discussion of these and related problems.

Youth and employment. If we study the history and conditions of the thirties, we will note that there was widespread unemployment. Then, ten years later, according to the census of 1940, 35 per cent of the unemployed were under twenty-five years of age, whereas only 22 per cent of the employable population were under that age. A youth study conducted in Maryland in the thirties revealed that economic insecurity was a widespread problem. Of the employable youth studied, all of whom had been out of school more than a year, 40 per cent had not obtained any sort of full-time employment, and the average time between leaving school and becoming employed for those dropping out of school before they were sixteen years of age was three and one-half years. The major cause of such a condition lies in our social and economic structure, rather than in faulty educational programs. However, this does not mean that our educational programs are adequate. A more genuine equality of educational opportunity would do much to alleviate problems of unemployment; a more realistic educational program would go a long way toward keeping youth in school until they have reached maturity.

Seemingly, people recognize that widespread unemployment among adults is a clear indication that something is wrong with the economic order of the time. However, in the case of widespread unemployment among youths, the blame is placed on the latter, or on those concerned with their education. There are some who point out the lack of vocational education, and others who say that too many people have been directed from the farms into commercial and industrial pursuits. There are some who go so far as to lay the blame on too much education, with its accompanying costs, and others who prescribe more education as the cure for the trouble. Although it cannot be denied that much can be done in a constructive manner for the education of youth, yet economic insecurity, uncertainty, and mounting periodic unemployment will remain until the source of the difficulty is cleared up.

The conditions that have brought about changes in the economic life and security of youth may be outlined as follows: The great resources of free land which the United States once possessed no longer exist. People have continued to migrate into cities until now almost sixty per cent of the population is urban. The industries and agriculture of the country have been mechanized, so that we may say that technological advancements have taken place at a more rapid pace than have social and economic changes in the ways of living. The depression brought full realization of a condition which had been evolving for some time in our social order, but had been, up to that time, only vaguely understood—namely, that proper provision for the induction of youth into adult life can no longer be made in the same way it was made in the past.

There are some who would associate unemployment among youth with economic depressions. However, there is good evidence that a serious gap exists between the inadequate preparation of young people and available opportunities for desirable jobs. This is clearly shown in data, presented in Figure 35 with respect to what boys and girls were doing in 1947. According to these data approximately one-half of the 12,630,000 boys and girls between fourteen and nineteen years of age were in school and were not working in October, 1947. The great prevalence of part-time employment among high school boys and girls, perhaps the most notable change in the volume of employment among high school youth, is closely associated with employment opportunities and trends. In order to determine what is happening to these out-of-school youth, The United States Department of Labor conducted a study of youth employment problems in Louisville, Kentucky. Of the 524 out-of-school youth interviewed a surprising number were unemployed.[2] In the age group 14–15, 46 per cent were unemployed; in the age group 16–17, 36 per cent were unemployed; and in the age group 18–19, 21 per cent were unemployed. The inability to find and keep jobs was keenly felt among these boys and girls. Two out of every three had been hunting a job for a month or longer, and one in five had been looking for a job six months or longer. One-half of the group had held their previous job for only three months or less.

[2] Elizabeth S. Johnson, "Teen-Agers at Work," *The Child*, October, 1948. See also, Elizabeth S. Johnson, "Employment Problems of Out-of-School Youth," *Monthly Labor Review*, 1947, Vol. 65, pp. 671–674.

The jobs these young people had been able to secure, especially those under 16 years of age, were in many cases undesirable from the point of view of hours and of working conditions. When young people leave school, most of them are in need of money and are anxious

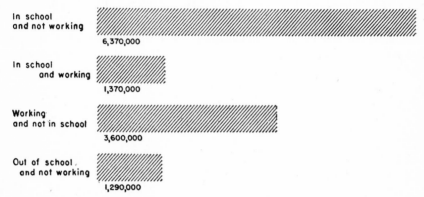

FIG. 35. *What Boys and Girls Ages 14–19 are Doing.* (Based on census estimates, October, 1947.)

to get a job. In trying to account for their failures to get jobs, most of them said: "Someone always gets there first." "Until you're 16 no one wants to hire you." "I can't get the kind of job I want without more education." To see the other side of the problem, employers were interviewed. Many of them agreed that young people under 18 were too immature and undependable. The employers expressed a preference for employing people with a high school education rather than those with better education.

Vocational needs of adolescents. The enormous increase in high school enrollment was pointed out in the previous chapter; however, a large number of adolescents graduate from high school or leave it before graduation with no vocational plans for the future, with a lack of understanding of their own potentialities, with little knowledge of occupational demands and opportunities, and with poor marketable skills. A challenge is presented to the educational and socializing agencies today to train adolescents to be productive and well adjusted citizens of the world of tomorrow. The vocational needs of adolescents should be thought of as all-inclusive. In their broadest interpretation they would include health, citizenship, social adjustments, mental hygiene, religious and moral development, and guidance.

In the study of the out-of-school youth of Louisville, the employers were interviewed regarding the type of preparation that they would recommend for young people desiring employment in their firms.[3] The majority of the 33 employers, who favored a minimum hiring age of eighteen years or older, considered school experiences more important than previous job experiences. To get some notion of the kind of educational preparation the employers considered most helpful, all 49 were asked what abilities they thought should be acquired by boys and girls in school as a preparation for employment. The replies seemed to fall into the three categories, as follows:

	Number of Employers
Development of good character traits	21
General knowledge and background	12
Specific vocational skills	5
No opinion or no report	3

The largest group of employers considered good character traits and sound habits of work and study the most important qualities for young people to acquire in preparation for employment. Specific vocational knowledge and skills were considered most important by five employers, representing stores, restaurants, hospitals, and a sewing-machine plant. It should be pointed out here, however, that these employers were not chiefly interested in seeking office skills, such as typing and stenography. Employers concerned with employing individuals for such positions would place a greater premium on the acquirement of specific vocational skills.

The fundamental differences between the problems of livelihood that confronted the primitive adolescent and those confronting the adolescent in civilized society today are: (1) Specialized training in highly developed fields of human endeavor has become essential for the pursuit of a livelihood. (2) The great differentiation of work brought about by specialization gives a diversity of advantages to different types of work and to different abilities. (3) Vocational prospects of girls have broadened to emphasize economic worth and individual ability in the main, rather than to reveal a sharp line of division on the basis of sex.

In the study by Doane [4] referred to in the previous chapter, the choice

---

[3] "Hunting a Career: A Study of Out-of-School Youth in Louisville, Kentucky," *United States Department of Labor Bulletin No. 115,* 1949.

[4] Donald C. Doane, "The Needs of Youth: An Evaluation for Curriculum Purposes," Teachers College, Columbia University, *Contributions to Education.* No. 848, 1942.

of a course of action was interpreted as indicating the existence of a psychobiological need in the area provoking the greatest intensity of response. More concern was noted in the area of *Vocational Choice and Placement* than in any other area. Among the ten topics within this area, those chosen most often were *How to apply for a job, Keeping a job,* and *Finding out what kind of work you are best suited for.* The topics related to training for a job were chosen less frequently than most of the vocational guidance items by both boys and girls in each age group. This seems to indicate that adolescents feel a greater need for vocational guidance than for vocational training. The feeling of need for such help was more evident in boys than in girls and reached its height as the time for leaving school drew near.

Interviews with youths will verify the notion that, if a student once drops out of school, he is not likely to go back again. Those fortunate enough to find reasonably good jobs, or even jobs that barely provide them with the necessities of life and some small amount of spending money, will hesitate to leave them. Youths without jobs are either unable to find enough money to return to school, or, most usually, have lost their ambition and desire to better themselves in the occupational world. It has constantly been noticed that one of the greatest evils of continued unemployment is its demoralizing effect. Studies conducted by the Division of Research of the Works Progress Administration reveal that many youths out of school "were not actively seeking work either because they were kept busy at home or because long unemployment had broken down their morale to the point where they had abandoned hope of getting work by their own efforts." [5]

The gulf is too wide between the program of the school and the vocational demands of life. Young people too often enter blind-alley jobs, or wander for years after they leave school from one job to another. The vocational needs of adolescents may be summarized as follows: (1) a better understanding of their own aptitudes and limitations, (2) occupational information—including occupational opportunities and job requirements, (3) vocational training, both in school and through work experiences, and (4) the opportunity to use their abilities once developed, i.e., the right to a job. If these needs are to be satisfied, there must be more vocational guidance in our schools. Guidance is based upon a recognition of the existence of individual differences and the philosophy of freedom of choice. The field of guidance has been

[5] "Urban Youth, Their Characteristics and Economic Problems," Division of Research, Works Progress Administration, Series 1, No. 24, 1939.

divided into six comprehensive areas. These are: (1) occupational information, (2) cumulative records, (3) counseling, (4) survey of training opportunities, (5) placement, and (6) follow-up.[6]

VOCATIONAL ASPIRATIONS AND OPPORTUNITIES

Vocational aspirations. There is a marked tendency on the part of high school students to aspire to a rather high goal in their vocational planning. More than one-third of them plan to enter the professions, which now include approximately 5 per cent of our adults. In addition to this, according to the *Fortune* Survey, over one-third of high school girls plan to enter business—mainly in clerical and secretarial positions. These vocational aspirations indicate that they are planning to enter occupations that will require training after high school. These findings are most important in relation to our present-day economy. There is a likelihood that as the semiskilled trades, farming, and other occupations chosen less frequently by high school students are made more attractive and provide a better income, more high school students will choose such vocations. More than half of the high school students who selected an occupation other than factory work or a skilled trade responded *yes* to this question put to them in the *Fortune* Survey:[7] "If you could be reasonably sure of earning as much money by being a skilled worker as at other things, would you then consider learning a skill or working in a factory or at a trade?" Of those who responded *no,* over one-half stated as the main reason for their refusal the fact that they had their minds set upon some other specific occupation. Less than five per cent, and these were mainly girls, stated that they would not do such work because it would lower their social prestige or position.

In the previous chapter it was pointed out that boys and girls become more concerned about occupational demands and about their own vocational aptitudes as they advance through high school. A study by Roeber and Garfield dealt with the differences in vocational preferences of students in grades 9 to 12.[8] Data from questionnaires concerning occupational interests were secured from 912 boys and 1,083

[6] R. E. Marshall, "Unifying the Guidance Program," *The Bulletin of the National Association of Secondary-School Principals,* October 1941, p. 78.

[7] "Fortune Survey," *Fortune,* December 1942, p. 9.

[8] E. Roeber and L. Garfield, "A Study of the Occupational Interests of High School Students in Terms of Grade Placement," *Journal of Educational Psychology,* 1943, Vol. 34, pp. 355–362.

girls in 22 different schools.   The communities covered by this survey varied in size from less than 2,000 to more than 15,000 population; 26.1 per cent of the fathers were classed as farmers.   The occupational choices of the boys and girls for grades 9 and 12 are given in Table LXII.

TABLE LXII

OCCUPATIONAL CHOICES OF BOYS AND GIRLS IN GRADES NINE AND TWELVE
(*After Roeber and Garfield*)

Boys			Girls		
Occupations	Percentages		Occupations	Percentages	
	IX	XII		IX	XII
Farmer	24.6	11.6	Stenographer-		
Engineer	9.1	10.1	Office worker	33.4	36.1
Aviator	8.6	4.3	Nurse	17.4	15.8
Mechanic	5.0	3.9	Teacher	8.0	12.3
Army-Navy	4.1	2.7	Beautician	4.2	3.2
Machinist	2.7	3.9	Commercial artist	3.8	2.8
Medicine-surgery	2.7	1.9	Housework	3.8	2.1
Carpenter	2.3	1.6	Journalism	2.7	1.8
Store clerk	1.8	2.7	Designer	2.3	2.5
Draftsman	1.8	3.9	Musician	2.3	2.1
Accountant	1.8	3.5	Factory worker	1.9	4.2
Lawyer	1.8	1.6	Waitress	1.5	3.2
Electrician	1.4	1.2	Store clerk	1.1	4.2
Musician	1.4	0.0	Telephone operator	0.4	2.5
Teacher	0.9	1.6	Actress	0.0	0.7
Factory worker	0.9	5.4			
Journalism	0.9	0.8			
Office clerk	0.5	3.1			
Commercial artist	0.5	1.2			
Tool and die maker	0.5	5.4			
Chemist	0.5	1.9			
Salesman	0.0	4.0			
Designer	0.0	2.7			

The results of this study are in agreement with those from other studies of this problem, with the exception of the larger number in the ninth grade choosing farming.   This may be explained by the large number of farm homes represented in the sample studied.   There is a pronounced tendency for students to become more realistic in their occupational choices as they advance from the ninth to the twelfth grade as evidenced by the increased number selecting occupations usually listed lower in the occupational scale. However, this study, as well as others, shows that choices are to a rather large degree concentrated in the professional and semiprofessional fields. This offers a problem and challenge to teachers and to guidance workers.

Mental ability and vocational aspirations. Many factors affect the vocational aspirations of adolescents. The vocational choices of 1,500 junior and senior high school students enrolled in some of the schools within and adjacent to the Philadelphia area were studied by Bradley.[9] The median IQ of these students was found to be 104.3, with a distribution and a range typical of that for an unselected group of high school students. The vocational choices of these students were classified into five categories: professional, business and clerical, skilled, military, semiskilled, and unskilled. The percentage of students from the various levels of intelligence expressing a vocational preference in each of the vocational groups is shown in Table LXIII. Although there is a continuous decline in the median IQ from 108.25 for those choosing the professions to 99.50 for those choosing unskilled types of work, there are individuals at varying mental levels choosing vocations from each of the occupational classifications. These data furnish evidence that mental ability is a factor related to vocational choice, but is only one of a number of factors.

Class status and vocational aspirations. This study by Bradley, supported by other studies, reveals that the social and economic status of the family is another factor affecting the vocational choice of adolescents. A study of the aspirations of Elmtown's youth showed that job opportunities are closely associated with the class position of the boy or girl seeking the job.[10] The family influence is such that class II boys and girls are given a preferable type of employment when they are employed at all; while class V and to a lesser extent class IV boys and girls must take what they are able to get with respect to full-time or part-time employment. The net result of this is that general office and clerical jobs are assigned to class II and class III boys and girls, while helping in the junk yards and doing menial odd tasks around the stores, factories, and other places of employment are left to boys and girls of class IV and of class V.

Each adolescent in the Elmtown study was asked to name the job or

---

[9] William A. Bradley, "Correlates of Vocational Preferences," *Genetic Psychology Monographs,* 1943, Vol. 28, pp. 99–169.

[10] Figure 36 is based on data reprinted by permission from *Elmtown's Youth* by A. B. Hollingshead, published by John Wiley & Sons, Inc., 1949.

TABLE LXIII

Mental Ability and Vocational Choice of High School Students (*After Bradley*)

Per Cent of Choice *

IQ's	Professional	Business and clerical	Skilled	Military	Semi-skilled	Unskilled	Total
140–149	.2	.0	.0	.0	.0	.0	.2
130–139	.6	.3	.1	.3	.0	.0	1.3
120–129	1.5	3.6	.7	1.7	.0	.0	7.5
110–119	5.5	7.2	3.2	5.3	.3	1.0	22.5
100–109	6.3	12.8	3.3	5.7	1.9	2.5	32.5
90– 99	3.0	9.0	2.7	8.3	.3	2.0	25.3
80– 89	.7	1.3	2.3	3.4	1.0	1.0	9.7
70– 79	.0	.3	.0	.0	.0	.7	1.0
Total	17.8	34.5	12.3	24.7	3.5	7.2	100.0
Median IQ	108.25	105.19	103.48	101.14	102.37	99.50	104.31

*The large number listing military was influenced by temporary conditions prevailing.

435

occupation he would like to follow when he has attained maturity. The results, presented in Figure 36, show the vocational choices by class for the different occupational groups. The adolescent boys of class I and class II wish first to be business and professional people (77 per cent) and second to be farmers. The girls wish to get married. Class III has somewhat similar desires, although only 36 per cent aspire to be professional and business people. The choice in the clerical area looms large with this group, a choice that no doubt reflects the influence of the

Courtesy, U. S. Department of Labor.

Adolescents need counseling and guidance as a part of their preparation for life.

adult vocational pattern that prevails among a large number of this class. The large increase between class II and class III in the undecided column indicates that many of these youngsters are unable to reconcile their aspirations with their abilities and opportunities. Many adolescents in class III aspire to a higher vocational level than that followed by their parents, but are not able to see their way clear, financially and otherwise, for securing the training needed to enter into such a vocation. There is a continued sharp increase among the undecided as we move from class III to a study of the choices in class IV and class V.

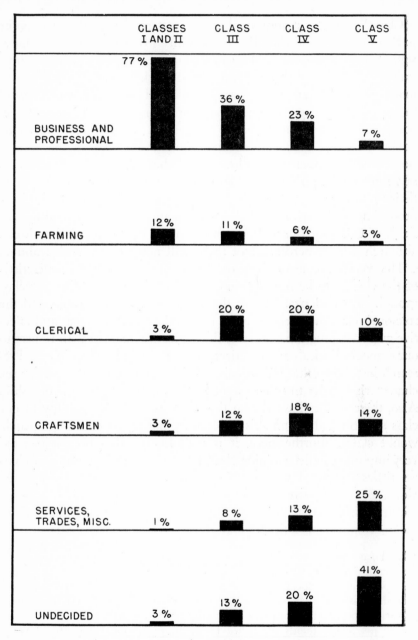

	CLASSES I AND II	CLASS III	CLASS IV	CLASS V
BUSINESS AND PROFESSIONAL	77 %	36 %	23 %	7 %
FARMING	12 %	11 %	6 %	3 %
CLERICAL	3 %	20 %	20 %	10 %
CRAFTSMEN	3 %	12 %	18 %	14 %
SERVICES, TRADES, MISC.	1 %	8 %	13 %	25 %
UNDECIDED	3 %	13 %	20 %	41 %

FIG. 36. *Vocational Aspirations of Elmtown's Youth by Class and Occupational Groups.* (After Hollingshead)

Class V presents a vocational choice pattern that is almost opposite to that of class I and class II. Uncertainty (41 per cent) stands out as significant in this group. Many miscellaneous vocations such as, animal trainer, juggler, and the like were listed by this group. In the craftsman group, containing 14 per cent, is included the largest percentage of those who have made specific plans for their future. Farming as an occupation shows little appeal to this group. A follow-up would no doubt show that a still closer relationship exists between the occupations actually chosen and followed and the class status. Needless to say, most of the girls will become housewives; although in this study their preferences were about equally divided between business and the professions and clerical work. The girls appear to be oriented toward fields of endeavor that will require some or much technical training, although a large percentage of them will never obtain such training.

The vocational aspirations of Elmtown's youth are characteristic of the choices to be found among other adolescents of different class groups. An understanding of the nature of our class groups and the role one's class membership plays is essential to a functional and effective guidance program. Despite the credos we may use with respect to our democratic ideals and practices, anthropological and sociological research has shown that individuals acquire certain anxieties and assume characteristic roles to some extent because of their class membership. Also, these studies have shown that there is a certain mobility in our class structure, although this is certainly not as great as most people would claim. Furthermore, it is well known that occupations are very important in determining class status, because of a number of characteristics associated with certain occupations, such as nature of the work, extent of power, nature of associates, training and educational requirements, income, security, and the like. Concerning their role in the development of attitudes and anxieties, Levin has stated:

In terms, therefore, of the relationships between given occupations and their common class status, certain attitudinal and belief requirements may be expected to be associated with the various occupations. It would not even be rash to assume that many of the emotional and personality requirements of various occupations are fundamentally based on class status factors and not on job requirements, as such. Thus, the professional is expected to appear, behave, feel and think quite differently than the skilled worker, and even more differently than the semi-skilled worker or unskilled worker. The stereotyped hierarchical classification of vocations is essentially a reflection of their class-conferring characters.

In a relatively mobile class society in which the vocations have class-conferring potency, it is obvious that ego-involvement with respect to occupational achievement would be high for many. Occupations must be selected, consciously or otherwise, in terms of their value in either maintaining the present class membership, if that is adequate to the individual's level of class aspiration, or in terms of their value in facilitating the individual's climb to the class considered higher, if he is motivated to do so.[11]

These findings and conclusions are further supported by the results of the study by Bradley presented in Table LXIII. We note that 17.8 per cent of the high school students expressed a choice in the professional group, while the data presented in Figure 37 show that less than 7 per cent of the employed in 1947 were classed as professional and semi-professional. Again, only 10.7 per cent listed skill or semiskilled, while 21.5 per cent of the gainfully employed were classified into the single grouping of operatives and kindred workers. These studies indicate a need for the schools to consider the problems and conditions actually existing and to provide an educational and guidance program harmonious with reality.

Other materials related to the vocational aspirations of adolescents are presented in Table LXIV.[12] There is a decided interest in securing a position offering some degree of security. Adolescents from low-income families are motivated by the desire for security on the job more than are adolescents from higher-income groups, those who have probably never experienced the need for a feeling of security.

Occupational choices and opportunities. Occupational choices have been so affected by the social factors that they are out of harmony with occupational demands. This becomes evident from a further study of choices commonly made by high school boys and girls. The influence of parents on the choice of the occupation of their children has been revealed in a number of general studies. Kroger and Louttit[13] give the results from a questionnaire study of 4,543 boys in four technical and academic high schools. About 90 per cent of the boys expressed vocational choices. A majority indicated choices higher than those of their fathers. When compared with census figures, 70 per cent of the boys indicated a preference for types of work engaged in by only 35 per cent of those gainfully employed today.

[11] M. M. Levin, "Status Anxiety and Occupational Choice," *Educational and Psychological Measurement,* 1949, Vol. 9, pp. 29–38.

[12] "Fortune Survey," *Fortune,* December 1942, pp. 8–9.

[13] R. Kroger and C. M. Louttit, "The Influence of Fathers' Occupations on Vocational Choices of High School Boys," *Journal of Applied Psychology,* 1935, Vol. 19, pp. 203–212.

Every teacher concerned with counseling students about their vocational aspirations should have information concerning occupational demands and trends.   As late as 1870 more than one-half of all American workers were engaged in agriculture.   According to the 1940

TABLE LXIV

HERE ARE THREE DIFFERENT KINDS OF JOBS. IF YOU HAD YOUR CHOICE, WHICH WOULD YOU PICK?

	All Students	Boys	Girls
A job which pays quite a low income, but which you are sure of keeping	47.0%	41.3%	52.9%
A job which pays a good income, but which you have a 50–50 chance of losing	29.5	30.2	28.8
A job which pays an extremely high income if you make the grade, but in which you lose almost everything if you don't make it	22.4	27.8	16.8
Don't know	1.1	.7	1.5

COMPARISON OF VARIOUS GROUPS WITH RESPECT TO PREFERENCE FOR JOBS THAT OFFER DIFFERENT DEGREES OF SECURITY

	Don't Know	Low-Income Security	Good Pay at 50–50	All or Nothing
Negroes	.9%	68.2%	13.1%	17.8%
Poor	1.3	60.3	24.0	14.4
Prosperous upper middle group	1.1	33.7	36.1	29.1
From laboring parents	1.3	58.6	23.8	16.3
From executive and professional parents	1.4	32.8	34.7	31.1
Uninformed	.6	60.0	22.9	16.5
Well informed	2.4	28.6	38.5	30.5

WHAT OCCUPATIONS ARE YOU PLANNING TO ENTER?

	All Students	Boys	Girls
The professions, in this order: engineering, nursing, teaching, arts, medicine, and law	35.8%	36.1%	35.5%
Business—mainly clerical and secretarial	21.1	8.0	34.4
Factory work, skilled trades, mechanics	8.6	14.4	3.1
Government work—mostly armed forces	4.5	8.5	.3
Farming	3.2	6.2	.2
Other	11.4	12.2	10.4
Don't know	15.4	14.6	16.1

census less than one-fifth of the workers were classified as farmers, farm managers, foremen, and farm laborers.   The continued mechanization of farming during the past decade has reduced the per cent of people engaging in farming; by 1947 the number was only 13.6 per cent.   The trends from 1940 to 1947 are shown in Figure 37.[14] This should serve as a guide in counseling and in training adolescents, although it is well recognized that community conditions and demands play an important role in the vocational aspirations and opportunities of adolescents. The school counselor should be realistic and honest in his relations with adolescent boys and girls. Vocational choices should in the final

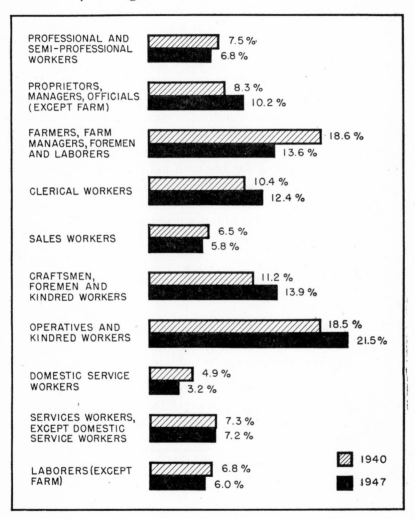

FIG. 37. *Occupational Distribution of Employed Workers in 1940 and 1947.*
(After Wool and Pearlman)

[14] H. Wool and L. M. Pearlman, "Recent Occupational Trends," *Monthly Labor Review,* 1947, Vol. 55, pp. 139–147.

analysis harmonize with the abilities of the individuals concerned and with the vocational opportunities that are likely to be present in the community or in the large urban area of which the community is in one sense an integral economic part.

VOCATIONAL NEEDS OF ADOLESCENTS

Educational demands. There is considerable evidence that the increased demand for universal education on the secondary level is not a result of the demands of various occupations and jobs. This is not to be interpreted to mean that there is no need for universal education on the secondary level; it means rather that such a need must be related to factors bearing upon increasing complexity of our social order, and not to actual demands of a vocational nature. However, as the average educational level of the young people entering the labor market is raised, those who leave high school without graduating will find themselves competing for jobs with individuals with more education. It has already been pointed out that employers consider many of these youth as too immature, undependable, or lacking in experience. A manager of several filling stations recently explained to the writer his reasons for hiring only helpers who had finished high school. He stated:

The high school graduates are more eager to make good.

They can be depended upon to be on the job at the time they are supposed to be there.

They are more intelligent and are thus able to assume responsibility when the need arises.

They are courteous to the customers and help to build up my trade.

It seems likely, that, while the high school students are learning science, mathematics, English, history, and other facts and principles in high school, they are also developing habits of orderliness, responsibility, dependability, cooperation, and ability to give and follow directions. These are habits and character traits considered by employers in the business, industrial, and professional occupational areas as most desirable.

During the summer and fall of 1938 more than 30,000 youths from seven widely separated cities, selected at random from lists of eighth grade graduates of 1929 and 1931, were interviewed by the Research

---

[15] *Disadvantaged Youth on the Labor Market.* Division of Research, Works Progress Administration, Series 1, No. 25, 1940.

Division of the WPA.[15] An analytic study was made of factors or conditions related to the "disadvantaged group." This group was made up of those youths who had been unemployed at least fifty per cent of their total time in the labor market and who had had, in addition, as the minimum amount, one year of unemployment. This disadvantaged group pointed out that the most common difficulty encountered was that of lack of experience. The question most frequently met by the individual seeking his first job related to *experience*. This is a serious problem. How is one to get experience without a job? It is, no doubt, one of the most important problems in the training of youths today for the labor market of tomorrow.

Vocational guidance. It has already been pointed out that the vocational aspirations of adolescents are not in harmony with occupational needs. There is also evidence that vocational interest patterns become fairly well stabilized by the time the individual reaches the eleventh grade, although circumstances and opportunities oftentimes affect vocational choices after this period. Taylor and Carter administered the Strong Vocational Interest Blank to 58 girls, first in the eleventh grade, and a year later in the senior year in high school.[16] Profiles of interests based upon the results obtained from these two tests revealed considerable stability in individual interest patterns during the year interval. This presents a challenge to those concerned with the guidance of adolescent boys and girls during the early high school years. In relation to this problem, Roeber and Garfield state:

> To the extent that students in the eleventh and twelfth grades tend to select occupations outside of the professions, we have evidence of a more mature and realistic point of view. The fact that such trends are limited implies that the secondary school has a definite responsibility for helping students look at occupations realistically. When we find that the choices of seniors differ only slightly from those of freshmen, and that the hopes of both groups are in great measure illusory, we can infer that the school has been relatively ineffective in the vocation guidance of its students.[17]

As we have suggested, when boys over fifteen or sixteen years of age who are interested in becoming lawyers, doctors, preachers, or members of some other profession are advised to enter some mechanical type of activity more in harmony with their abilities, they may often resist

---

[16] K. V. F. Taylor and H. D. Carter, "Retest Consistency of Vocational Interest-Patterns of High-School Girls," *Journal of Consulting Psychology,* 1942, Vol. 6, pp. 95–101.

[17] *Op. cit.,* pp. 361–362.

such advice. Furthermore, high school students have too often been attracted to professional and so-called "white collar" jobs under the influence of the example of some outstanding individual. Perhaps it is a successful uncle who studied law and is now a superior-court judge, or it may be an acquaintance who has made much money through selling life insurance. Now it is at this point that a great deal of vocational advice fails. The individual tends to picture his chosen work from the viewpoint of those who are highly successful in the work; he does not see its failures and hardships. The professions are held up as being clean, honorable, easy occupations offering good pay and considerable social prestige; the more mechanical activities are conceived to be laborious and dirty, unskilled, inferior in social status and in pay. This division, which sprang up before the day of the trained engineer and farmer, is not as well defined today; but the line of demarcation has been set up in part in the minds of the majority of boys and girls. One of the most prominent developments in recent years, in connection with the expansion of education, is the raising of so many other lines of human endeavor to a level close to that of the professions.

Not only is the discussion presented here true for young men, but similar factors are also valid for high school girls. In the past vocational guidance has been meager, and almost exclusively for men, but today it is reaching into the lives of high school girls. Observe the large number of girls occupied in various pursuits; notice the many lines of endeavor in which women engage. With the inclusion of so many activities in woman's domain has come their acceptance as desirable social positions, and thus they have become generally occupied by young women. Vocational guidance, then, has a prominent appeal for the young women of tomorrow.

Changes in the life activities, and particularly in the vocational activities, of women during the past century have been accompanied by innumerable problems for the adolescent girl. Cultural norms relating to the nature of the feminine role have been established. These norms have been passed down as part of our social heritage. Specifically, the male expects the female to play this feminine role.

The other role, introduced as a result of the induction of women into many vocational fields, tends to eliminate the factor of sex and the sex role. Here the girl is encouraged to study hard in school so that she can enroll in a certain school for nurses. The father and mother may point with pride to their daughter, who is now a laboratory technician at some reputable hospital. These contradictions of roles present difficult problems for adolescent and post-adolescent girls.   It is through her family and through boy friends in particular that she meets these contradictions and inconsistencies. The writer observed this in a study of the problems of adolescents, referred to in Chapter II. One of the most common problems added by girls to the check list was wondering whether or not I will ever get married.   In discussing these inconsistencies in relation to the girl's adjustments, Komarowsky stated:

. . . Generally speaking, it would seem that it is the girl with a 'middle-of-the-road personality' who is not happily adjusted to the present historical moment. She is not a perfect incarnation of either role but is flexible enough to play both. She is a girl who is intelligent enough to do well in school, but not so brilliant as to 'get all A's'; informed and alert but not consumed by an intellectual passion; capable but not talented in areas relatively new to women; able to stand on her own feet and to earn a living but not so good a living as to compete with men; capable of doing some job well (in case she does not marry or otherwise has to work) but not so identified with a profession as to need it for her happiness.[18]

Vocational training in the schools. The ever-increasing complexity of our industrial order emphasizes the necessity for vocational guidance and counseling, as well as for the provision of certain sample types of industrial training for boys and girls, who will later perform similar productive activities in their economic life.   Traditionally, young people have met this problem by adjusting themselves to vocational requirements outside of school, through actual employment as beginners. Abrupt transitions accompanied by emotional upsets, "rougher" factory conditions, and the danger of exploitation were some of the results due to the limitations of a system that failed to integrate school with postschool experiences. Progressive schools today are committed to the principle of bridging the gap between the life of the community and the activities within an educational institution.

There are some skills that are basic to broad fields of work; and there is knowledge of a vocational nature that is closely related to a large group or class of occupations or vocations. For example, agriculture is a term that is used in connection with many occupations— even with those of the proprietor of the country grocery store and the man who distributes farm goods and machinery. There is knowledge about the motor car and about the engine that pulls it through space that is important not only for the manufacture, repair, and operation

[18] M. Komarowsky, "Cultural Contradictions and Sex Roles," *The American Journal of Sociology,* 1946, Vol. 52, p. 189.

of the motor car, but for many related occupations as well. Thus, training in agriculture, home economics, industrial arts, and business education tends to function in a manner sufficiently extensive to justify its existence without impairing training in other principles necessary to the wholesome development of the individual student. "Furthermore," as Bent and Kronenberg have pointed out, "these fields, viz., home economics, industrial arts, business education, and agriculture have social and personal value as well as vocational, which is an additional reason why they can be included." [19]

However, the education of adolescents who give themselves wholeheartedly to the academic pursuits of high school on a full-time basis is destined to become a less important function of the public secondary school. The idea of "liberal" (academic) education for the gifted and "vocational" education for those less well endowed academically seems to some to be an undemocratic caste-form of school program. In recent years this attitude has been changing and, beginning with the Smith-Hughes Law of 1917, the federal government has constantly sought to encourage and stimulate the development of vocational education. Substantial federal funds (available to the states on the dollar-for-dollar matching principle) for this work are in large measure responsible for the great variety of courses offered and the large number of students now enrolled in them.

In the light of the findings of modern psychology relative to generalized experience and socio-industrial developments making for rapid changes in occupations, vocational programs need to be organized for broader skills and knowledge rather than around narrow skills and highly specialized knowledge. It has been pointed out that such programs should consist of the following phases: "(a) practical, concrete activities and experiences of the occupations represented, (b) related science and technical information, and (c) the social and economic understandings and appropriate attitudes." [20] The school program provides largely for the second of these and to a lesser degree for the third. In order to provide for the first, it is necessary to readjust the educational and vocational programs.

In connection with the organization and planning of vocational courses, it has been pointed out that, "there is increasing recognition

[19] Rudyard K. Bent and Henry H. Kronenberg, *Principles of Secondary Education.* New York: McGraw-Hill Book Co., 1941, p. 359.

[20] L. V. Koos, J. M. Hughes, P. W. Hutson, and W. R. Reavis, *Administering the Secondary School.* New York: American Book Co., 1940, p. 75.

that highly specialized types of vocational education should be reserved until the period immediately prior to the time when the pupil leaves the full-time school, and that in many cases the young person can better come back for specialized vocational courses after he has made a beginning in some suitable occupation." [21]   It has been proposed that the reorganized secondary school curriculum should involve the specific activities that high school youth is now experiencing, subject to guidance and integration of these experiences with materials related to different aspects of the school program.   There are two types of such activities found in most communities that can be used for this purpose. One of these is made up of the gainful occupations; the other involves civic-social participation.   According to Julius L. Meriam, twelve per cent of all high school pupils are wage earners on a part-time basis.[22] Many of these students work because of economic necessity; others are motivated by special interests and the desire to be usefully and actively employed.

A survey of the work opportunities in almost any community, under normal conditions, will reveal a great number of jobs well within the capacities of high school youth, and, at the same time, educationally valuable, if conducted under the supervision and guidance of competent school authorities.   There is perhaps no limit to what can be done in this connection.   However, it is obvious that difficulties may arise relative to child-labor legislation or to conflict with labor organizations. However, if this work is conceived to be an educational adventure and if precautions are taken to safeguard the pupils from exploitation, one part of the difficulty will have been overcome.   Furthermore, if the public recognizes that the work is done as part of the school program, and is not a procedure for supplying cheap labor, the second problem can be overcome as well.

The cooperative plan of education, whereby boys and girls of senior high school age spend part of their time at work and part in school, has many advantages.   Such plans have been used successfully in various types of school programs, and have received the sanction of a number of educational agencies.   There are, however, definite precautions to be taken, for the dangers inherent in part-time work should be eliminated as far as possible.   Experience in certain types of work,

[21] Floyd W. Reeves, *Youth and the Future*.   Washington: American Council on Education, 1942, p. 139.

[22] Julius L. Meriam, "The High-School Curriculum," *Phi Delta Kappan,* 1942, Vol. 25, pp. 13–16.

it has been pointed out, is oftentimes available in the home. Qvol Spafford has stated:

> Young people, who wish to do so, may secure home employment for almost any kind of work in which they are skilful, for as much time as they wish to be so occupied. Home economics teachers have a real responsibility for making this employment an educational experience for those who undertake it.[23]

The transition from school to work. Each year about a million boys and girls begin full-time work. For many of them this is an abrupt change from the sheltered life of the home and of the school to employment in the factory, in the office, at the counter, or on the assembly line. One day they are in school, an institution operated for the guidance and education of the boys and girls into useful citizens. The next day they are on their own, a cog in the industrial machinery or in the business enterprise. Their particular place in this vast enterprise is determined largely by the needs and demands of the particular time. A transition so rapid as this is sure to produce job dissatisfaction and vocational maladjustments.

The first job is a milestone in the individual's life. It appears as the first real step toward adulthood and independence. A wrong start— getting fired, finding the job beyond one's abilities, finding that the job fails to offer the things the individual was aspiring toward, or finding management unreasonable with him in his early efforts—is a frustrating experience for the young worker and adversely conditions him toward a working life. Thus, there is a definite need for preparing young people to make satisfactory vocational adjustments. Karl Smith's experience indicates some of the problems faced by a boy who drops out of school with no vocational plans, very little vocational training, and apparently no vocational guidance.

> Karl Smith, just past his sixteenth birthday, had worked just one week in the year since he left school. He dropped out of school for a combination of reasons after 4 months in the sheet metal course of the local trade school. He had serious financial problems at home. He was disappointed at not being admitted to the auto mechanic course and he felt stigmatized at school because he had been called into court for truancy. His one brief job had been in an army supply store where he worked as stock boy. Although really interested in finding steady work he had not registered at the employment service, because he felt sure that his juvenile court record would

---

[23] Qvol Spafford, "Adjusting Home Economics to Wartime Needs," *The School Review,* 1943, Vol. 51, p. 36.

'queer' his chances there. Confused and discouraged by his failure to find a job, he was following the 'help wanted' ads in the newspapers. Karl had no plan for the future, since his interest in learning to be an auto mechanic had been frustrated.[24]

The problems faced by Karl reveal the need for assistance in general vocational orientation, placement counseling, and assistance in finding a job.

The story of Jean Black indicates, however, that graduation from high school does not necessarily assure one of a job or even of a satisfactory vocational orientation.

Jean was an attractive girl of 18 with intelligence, poise, and considerable musical ability. She not only completed high school but spent 6 months in college. She had 2 months' experience in sales work before going to college, and her parents gave their approval and financial support to her education.

Yet Jean was 'in a quandary as to where to turn.' She did not like her brief experience in selling. But she was also dissatisfied with the music course she took in college, because she considered it would not lead to practical employment. Jean had no interest in returning to school for a business course, however. She was marking time with a Saturday job in a downtown department store, and would have welcomed counseling from any source that could have helped her get a sense of direction.[25]

Values of work experience. For the majority of youth the need is for work experience rather than for training for some specific job. As a result of studies dealing with this problem, Jacobson has stated: "What they need is work experience, vocational guidance, placement, and an understanding of vocations and their possibilities."[26] Materials presented in the study by Bell[27] lend support to this notion. According to the opinion of employers in occupations believed to employ seventy per cent of all workers, more than two-thirds of those workers could be trained to reach full production in one week or less, and less than ten per cent of them would need to be trained for a period of six months or more.

Some of the work experiences that might well be offered to youth

[24] "Hunting a Career: A Study of Out-of-School Youth in Louisville, Kentucky," *United States Department of Labor Bulletin No. 115,* 1949, p. 90.

[25] *Ibid.,* p. 90.

[26] North Central Association of Colleges and Secondary Schools: *General Education in the American High School.* New York: Scott, Foresman and Co., 1942, p. 272.

[27] Howard M. Bell, *Matching Youth and Jobs.* Washington: American Council on Education, 1940, p. 58.

are those involving community beautification and betterment. Certainly, the care of wild life, the conservation of natural beauty as well as of natural resources, and the elimination of rubbish and other unpleasant elements from the physical environment will, in the end, make for community betterment as well as for immediate beautification. The extent to which such enterprises are carried out will depend in a large measure upon the personnel concerned with the initiation of the program. Problems connected with sanitation, recreational projects, and social needs can well be solved through the activity of youth groups, provided help, encouragement, and leadership are given by schools and civic groups in initiating the program. Some communities are providing work experiences in various types of community surveys and in certain types of research activities relative to community conditions, needs, and problems. These and many other more or less related types of work are among the things that can be done by youth groups. Such work experiences may thus be valuable for the improvement of the community at the same time that they are furnishing experiences of an educational nature to the youths involved in them.

Charles P. Schwartz, Jr., of the University High School, Chicago, Illinois, spent eleven weeks during the summer of 1942 on a farm in Wisconsin. His description of some of his activities there reveal the values that such work experiences may have for an urban boy. He writes as follows:

Before we left we had quite a few meetings at school, learning, after a fashion, about farm life. We also sent to the Department of Agriculture for some pamphlets, but we found that we needed some practical experience before they would do us much good.

Finally the day arrived. It was Saturday, June 20th. We left the city at noon and arrived at the farm about two o'clock.

That evening I did my first chores, throwing down silage, feeding horses and pigs, and a few other miscellaneous chores. These were to increase as the summer went on, but then they were all I could handle. Never having been on a farm before, I learned quite a few facts my first day: that milk was warm when it came out of a cow, that silage was warm, that pigs really deserved the name of hog. . . .

Among the things which I thought were unusual during my first few days were: how horses could eat so much hay, and how much housekeeping horses and cows required; I had to clean the barns. . . .

One of my biggest jobs during the summer was the shoveling of grain during combining. After combining was over I wondered how I could ever have shoveled so much grain but I guess I did. . . .

I had to go home on the fifth of September, but as I look back on the summer I feel that it was about the best I ever had and that if I'm needed I would like to go back. I acquired many new skills and saw the rural way of life and its outlook.[28]

As schools come to utilize the work experience of pupils to the greatest degree, it becomes imperative for the guidance program to be broadened in scope. This broader program will encompass the needs of individual boys and girls in terms of their background of experiences as these relate to their future needs. This changing concept of guidance has been described in the following manner:

The conclusion seems inescapable that elements of a guidance program which is to serve even purely vocational objectives must begin in the elementary school. Factors of organization will radically alter the content and method of such a program, but principles and objectives will remain the same. The school will accumulate an inventory of pupils' traits; teachers will seek to individualize their instruction, and to bring about reasonable adjustments of pupils to their environment; and the school as a whole will aid pupils in making choices wisely and in viewing the work-world in the light of facts. Most important of all, the individual may be accepted for vocational training with an assurance that he feels he is undertaking the next step in a logical and continuous educative process.[29]

The American Youth Commission gave a great deal of consideration to the problems involving the place of work experience in an educational program. In an effort to clarify its findings, it formulated certain general principles, two of which are:

1. Appropriate amounts of useful work are desirable elements in the experience of children and youth of all ages. During the years of compulsory school attendance, such work should be subordinated to the requirements of schooling. In many instances, productive manual labor and other forms of useful work should be introduced into the school program as an element on a par with other major elements of a well-rounded curriculum.

2. In the personal development of every young person there comes a time when, in his or her own interest and in the interests of society, employment should replace school attendance as his or her major occupation. For many young persons this time comes at the age of 16, the age up to which school attendance should be compulsory. . . .[30]

---

[28] Charles P. Schwartz, "Eleven Weeks on a Farm," *Progressive Education,* December 1942, pp. 432–433.

[29] Committee to Study Postwar Problems in Vocational Education, *Vocational Education in the Years Ahead*. Washington: U. S. Office of Education, 1945, p. 320.

[30] The General Report of the American Youth Commission, *Youth and the Future*. Washington: American Council on Education, 1942, p. 58.

SUMMARY

Vocational training is a function, a resultant of technological developments, economic changes, and social forces. It is based upon a recognition of the value of the individual as a member of a social group. Within the social group is variety, caused by the specialization of labor, and this calls for vocational guidance and training to the end that each individual may be successful and adjusted in his place in the world of work.

There is no unanimity of opinion concerning the effects technological developments are having upon the concept of the function of schooling. In many cases, such as in that of the specialization found on the assembly line, the need for training in specific vocational skills has decreased. It appears that there is now a need for a broader conception of vocational training, which will include character and personality development. Furthermore, the increased complexity of our social and economic structure has increased the need for considering training in citizenship as part of the vocational training program.

Scientific research has shown the following general facts to be characteristic of adolescent vocational interest:

1. Pupils often make vocational choices on the basis of some single momentary factor, such as: social approval, some friend or kin engaged in the activity, an enthusiastic lecture, recency of contact with some animating personality, and so forth.

2. Even choices that are made momentarily are usually supported by some fairly well developed interest. The choices made are somewhat in harmony with earlier life bents.

3. Pupils of both mediocre and superior mental ability have, to a very large extent, vocational preferences.

4. Vocational interests and ambitions of high school pupils are not in harmony with the actual possibilities of employment.

THOUGHT PROBLEMS

1. Account for the fact that the youth group is the one most affected by periods of unemployment.

2. What are some of the major causes of occupational maladjustments? Give some remedies.

3. What are the values of *work experience* for high school pupils? What are some dangers to be avoided?

4. List some of the basic factors to be considered in a sound vocational guidance program.

5. How has technological development affected the nature of vocational training? Illustrate your answer by reference to some jobs with which you are acquainted.

6. Using any data available relative to the number of people employed in various types of work in some nearby community (see the Census), outline the type of educational program needed for such a community.

7. What occupational trends are suggested in Figure 37? What are the implications of these trends to education? To vocational guidance?

8. How would you account for the large number of high school boys and girls undecided with respect to a vocation? What generalizations would you make from the results presented in Figure 36?

SELECTED REFERENCES

Activities of the American Youth Commission. Washington: American Council on Education, January, 1937; second edition, March, 1937.

Averill, L. A., *Adolescence.* Boston: Houghton Mifflin Co., 1936, Chap. XIII.

Bell, H. M., *Matching Youth and Jobs,* Part 1. Washington: American Council on Education, 1940.

Carter, Harold D., "The Development of Interest in Vocations," *Forty-third Yearbook of the National Society for the Study of Education,* 1944, Part 1.

Cole, Luella, *The Psychology of Adolescence* (Third Edition). New York: Rinehart Co., 1948, Chap. XVI.

Chamberlain, L. M., and Kindred, L. W., *The Teacher and School Organization.* New York: Prentice-Hall, Inc., 1949, Chaps. XIII and XIV.

Committee of the National Society for the Study of Education (F. J. Keller, Chairman), "Vocational Education," *Forty-second Yearbook of the National Society for the Study of Education,* Part 1, 1943.

Garretson, P. K., "Relation between Expressed Preference and Curricular Abilities of Ninth Grade Boys," Teachers College, Columbia University, *Contributions to Education,* 1930, No. 396.

Hollingshead, A. P., *Elmtown's Youth.* New York: John Wiley & Sons, Inc., 1949, Chap. XI.

Hoppock, R., *Group Guidance: Principles, Techniques, and Evaluation.* New York: McGraw-Hill Book Co., 1949.

Hutson, P. W., "Selected References on Guidance," *The School Review,* 1948, Vol. 56, pp. 421–425. This is an annotated bibliography of selected writings appearing in the last half of 1947 and in the first half of 1948.

Landis, Paul H., *Adolescence and Youth.* New York: McGraw-Hill Book Co., 1945, Chaps. XV, XVI, and XVII.

Myers, George E., *Principles and Techniques of Vocational Guidance.* New York: McGraw-Hill Book Co., 1941.

Trout, D. M., "Academic Achievement in Relation to Subsequent Success in Life," in E. G. Williamson (Editor), *Trends in Student Personnel Work*. Minneapolis: University of Minnesota Press, 1949, pp. 201–217.

Warters, Jane, *High School Personnel Work*. New York: McGraw-Hill Book Co., 1946.

For a complete review of the literature for the three-year period ending October 1, 1947 see "Counseling, Guidance, and Personnel Work," *Review of Educational Research*, 1948, Vol. 18, No. 2.

XXI

Adolescents and Democracy

THE SUPERIOR MAN IS LIBERAL TOWARDS OTHERS' OPINIONS, BUT DOES NOT
COMPLETELY AGREE WITH THEM; THE INFERIOR MAN AGREES WITH OTHERS'
OPINIONS, BUT IS NOT LIBERAL WITH THEM. *Confucius*

Human resources. The possibilities of a nation are limited by its
resources. The boys and girls enrolled in our schools today are the
human resources upon which the future of our country is largely dependent.[1] The development of these resources represents the major
task of our schools. These youths cannot be neglected without disas-
trous consequences. In times past the rearing and induction of boys
and girls into adult life were functions of the family, the church, the
schools, and privately controlled business and commercial concerns.
These agencies worked harmoniously, according to the general philosophy of individualism, in bringing individuals to independent and
self-sustaining adulthood. However, changed social and economic
conditions have so altered conditions that many adolescents and youths
are faced with critical problems relative to education, health, employment, and social and recreational needs. If the human resources
present in growing boys and girls are to have the fullest and most
complete utilization, those boys and girls must be guided in their
maturation and development.

The previous chapters have pointed out the influences of various
forces and conditions on the development of the adolescent's philosophy
of life.[2] The question of the relative influence of different institutions
has been studied by many students of educational psychology. The
Institute of Student Opinion Poll, sponsored by *Scholastic Magazine,*
conducted a number of studies of adolescent opinion. Members of

[1] See Chapter I for the distribution of adolescents.
[2] See the *National Parent Teacher* for April, 1946 for a more complete discussion of this and its implications for the home.

455

the Institute are 1,555 high schools broadly representative of the area of American secondary education.   One of its surveys asked this question: "In your opinion which of the following influences your thinking to the greatest extent: parents and family, schoolteachers, close friends ('the gang'), community and student opinion, magazines and newspapers, radio, movies, schoolbooks, or church?"   More than 101,500 students responded and gave frank opinions with the candor that parents and teachers learn to expect from high school boys and girls. The answers given to this question are presented in Table LXV.

The large vote in favor of parents show that high school youth still rely upon the home to provide a basis for their attitudes and opinions. They express the notion that their parents know them better, that they are able to argue more freely with their parents, or that their parents are more reasonable with them when they need help.   The

TABLE LXV

THE RESPONSE OF HIGH SCHOOL STUDENTS TO THE QUESTION: *Which of the following influences your thinking to the greatest extent: parents and family, schoolteachers, close friends ("the gang"), community or student opinion, magazines and newspapers, radio, movies, schoolbooks, or church.*

Family	38%
Magazines and newspapers	17%
Close friends ("the gang")	11.5%
Radio	10.5%
Community or student opinion	6%
Schoolteachers	5%
Church	5%
Movies	3%
Schoolbooks	1%
Don't know	3%

function of the school cannot be wholly ignored; however it would appear that the school functions indirectly as well as directly by preparing students for home membership and by training them to become parents of a future generation, better prepared to guide their children in the solution of problems.

Autocratic, democratic, and laissez-faire controls.   The effects of autocratic and democratic control on the home and on community life have been discussed in earlier chapters.   One of the most important means of preparing boys and girls for participation in democratic ways of life is to provide them with opportunities for participating in the various institutions concerned with their guidance and development. However, the relationship between autocracy, democracy, and complete individual freedom (laissez-faire) is poorly understood by the average counselor, teacher, and parent. The relationship is usually thought of as following a linear scale, at one end of which is laissez-faire and at the other autocracy. Democracy is then thought of as a form of control falling around the mid-point between these extremes. That such a notion is incorrect has been well illustrated by Kurt Lewin.[3] According to his concept, these forms of control should be perceived as a triangle. Since both democracy and autocracy are types of leadership involving controls, they are somewhat similar. These, then, can be perceived of as being on a straight line. Autocracy presents that type of leadership in which all the controls are highly centralized. In the case of complete democracy these controls lie in the voice and action of the people. The line between autocracy and democracy represents a continuum—showing all degrees from complete leadership responsibility and control to complete group responsibility and control. This relationship is illustrated in the triangle of Figure 38. Autocracy and democracy diminish and converge at *LF,* which represents laissez-faire. At this point there is an absence of both democracy and autocracy; there is complete individual freedom and perhaps chaos.

Adolescent boys and girls need to be taught the true meaning of democratic controls through both precepts and example. The best protection against the encroachment of autocratic controls in our national life is for our homes, clubs, schools, churches, and other institutions to operate in a democratic manner. Adolescents should learn that with freedom goes responsibility, that controls are essential if a harmonious social order is to exist, and that if they desire to make their own choices they must be willing to accept responsibilities involved in such choices. In a democracy controls must be established in the hearts of men. Thus, habits of control must begin during the early years of life and become an integral part of *the self* as the individual grows into maturity.

Essentials for growth in self-control. There is a need for the home, school, church, and other agencies, concerned with the guidance and direction of boys and girls, to cooperate in their programs, which affect the lives and activities of adolescents. It was suggested in Chapter XI that adolescent boys and girls use various means to bring about their independence of parental control. It is difficult for many parents to

[3] K. Lewin, "The Dynamics of Group Action," *Educational Leadership,* 1944, Vol. 1, pp. 195–200.

relinquish their authority. The school and other agencies also find themselves playing an authoritarian role, and thus fail to provide for the adolescent's need for achieving independence and maturity. The problem of how to get more experience in self-discipline and

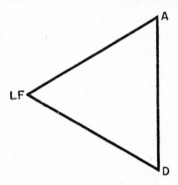

FIG. 38. *The Relationship of Autocracy* (A), *Democracy* (D), *and Laissez Faire* (LF). (After Lewin)

group cooperation (essentials for effective citizenship in a democratic society) in responsible and significant activities is thus made a difficult one. A summary of the needs of the established social forces as they affect the adolescent include:

1. A better understanding for parents of the importance of late childhood and early adolescence as a basic period for training in responsibility and self-discipline.

2. A recognition by parents of the characteristics and needs of adolescents as a basis for providing for their achievement of independence and status, and for preparation for adult responsibilities.

3. High schools of a democratic nature concerned with providing worthwhile learning experiences in which the students shall have a responsible share in planning.

4. High schools concerned with the development of cooperative, responsible citizens, schools that will furnish genuine experiences in self-government in matters that directly concern the lives of the individual students.

5. Community enterprises that will furnish opportunities for adolescents to plan and execute activities of a wholesome nature which provide for the moral, social, and personal growth of adolescents. (It is here that the church can function effectively in providing opportunities and guidance in recreational, social, and spiritual development of adolescents.)

6. The cooperation of the industrial and business life of the community in providing rich and extensive opportunities for adolescents to explore the

world of work. This would provide these boys and girls with a basis for making sounder decisions relative to their educational and vocational plans.

Courtesy, Vocational Agriculture, University of Georgia.

Meaningful and Stimulating Experiences provide the Best Preparation for Life.

FROM ADOLESCENCE TO MATURITY

The maturing adolescent. Throughout preceding chapters it has been emphasized that as the child grows into the period of adolescence, following that of childhood, he is truly entering upon a new sphere of activity. He is reaching into a new social atmosphere, his maturing physiological nature is asserting itself along new channels, and new impulses are arising. It has furthermore been pointed out that behavior is not explicable wholly in terms of the stimulus-response hypothesis but rather in terms of the individual as a whole. This means, in the case of the adolescent, his biological and sociological past, as well as the momentous present.

The adolescent, with his rapid physiological changes, with his new type of physical potency, with his increased physical strength and vigor, with his growing impulses relative to others, is not the same organism that responded to various stimuli during infancy and childhood; because of his organic changes his responses to various stimuli are quite different from what they were just a few years ago. At four years of age Tom will call to Mary, a neighbor's child, to climb over the fence and play in the sand pile with him. At the age of seventeen, Tom will likely be calling over the phone rather than over the back fence. This time the call will be for an automobile ride, a dinner dance, or a swim in the lake. The impulses prompting the call over the back fence to play in the sand and those prompting the call over the telephone to go for a ride or a stroll are different; the interests of a maturing organism have replaced those of the playful child.

Thus behavior changes somewhat in harmony with the physiological changes that are taking place at this period; also, with such changes in behavior activities Tom and Mary face increasing responsibilities and increasing needs for adequate adjustments to a changing condition in their own life and environment. To express it analytically, the drives of both Tom and Mary have undergone pronounced changes. This is a clear illustration of the development of heterosexuality, referred to in Chapter X.

So that their responses may be adjusted in harmony with the changed physiological self, adolescents should be prepared to know and understand the nature and significance of the changes and activities that take place during adolescence. It is not enough that the child be informed concerning changes after they have begun; in fact, if those from whom he should have received information have not given it to him by this time, it is quite likely that he will have gained it from unguided and often ill-informed sources. Canivet[4] found from a questionnaire given 697 men and 153 women that they had received information about sex between the ages of 8 and 12 years, the pre-adolescent stage of life. A child who is dealt with frankly, positively, and honestly by parents or other advisers will not, to satisfy a naïve curiosity, find it necessary to seek information through misinformed or otherwise questionable channels.[5]

Margaret Mead[6] found, during a year's stay with the Samoan people, an almost total lack of mental-hygiene problems. This she attributed in the main to frankness in dealing with problems of childhood and the open attitude toward physiological processes, functions, and changes.

[4] N. Canivet, "Enquête sur l'initiation sexuelle" (A study of sexual enlightenment), *Archives de Psychologie,* 1932, Vol. 23, pp. 239–278.

[5] For a rather comprehensive discussion and guide to sex education developed for use in the schools, see F. B. Strain, *Sex Guidance in Family Life Education; A Handbook for Schools.* New York: The Macmillan Company, 1942.

[6] Margaret Mead, *Coming of Age in Samoa.* New York: William Morrow and Co., 1928.

The consensus among physicians, biologists, psychologists, educators, and other students of human nature is that information should be given to the child in harmony with his curiosity and ability to understand. As Bigelow points out:

The accumulating evidence is pointing towards the conclusion that the 'critical' aspect of human puberty in highly civilized countries is probably due very largely to unhygienic conditions, most of which are preventable or correctable in childhood and adolescence.[7]

Changes of needs and goals with age. New needs or a change in existing needs may arise as a result of a variety of circumstances and conditions. During the period of growth and development, new needs are continually appearing in the child's life, while certain childhood needs related to care and protection disappear or lose their potency. The development into adolescence introduces a different self and different concepts of the self. This was emphasized in earlier chapters. Likewise, old goals are reorganized and new goals are introduced. The individual begins to give more consideration to the things that he will be doing a few years hence, when he is more completely independent of his family. Educational problems, financial difficulties, job opportunities, vocational decisions, and the ultimate establishment of a home of his own appear as more realistic goals than they had seemed during the early childhood years.

Parents and teachers should be aware of the fact that the needs and goals of the adolescent are indeed different from those of the younger child, but are not identical with those of the mature adult. Each period of life may be said to present a different individual with different needs, although life itself is a continuum and the individual life span is a continuous process of development and change. Concerning these conditions Kurt Lewin has stated:

Generally speaking, needs may be changed by changes in any part of the psychological environment, by changes of the inner-personal regions, by changes on the reality level as well as on the irreality level (for instance, by a change in hope), and by changes in the cognitive structure of the psychological future and of the psychological past. This is well in line with the fact that the total life space of a person has to be considered as one connected field.[8]

[7] M. A. Bigelow, *Adolescence: Educational and Hygienic Problems.* New York: Funk and Wagnalls, 1924, p. 33.

[8] Reprinted by permission from K. Lewin, "Behavior and Development as a Function of the Total Situation," in *Manual of Child Psychology* (L. Carmichael, Editor), published by John Wiley & Sons, Inc., 1946, p. 824.

Youth and marriage. Studies cited in Chapter II indicated that love and marriage loom large in the lives of adolescents. There is some evidence that the sex drive reaches its maximum during this period; however, the effects of technology and the demands for increased schooling have tended to prolong the period of adolescence and youth so that the individual is unable to assume the responsibilities of family life during the post-adolescent period. This has had a profound effect upon our social and moral structure, although many parents, teachers, religious workers, and others would like to remain blind to the changed conditions. This has no doubt so affected the sexual lives and practices of the present generation that many activities that were seriously frowned upon by those of a few generations ago are quite widely accepted today. It appears that a democratic society is faced with one of two choices in this connection. Either we must recognize the fact that the sex drive is powerful during this period and that guidance rather than repression of adolescents in their social activities must be followed, and that increased sexual activities will be found where such a program is not instituted, or we must provide in some way for earlier marriages. Although the latter situation may appear to many to be the ideal it seems unlikely that it can be followed with our present social structure.

The social effects of technology. The adolescent of today is forced to adjust to a culture characterized by change, which brings with it confusion and conflict. This period of change is a result of the innovative forces of science, invention, and discovery, which have been termed, in recent years, *technology.* Gradually the entire world is being affected by this technological revolution. Work has become more highly specialized; individuals have become more dependent upon one another; natural boundaries have, to a large degree, disappeared; distances have been reduced from weeks and months, in terms of time, to hours; and nations have been brought closer together. Not only are the consequences of this revolution apparent in the changing material structure of society and the world of nations, but it is having an increasing effect upon the social and spiritual life of individuals and groups; a fact especially evidenced by the habits, attitudes, and values of adolescents today as contrasted with those of adolescents at the beginning of the present century.

The facts presented relative to economic trends lead one to the obvious conclusion that professional administrative and engineering positions cannot be the goal of all adolescents. Our economic system operates in many ways to limit social mobility, although social mobility —the opportunity to climb the economic and social ladder—has always been the dream of those coming to America from other lands. Youth should no longer be deluded by wishful thinking into reaching beyond its grasp. Just because a boy or girl does not choose the professions or some advanced technical field for his life work is no proof that he is a failure. Every individual should be brought to realize his responsibility to himself and to society, and to use his initiative in adjusting to the social-economic order in which he lives.

Adolescent boys and girls must be taught what the past several generations were not taught: namely, to realize that technological advancements have brought with them new problems and added responsibilities. The printing press, the radio, the movie, and the airplane have, or should have, revealed to us that technological advancements may serve as evil and destructive forces as well as beneficial ones. All too often we have hailed the benefits of science without considering the price that is being paid for them in terms of their effect upon society. H. L. Shapiro forcibly pointed out the price tag when he wrote:

In no aspect of our lives as members of a complex industrial community, or as a nation in the modern world, has technology brought greater responsibilities than in our attitudes toward the various groups that make up our society, or toward the peoples that constitute mankind. It is a commonly observed truism that the world grows more interdependent, and that our society demands increased cooperation from all of its members, as mechanization progresses.[9]

In a society that is being transformed by technological advancements as rapidly as is ours, capacity for adaptation and adjustment is the one quality that will be most needed for effective and successful living. These transformations are destined to affect all phases of our lives and all our institutions. Although profound changes have appeared, the full effects of technology are yet to come. This will call forth, on the part of youth, the capacity to learn, the motivation for learning, the willingness to assimilate, and the readiness to make adjustments required for a dynamic democratic society. Old experiences will be no handicap to youth, although many are handicapped by outworked or outmoded concepts handed down to them largely through the family and through the class structure. In this connection John G. Rockwell stated:

[9] Harry L. Shapiro, "Anthropology's Contribution to Interracial Understanding," *Science,* May 12, 1944, p. 373.

Seeing the sights in the nation's capital. Finalists in the Science Talent Search conducted by the Westinghouse Educational Foundation.

The role of learning in biological adjustment seems to be strategic. The sensitivity of this mechanism to the various conditioning agents suggests that its usefulness—if one may speak in such terms—lies in the fact that this gives to the animal a potentiality for variation and a plasticity nicely adapted to meet the needs of a constantly changing environment.[10]

Youth and the socio-economic outlook. Public opinion polls reveal that some interesting changes have taken place in the attitudes of the public during the past several decades, all of which is indicative of the socio-economic revolution through which we are passing. These attitudes have been reflected in our policies in international affairs, trade, monetary standards, social welfare, security, educational opportunities, health, and unemployment. These changes have especially affected individuals in the lower occupational scale, and have been enhanced by universal education, the press, the radio, and labor unions. Today the average individual is better informed than at any other period in American history. All this means that there has come about a change in the outlook, aspirations, and values of the great mass of American citizens. These changes are without doubt reflected in changed attitudes of adolescents toward things wanted relative to employment.

These changes are reflected in the attitudes of the American worker as revealed by a comparison of the results of the *Fortune* magazine survey of 1940 with those obtained from the survey seven years later.[11] In 1940 there were from nine to ten million unemployed, whereas seven years later we were in a postwar inflation with unemployment at about one-fifth of that of the earlier period. Yet the prewar citizen was less anxious to have security and a steady job with low pay than was the postwar citizen. The adolescent in our schools today is desirous of finding employment, with some opportunity for advancement, but with a fair degree of security. Most of them are too realistic to be lulled into daydreaming of prestige, power, and fortune. True, many adolescents have developed aspirations far beyond that of possible fulfillment. However, changing economic conditions are affecting the attitudes of adolescents toward social-economic policies and legislation.

As individuals move from the sheltered life of early and late childhood into the more responsible world of adulthood, these transforma-

[10] J. G. Rockwell, "How We Learn—Some Physiological Factors," *Mental Health in the Classroom.* Thirteenth Yearbook of the Department of Supervisors and Directors of Instruction, National Education Association, Washington, D. C., 1941, p. 88.

[11] "The Fortune Survey," *Fortune,* January 1947, Vol. 35, p. 10.

tions become more realistic and evident. That postadolescents are giving consideration to the social and economic problems of today is evidenced by answers received from college students to the questions listed in Table LXVI.[12] These results indicate that the more advanced college students are more favorable to the United Nations organization and to the idea of an international police force and world court than are the younger students.

In the midst of cultural confusion, variances in values, and socio-economic changes, it is not surprising that many adolescents meet with conflicts and frustrations. Along with the freedoms that are provided for youth in a democratic land, there is a need for freedom from fears related to unemployment. This is a challenge that faces the present and the future generations. This is a problem our basic institutions can ill afford to ignore. In discussing the function of the schools in meeting this challenge, Newton Edwards stated:

> Indeed, it may not be too much to say that for this and the next generation the most important obligation of the American educational system is to prepare youth to pass sound judgment on fundamental patterns of public and social policy, to equip them with the values, motivations, intelligence, and knowledge they will need in working out co-operatively the design of a new society.[13]

YOUTH AS CITIZENS OF A DEMOCRACY

Youth in politics. According to the 1940 census, over eight million people who had never before voted for a President were qualified to participate in the election of that year. Young people invariably follow in the footsteps of their parents in their political affiliations. However, the young people of today have certain characteristics distinguishing them from their counterparts of previous decades. Today youth is likely to be better educated than it was forty or fifty years ago. Secondly, young people now do not recognize party loyalty to the same degree as did their fathers of old. "Stand by the party no matter

---

[12] Karl C. Garrison, "A Comparative Study of the Attitudes of College Students Toward Certain Domestic and World Problems," to be published in *Journal of Social Psychology*. See also Benjamin Shimberg, "Information and Attitude Toward World Affairs, *Journal of Educational Psychology*, 1948, Vol. 19, pp. 206–222. Some of the items used in the study conducted by the writer were taken from this study. The study reported by Shimberg was based upon results obtained from a nationwide sample of approximately 10,000 high school students.

[13] N. Edwards, "The Adolescent in Technological Society," *Forty-third Yearbook of the National Society for the Study of Education*, Part 1, 1944, p. 196.

TABLE LXVI

RESPONSES OF COLLEGE STUDENTS TO CERTAIN PROBLEMS INVOLVING
DOMESTIC AND INTERNATIONAL RELATIONS

Should the development of large power projects for flood control be carried on by private industry or by the government?	PRIVATE INDUSTRY	?	GOVERN-MENT
Freshmen	17%	2%	81%
Sophomores	8%	14%	78%
Juniors	12%	5%	83%
Seniors	13%	6%	81%
Should the large forest areas be controlled for conservation purposes by private enterprise or by the government?			
Freshmen	22%	5%	73%
Sophomores	7%	10%	83%
Juniors	11%	2%	87%
Seniors	13%	0%	87%
Should free medical services and an annual examination be provided for all the citizens?	No	?	YES
Freshmen	22%	5%	73%
Sophomores	24%	15%	61%
Juniors	27%	17%	56%
Seniors	37%	21%	42%
Do you or do you not think that the United Nations should be strengthened to make it a world government with power to control the armed forces of all nations, including the United States?	Do	?	Do NOT
Freshmen	58%	10%	32%
Sophomores	53%	10%	37%
Juniors	53%	18%	29%
Seniors	54%	25%	21%
Do you or do you not believe that all nations should form a world organization with power to use an international police force against any nation, including the United States, which tried to start a war?	Do	?	Do NOT
Freshmen	73%	17%	10%
Sophomores	76%	14%	10%
Juniors	77%	13%	10%
Seniors	86%	8%	6%

what they espouse nor whom they elect," has lost its appeal. Furthermore, the typically American, pervasive interest in experimentation has entered the political field, bringing with it the hint to youth that it is entitled to propose a better design even though that design may be

opposed to past experience.   It should be pointed out again that young
people need guidance in these activities, in order that democratic ideals
and democratic ways of behaving may become a part of them, a point
emphasized, with a caution against overdoing it, by Wilma Lloyd:

> In our society, under the dominance of our democratic tradition, the neces-
> sity of self-determination in its relation to social co-operation is clearly rec-
> ognized.   Continuing development of democratic ideals depends upon it.
> But by and large we have fallen, in practice, between two poles in the edu-
> cation of young people.   We give them too much direction, stultifying the
> capacity for self-determination, or we give them too little direction, thus
> throwing on the child a responsibility beyond his capacity to bear. . . .
> Recognizing the importance of self-determination demands a change in
> our attitude and behavior.   Character is a way of living, motivated by a
> system of values which constitutes the individual himself.   We cannot give
> these values; they are won through the process of self-discovery in evaluat-
> ing both self and world in the ongoing activity of living. . . .[14]

Youth and the freedoms.   Freedom is an American tradition.   The
cry of "freedom" has played an important role throughout our history
—in both war and in peace.   The results of the 1942 *Fortune* Survey of
the opinions of American high school youth are most encouraging to
those who believe in the principles of our democratic government.[15]
They reveal that youth has an ardent devotion to liberty, and indicate
quite definitely that its devotion to liberty's ideals is more related to the
things for which our forefathers fought than to conditions that have
evolved, within recent times, from our industrial and technological
developments.   According to the results of the Survey, presented in
Table LXVII, 82.5 per cent would be least willing, of all the freedoms,
to give up that of speech or of worship.   It may be said by some that
these results are conditioned by the fact that high school youths never
see the possibility of earning $3,000 per year, and that thus it is easy
for them to give up the right to do so, but this is not the case, as is
shown by their answers to a subsequent question put to them.   The
average high school boy expects to be earning more than $3,000 per
year ten years after he is out of school.

The establishment of attitudes of cooperation, tolerance, loyalty, and
honesty cannot be left wholly to the home; neither can the teaching of
reverence, fidelity, and worship be left wholly to the church.   The
school must not be regarded simply as an institution for the dissemina-

[14] Wilma Lloyd, "Adolescence—A Quest for Selfhood," *Progressive Education,*
April, 1939, p. 245.
[15] "The Fortune Survey," *Fortune,* November 1942, p. 8.

tion of skill in the three R's, and of certain organized bits of knowledge. A serious obstacle to the improvement of educational methods, however, is the present uncertainty as to what are the most desirable attitudes and ideals to inculcate. Ideals evolve progressively with the ex-

TABLE LXVII

RESPONSES OF HIGH SCHOOL STUDENTS TO THE QUESTION: "IF YOU HAD TO GIVE UP ONE OF THESE THINGS, WHICH WOULD YOU BE LEAST WILLING TO GIVE UP? WHICH ONE WOULD YOU BE MOST WILLING TO GIVE UP?"

	Least Willing	Most Willing
Freedom of speech	46.0%	.9%
Freedom of religion	36.5	1.8
The right to vote	5.2	6.4
Trial by jury	3.8	3.9
The right to change jobs if you want to	3.0	20.8
The right to earn more than $3,000 a year if you can	2.3	59.8
Don't know	3.2	6.4

periences that create them. Hence, the ideals of our democratic society must be restated from time to time, to accord with our efforts toward and experiences in better living. Again, persistence in a purely academic approach, rather than adoption of one centered in the interests and needs of adolescents, presents a barrier to the development of vitalized, meaningful attitudes and ideals. A factor that must be recognized today is the effect upon our materialistic order of the pressure exerted by institutions and forces that are primarily interested not in educational and moral values, but only in individual or group aggrandizement. It has been pointed out by Isaiah Bowman, President of Johns Hopkins University, that:

We can avoid the authoritarian road only by choosing the free road. A body of prescribed doctrine, approved books, dialectical smartness and *a priori* opinion masquerading as 'wisdom' can give 'unity' also, as authoritarian governments have amply demonstrated. The strength of our American unity is in our free way of uniting. Our purpose is to train that freedom into responsibility. The growth of our educational program is due to the demonstrated need for the trained and responsible men and women we help to produce, with all faults of training and learning fully and freely admitted.[16]

[16] From *Science,* January 30, 1942, p. 126.

If our youth is to assume a positive rather than a neutral or perhaps negative attitude toward democracy, it must be given a function to perform and responsibilities to fulfill, gifts that would make come true the dream of many to grant to youth the opportunity for education, for health, for the satisfaction of social and recreational needs, and for participation in democratic living. The fathers of this nation have left to us the heritage of freedom of action, but with that heritage goes the responsibility of preserving it for future generations. Indeed, to the youth of every age is given the responsibility of preserving and enlarging the significance of the principles underlying the American way of life, so that future generations may find those ideals more firmly established, and the visions of those who sacrificed their lives and fortunes to establish them more completely fulfilled.

Education for world citizenship. Education for world citizenship can only be brought about when world citizenship becomes part of the goal of society, and particularly the goal of teachers, parents, and others concerned with the guidance of youth, for both now and for the future. Training for world citizenship, like training in democratic ways of living, cannot be relegated to some special department of the school; it will require the attention and consideration of every teacher of every subject, coordinated with ideals and practices in the home, on the playground, at church, and in other community activities.

American boys and girls must be trained to detect the economic fallacies that led us toward imperialism at one decade and into isolation a generation later. They must come to understand that favorable conditions for trade will provide opportunities for other nations to pay for our goods by selling us their goods in exchange. They must be given an understanding of how our technology operates, and how it has affected our way of living. They must come to see that the various agencies of production are so complex and interrelated that, without planning and regulation, economic chaos would result. They must be shown the possibility and necessity of a full life for all members of our society. Relative to these needs, Douglass has pointed out:

The fundamentals of American education must prepare the American people to resist efforts of any propaganda group, British or otherwise, to maintain imperialism and the exploitation of weaker nations in Asia, Africa or elsewhere. . . . It must enable him to understand and to discourage the exploitation of less well developed peoples by powerful and unscrupulous American business organizations such as has prevailed for more than three quarters of a century in Mexico, all South American countries, Cuba, Puerto Rico, Hawaii and the Philippines.

The fundamentals of American education must include a deep inculcation of American ideals, the ideals of democracy and equal opportunity. They must include inculcation of ideals which make for co-operation and the general welfare, ideals of fair and upright dealings with all others. . . .

The fundamentals of American education must prepare the American citizen to think for himself, so that he will not be herded about by newspapers and radio demagogues. . . . Habits and abilities of clear thinking which will pierce the surface of clever propaganda and free the American citizen from the ring in his nose by which he has been led in recent decades must be included.[17]

SUMMARY: FROM ADOLESCENCE TO ADULTHOOD

The boys and girls now in school are the potential citizens of tomorrow. If these boys and girls are to function effectively as responsible and as well adjusted citizens in a democratic society, certain imperative needs must be met. These needs have been stated as follows:

1. All youth need to develop salable skills and those understandings and attitudes that make the worker an intelligent and productive participant in economic life. To this end, most youth need supervised work experience as well as education in the skills and knowledge of their occupations.

2. All youth need to develop and maintain good health and physical fitness.

3. All youth need to understand the rights and duties of a citizen of a democratic society, and to be diligent and competent in the performance of their obligations as members of the community and citizens of the state and nation and of the world.

4. All youth need to understand the significance of the family for the individual and society and the conditions conducive to successful family life.

5. All youth need to know how to purchase and use goods and services intelligently; understanding both the values received by the consumer and the economic consequences of their acts.

6. All youth need to understand the methods of science, the influence of science on human life, and the main scientific facts concerning the nature of the world and of man.

7. All youth need opportunity to develop their capacities to appreciate beauty in literature, art, music, and nature.

8. All youth need to be able to use their leisure time well and to budget it wisely, balancing activities that yield satisfactions to the individual with those that are socially useful.

[17] Harl R. Douglass, "Essentials of a Post-War Educational Program," *The Educational Forum*, 1943, Vol. 7, pp. 370–372.

9. All youth need to develop respect for other persons, to grow in their insight into ethical values and principles, and to be able to live and work cooperatively with others.

10. All youth need to grow in their ability to think rationally, to express their thoughts clearly, and to read and listen with understanding.[18]

Technology has brought with it many problems in our cultural, social, and economic order. The effects of science and invention are being felt in almost all parts of the world, and are creating world, rather than individual, problems. The education of boys and girls demands that consideration be given to these problems, and that training for *world citizenship* be one of its goals. Again, if democracy is to survive and function effectively for the well-being of all the people, the school must accept its challenge, and provide students with the information, skills, and attitudes that will equip them to meet the problems of to-morrow. Youth must be trained to pass sound judgment on national and international issues and policies; and it must be equipped with values consistent with a philosophy of world-wide brotherhood.

THOUGHT PROBLEMS

1. Discuss the importance of adolescents as resources in a democratic society.

2. Evaluate the three major growth-steps, which adolescents in our culture are expected to take, presented in this chapter.

3. What is the significance to you of the findings relative to youth and the freedoms presented in Table LXVII?

4. What do you understand the term *world citizenship* to mean? How is this problem related to education?

5. What are the major social changes that have resulted from technology? What are the implications of these changes to education?

6. What do you believe to be the function of the school in relation to problems of courtship and marriage? Give some specific things that the school might be able to do in fulfilling this function.

SELECTED REFERENCES

American Youth Commission, General Report, *Youth and the Future*. Washington: American Council on Education, 1942.

Chambers, M. M., and Exton, Elaine, *Youth—Key to American Future: An Annotated Bibliography*. Washington: American Council on Education, 1949.

Cole, L., *Attaining Maturity*. New York: Rinehart & Company, 1944.

Edwards, Newton, "The Adolescent in Technological Society," *Adolescence*. Forty-third Yearbook of the National Society for the Study of Edu-

[18] "The Imperative Needs of Youth," *Planning for American Youth*. Published by the National Association of Secondary-School Principals.

cation, Part 1. Chicago: Department of Education, University of Chicago, 1944.

Edwards, Newton (Editor), *Education in a Democracy.* Chicago: University of Chicago, 1941.

McClusky, H. V., "The Changing Needs of Young Adults," in E. G. Williamson (Editor), *Trends in Student Personnel Work.* Minneapolis: University of Minnesota Press, 1949, pp. 40–51.

Pierce, W. C., *Youth Comes of Age.* New York: McGraw-Hill Book Co., 1948.

Warters, Jane, *Achieving Maturity.* New York: McGraw-Hill Book Co., 1949.

APPENDIXES

APPENDIX A

Selected Bibliography

Abel, Theodore, and Kinder, Elaine F., *The Subnormal Adolescent Girl.* New York: Columbia University Press, 1942.

The problems of subnormal adolescent girls are presented in a systematic and comprehensive manner. Special difficulties encountered by the subnormal adolescent girl in her home, in school, in the factory, and in the community at large are given special consideration.

Arlitt, Ada H., *Adolescent Psychology,* with a preface by Cyril Burt. London: Allen, 1937.

A valuable view of adolescent growth; both theoretical and practical discussions are presented.

Blatz, William E., *The Five Sisters.* New York: William Morrow and Co., 1938.

A report on the growth and development of the five Dionne sisters which points to the importance of environment and education in the development of the individual.

Blos, Peter, *The Adolescent Personality; A Study of Individual Behavior.* New York: Appleton-Century-Crofts, Inc., 1941.

This volume makes use of the case-history method of studying the adolescent. The influence of various conditions on the development of the personality is revealed through the case-study procedure.

Boynton, Bernice, "The Physical Growth of Girls. A Study of the Rhythm of Physical Growth from Anthropometric Measurements on Girls between Birth and Eighteen Years." *University of Iowa Studies in Child Welfare,* 1936, Vol. 13, No. 4.

A tabular and graphic presentation of results obtained from 22 physical and anthropometric measurements of girls between birth and 18 years.

Brooks, Fowler D., *Psychology of Adolescence.* New York: Houghton Mifflin Co., 1929.

An exhaustive and more or less encyclopedic treatment of adolescent growth. Attention is also given to problems involved in personality growth and adjustment.

Cole, Luella, *Psychology of Adolescence* (revised edition). New York: Farrar and Rinehart, Inc., 1948.

A readable presentation of the development of adolescents. Many brief case studies and pictorial graphs are introduced to illustrate and clarify different problems.

Crow, L. D., and Crow, Alice, *Our Teen-Age Boys and Girls*. New York: McGraw-Hill Book Co., 1945.

The authors have presented a constructive and functional approach to the problems of teen-age boys and girls. Included are pertinent questions and illustrative stories taken from the lives of these young people.

Educational Policies Commission, National Education Association, *Education for All American Youth*. Washington: National Education Association, 1944.

A plan projected into the future designed to provide for the education of all the American youth.

Fleming, C. M., *Adolescence*. New York: International Universities Press, 1949.

A selection of anecdotes is here presented in an interesting style providing useful materials for interpreting the adjustment problems of the adolescent. A useful bibliography accompanies each chapter.

Fisher, D. Canfield, *Our Young Folks*. New York: Harcourt, Brace and Co., 1943.

In this volume the writer presents an informal report on problems faced by the great body of American youths who are maladjusted as a result of the narrow academic training provided in most schools. Mrs. Fisher points out the need for the use of aptitude tests in the guidance of students to avert vocational maladjustment. She also pleads for a more functional education to prepare youths to meet the problems of today.

Garrison, Karl C., *Psychology of Adolescence* (fourth edition). New York: Prentice-Hall, Inc., 1951.

A presentation of materials from recent studies bearing on the physical, mental, emotional, social, and spiritual development of the adolescent. The last half of the book is devoted to personality development.

Gleming, C. M., *Adolescence: Its Social Psychology*. New York: International Universities Press, Inc., 1949.

An introduction to recent findings from the fields of anthropology, physiology, psychometrics, and sociometry that bear on the growth and behavior of adolescents.

Harris, Erdman, *Introduction to Youth*. New York: The Macmillan Co., 1940.

This is a nontechnical presentation of information and suggestions on the guidance of youth. It is written for the layman as well as for the student of adolescent psychology. There is special emphasis upon the need for guiding youths in the formulation of a philosophy of life.

Havighurst, R. J., and Taba, Hilda, *Adolescent Character and Personality*. New York: John Wiley & Sons, Inc., 1949.

A preliminary report of studies made on all youths who were sixteen years old in 1942 from a midwestern town referred to as "Prairie City." A variety of sociological and psychological techniques were used in gathering data on these youths and their environment.

Hollingworth, Leta S., *The Psychology of the Adolescent.* New York: Appleton-Century-Crofts, Inc., 1928.

A most valuable book on the growth, sex, weaning, and maturity problems of adolescents. Students, parents, and teachers will find this material of interest and applicable to the adolescent stage of life.

Hurlock, Elizabeth B., *Adolescent Development.* New York: McGraw-Hill Book Co., 1949.

A comprehensive treatment of adolescent development is presented in this volume. The reading materials are based upon a wide selection of research studies.

Jones, Harold E., and others, *Adolescence,* Forty-third Yearbook of the National Society for the Study of Education, Part 1. Chicago: School of Education, University of Chicago, 1944.

An important addition to the literature on the psychology of adolescence, based on a summary of investigations in the fields of physiology, physical measurements, psychology, and sociology. A large group of well-known students of adolescent psychology have contributed to the findings and interpretations presented in this volume.

Keliher, Alice, *Life and Growth.* New York: Appleton-Century-Crofts, 1938.

Many questions of youth related to physical and personality development are proposed and answered by Keliher. Graphical illustrations aid in making the data more meaningful.

Knoebber, Sister Mary M., *Self-Revelation of the Adolescent.* New York: Bruce Publishing Company, 1936.

A presentation of the feelings, ideals, and personality problems of adolescent girls as viewed by the adolescent girl herself.

Kunkel, Fritz, *What It Means to Grow Up.* Translated from the German by Barbara Keppel-Compton and Hilda Niebuhr. New York: Charles Scribner's Sons, 1936.

A translation of a German book giving considerable attention to the nature and significance of adolescent development.

Landis, Paul H., *Adolescence and Youth.* New York: McGraw-Hill Book Co., 1945.

Special emphasis is given to the social structure affecting the individual during the developmental stage. Problems concerned with the attainment of maturity in the moral, marital, and economic fields are given considerable attention.

McKown, H. C., and Lebron, Marion, *A Boy Grows Up.* New York: McGraw-Hill Book Co., 1940.

A practical handbook of guidance for boys which should be useful in connection with problems of how to get along with each other, with the family, and with others. It also deals with questions of managing money, securing a job, and the like.

Mead, M. J., *From the South Seas: Studies of Adolescence and Sex in Primitive Societies.* New York: William Morrow and Co., 1939.

A one-volume edition of three anthropological works: "Coming of Age in Samoa," "Growing up in New Guinea," and "Sex and Temperament."

Meek, Lois H., and others, *The Personal-Social Development of Boys and Girls with Implications for Secondary Education*. New York: Progressive Education Association, 1940.

This is a report on which the thinking and experience of a number of people, including teachers, counselors, specialists, administrators, and research workers have been brought together. The report is especially concerned with the guidance and direction of adolescents in their personal-social relations.

Pryor, Helen Brenton, *As The Child Grows*. New York: Silver, Burdett & Co., 1943.

Part I of this volume deals with the various factors related to growth including differences in body build. Part II treats of growth at the various age levels, preschool, primary, pre-adolescence, adolescence, and post-adolescence.

Shock, N. M., "Physiological Factors in Development," *Review of Educational Research,* 1947, Vol. 27, pp. 362–370.

The author presents a brief summary of research on dietary and nutritional influences on development and the effects of disease on development.

Strain, Frances Bruce, *Love at the Threshold*. New York: Appleton-Century-Crofts, Inc., 1939.

This volume has been developed to aid parents, teachers, and counselors of adolescents. Beginning with a presentation of questions asked by adolescents and post-adolescents about dating, a splendid discussion of the various problems related to love attachments during adolescence is presented.

Symonds, Percival M., *Adolescent Fantasy*. New York: Columbia University Press, 1949.

The objective and dynamic approaches to the study of the adolescent's personality is here presented. The picture-story technique and other procedures are used in arriving at a better understanding of adolescent needs, problems, and fantasies. A rather complete case study of a maladjusted boy and one of a well adjusted boy are included.

Taylor, Katherine W., *Do Adolescents Need Parents?* New York: Appleton-Century-Crofts, Inc., 1938.

This is designed as a guide to parents in dealing with the problems that emerge as their children grow into and through adolescence.

Warters, Jane, *Achieving Maturity*. New York: McGraw-Hill Book Co., 1949.

This is a nontechnical presentation of problems related to growing up. There is a practical and realistic approach to problems of adolescents and youth, based in part upon some of the recent research studies in this area.

Wile, I. S., *The Challenge of Adolescence*. New York: Greenberg, 1939.

Taking data from the fields of biology, psychology, sociology, religions, and ethics, the author presents an interpretation of adolescence.

Zachry, Caroline B., in collaboration with Lighty, Margaret, *Emotion and Conduct in Adolescence*. New York: Appleton-Century-Crofts, Inc., 1940.

This is a presentation of the findings of the study of adolescents by the Commission on Secondary-School Curriculum of the Progressive Education Association, begun in 1934 and concluded in 1939. There are three parts to the volume. Part I describes the adjustment of the adolescent to himself as a developing person; Part II deals with his changing personal relationships; and Part III shows the development and change of his attitudes toward basic social institutions.

APPENDIX B

Diagnostic Child Study Record[1]

I. Identifying Data
   Name..Date of Birth...............
   Place of Birth..Age.................
   Present Address...Sex................
   Father's Name..................Mother's Name.......................
   Ages of Brothers..................Ages of Sisters...................
   School ....Grade ....Date ....

II. Developmental History
   Write a brief biography of your life.
   ....
   ....

III. Recreational Activities and Interests
   1. When you have some free time, what do you like best to do? ....
   ....
   ....

   2. What do you usually do:
      a) Immediately after school?...
      b) In the evening?...
      c) On Saturdays?...
      d) On Sundays?...
   3. What games do you play most?..
   ....
   What games do you prefer?..
   ....

   4. Write the full names and ages of three close friends.
   ....
   ....
   ....

[1] Reproduced by permission of the publishers from Karl C. Garrison, *Psychology of Exceptional Children,* Revised Edition, pp. 487–489. Copyright 1950 by The Ronald Press Company.

482

5. Do you have many friends or few?...

6. Do you prefer to be alone most of the time or with others?...................

....

7. Do you prefer activities with boys, girls, or mixed groups?.....................

8. How often do you go to the movies?...
   With whom usually?..
   Underline the kind of picture you like best: comedy—western—love—
   mystery—serial—gangster—educational—cartoons—news—others— ....

9. Do you enjoy reading?...................................Name any books which you
   have read and enjoyed during the past year. ....

....

10. What are your favorite radio programs?   First choice...............................
    ....Second choice.......................................
    ....Third choice...

11. Do you have a pet?................What? ....

12. Name any hobbies you may have. ....

....

IV. Attitudes Toward Home and Family Life

1. Do you have any preferences among the members of your family?
   .... Who?...

2. What recreational facilities are there in the home?...................................

....

3. Name things the family tends to do together. (For example, go to
   picnics, go to church, etc.)..

4. At what time do you usually go to bed?...
   When do you usually get up?...

5. What is the attitude of your father and mother toward your friends?

....

6. Who in the family advises with you about your actions and problems?

....
   Is there anyone whom you feel free to go to about almost any prob-
   lem?................................. Whom...

7. Do you feel that you are understood by the other members of your
   family? ....

8. Do you have an allowance?.....................Have you earned any money
   during the past year?.....................How?...

V. Attitudes Toward School

1. Do you like school?...............................What are your favorite subjects?
   First choice..Second choice...........................
   ....Third choice......................................
   What subjects do you dislike?...

2. What clubs do you belong to at school?...

....

3. What is your biggest school problem?...

....

4. Are your teachers in general fair with you?..

....

5. Do you enjoy doing things with the members of your class?.................

....

6. In what ways would you like to have school different from what it is?

....

....

VI. Aspirations and Ambitions

1. If you could have three wishes come true, for what would you wish?
First wish...
Second wish...
Third wish..

2. What things do you oftentimes wonder about?.....................................

....

3. What occupations attract you most? First choice.................................
....Second choice..
Why? ....

4. What are some things which you would like to do in the course of the
next ten years?...

....

5. What are your goals for ten years hence? (For example: What
would you like to be doing? What would you like to have? Where
would you like to be? Who would you like to be?).................................

....

....

APPENDIX C

Annotated Bibliography of Popular Literature Related to the Adolescent Age[1]

Popular literature touching upon adolescence presents in a vivid manner the problems faced by real adolescent boys and girls as they grow toward maturity, although the stories themselves may be pure fiction. Since such books give a detailed and perhaps more realistic interpretation than scientific compilations of facts in a textbook, they should be of value to the reader in helping him to understand the significance of the adolescent period.

There are many books dealing with this period of life. The following list is by no means complete. In its compilation, the writers made an extensive survey of the field, and their list is the result of a selective process. Some of the books to be found in it deal with *growing up* in general. Others deal with the adolescent over a short period of time, and still others are concerned with some special problem of the adolescent years. The extent to which the social setting of the adolescent is introduced varies considerably—depending upon the author's points of view, interests, and purposes. From this bibliography, presented in annotated form, the reader will be able to find materials relative to adolescence that will be of interest and value to him.

Aldis, Dorothy, *All Year Round*. Boston: Houghton Mifflin Co., 1938.
     The influence of a mother's triumphs and mistakes upon her three children, the nervous condition of a four-year-old, and the skillful treatment of an adolescent daughter are among the many interesting things sympathetically treated by the author.
Armstrong, Margaret, *Fanny Kemble*. New York: The Macmillan Co., 1938.

---

[1] This is a revision of a list devised by Richard L. Wampler and Karl C. Garrison and has been used in classes in child and adolescent psychology.

This book furnishes a vivid picture of Victorian child psychology, revealing its differences from modern views. When Fanny's family can do no more with her, they turn her over to a boarding school. Her treatment there contradicts all modern practices.

Baker, Mrs. D. D., *Young Man with a Horn*. New York: Houghton Mifflin Co., 1938.

Here is an example of innate ability strong enough to carry a youth from the confines of a poor environment to the heights of success. His interest in books and music helped him to ignore hardships.

Benney, Mark, *Angels in Undress*. New York: Random House, 1937.

A boy's unsuccessful struggles against the evils of postwar society and his attempts to compensate for the disadvantages of his social environment result in harm to himself and others.

Benson, Sally, *Junior Miss*. New York: Random House, Inc., 1941.

A collection of stories of Judy Graves and her family has been brought together in this volume. The tales of Judy, a twelve-year-old pupil at a private school for girls in New York City, give a good insight into child nature and life in New York.

Carter, John Franklin, *The Rectory Family*. New York: Coward-McCann, Inc., 1937.

This whole book is a picture of normal family life in New England in the period just before World War I. The sensitiveness of the children to ridicule, embarrassment, and failure is sympathetically presented.

Childs, Marquis W., *The Cabin*. New York: Harper and Brothers, 1944.

This is an understanding story of the life of a thirteen-year-old boy one summer on a Middlewestern corn farm. Tragedies unfold and are successfully met.

Cormack, M. B., and Bytovetzeski, P. L., *Swamp Boy*. New York: McKay, 1948.

The story revolves around the life and experiences of sixteen-year-old Clint Sheppard of the Okefinokee swamp of Georgia. Clint has been taught the love of the swampland by Tom, a Seminole Indian, who is the recognized leader of the community. Clint's problems of adjusting to town ways of living as well as those of the swamp environment is interesting as well as amusing.

Cronin, A. J., *The Green Years*. Boston: Little, Brown and Co., 1944.

Green Years is a stirring story of Robert Shannon from his eighth year to his eighteenth. Robert may be classified as a waifish little boy, depending upon his grandfather for affection and security.

Curie, Eve, *Madame Curie*. New York: Doubleday, Doran and Co., 1933.

A convincingly accurate, though brief, record of the super-scientist's childhood and the conditions and reverses that made her shy, nervous, overemotional, and mature for her age but which could not squelch her genius.

Daphne, Athas, *Weather of the Heart*. New York: Appleton-Century-Crofts, 1947.

The problems and difficulties of two adolescent girls living at Kittery Point are presented in a very fascinating manner.

Delafield, E., *Nothing Is Safe*. New York: Harper and Brothers, 1937.

Because of unhappiness caused by a split home, a brother and sister became very dependent upon each other. This dependence is blamed for a neurotic condition which the boy develops and the two children are separated. Owing to lack of understanding of the real difficulty it was never remedied and the children continued to suffer.

De La Roche, Mazo, *Growth of a Man*. Boston: Little, Brown and Co., 1938.

A complete misunderstanding of his behavior by his grandmother leads an eight-year-old boy to overvalue success in school. Weaknesses of nineteenth century teaching are evident. Strict discipline at home and overwork do not lessen Shaw's determination to succeed.

De La Roche, Mazo, *The Very House*. Boston: Little, Brown and Co., 1937.

In the form of fiction the author manages to set forth the principles of modern child psychology so that the lay reader can understand and perhaps transfer them to his own treatment of children.

Dell, Floyd, *Homecoming*. Boston: Little, Brown and Co., 1937.

Even the love of parents cannot protect Floyd from the effects of poverty. Lack of friends leads him to seek refuge in books.

Doyle, Helen M., *A Child Went Forth*. New York: Gotham House, Inc., 1934.

During her whole childhood Helen adjusts to situations from which there is no escape. The strength of character she developed by overcoming these obstacles enriched her whole life.

Farrell, James T., *Father and Son*. New York: Vanguard Press, Inc., 1940.

A lengthy but impressive story of the development of Danny O'Neill from adolescence up to the age of nineteen. There is a sort of realism presented which reveals the author's understanding of human nature.

Farrell, James T., *No Star Is Lost*. New York: Vanguard Press, Inc., 1938.

Like *Studs Lonigan* by the same author, this book portrays a boy's struggles with and defeat by his environment. The lack of adjustment of the school to the needs of the community it serves is also evident.

Farrell, James T., *Studs Lonigan*. New York: Vanguard Press, Inc., 1932.

This story presents a sociological study of the influences of a vigorous but often unfortunate environment upon Studs Lonigan, a son of middle-class Chicago.

Field, Isobel, *This Life I've Loved*. New York: Longmans, Green and Co., 1937.

Isobel Field's story furnishes a study of the effect a change of environment had upon her childhood and the way her experiences in a mining camp influenced her life in a large city.

Gibbs, A. Hamilton, *The Need We Have*. Boston: Little, Brown, and Co., 1936.

Denny at fourteen impresses one as the impish, young-for-his-age result of an oversheltered childhood and then as a prodigiously mature mind of adult level with insight, judgment, and subtlety.

Gosta of Gerjerstam (Translated by Birkeland), *Northern Summer.* New York: E. P. Dutton and Co., Inc., 1938.

The author has succeeded in picturing the happy development of children in a family living on a Norwegian island. Free from social conventions, and aided by the enthusiasm and guidance of their parents, they live naturally and happily together.

Gunnarsson, Gunnar, *Ships in the Sky.* New York: Bobbs-Merrill Co., 1938.

Although the life of a child in Iceland must of necessity be quite different from that of children in our country, this book presents a valuable picture of how a child develops in lonely regions where the family is his whole society.

Hagedorn, Hermann, *Edwin Arlington Robinson.* New York: The Macmillan Co., 1938.

As the youngest child in the family, Edwin Arlington Robinson was neglected, but was probably saved from maladjustment by his play-life with the neighborhood gang. Although not encouraged by others, his poetic talents developed naturally.

Harriman, John, *Winter Term.* New York: Howell, Soskin Publishers, Inc., 1940.

The author presents an unsentimental story of American schoolboy life; and in this he shows unusual understanding of how boys think and feel.

Havighurst, Walter, and Boyd, M. M., *Song of the Pines.* Philadelphia: John C. Winston Co., 1949.

Nils Thorsen, a Norwegian boy, joined a group of pioneer settlers at the age of fourteen. His inventive genius was a source of help to these Wisconsin pioneers in their lumbering activities.

Herbert, F. H., *Meet Corliss Archer.* New York: Random House, Inc., 1944.

Corliss, a subdeb daughter of a well-to-do lawyer in an American city, is well known to many through the Good Housekeeping stories. The author presents some interesting experiences in Corliss' home and social life.

Hulme, Katheryn, *We Lived As Children.* New York: Alfred A. Knopf, Inc., 1938.

Although often unhappy because their parents were divorced, this family of children lived a fairly satisfactory life. Many of the characteristic periods of childhood including gang life, collecting, etc., are included.

Kehoe, William, *A Sweep of Dusk.* New York: E. P. Dutton and Co., 1945.

The materials of this volume revolve around the problems of an oversensitive adolescent reared by an overbearing mother. The materials are drawn from the author's own experiences.

L'Engle, Madeleine, *The Small Rain.* New York: The Vanguard Press, Inc., 1945.

This is a touching story of the problems encountered by a young and talented artist during the adolescent years. Her disillusionments are characteristic of the life of many adolescents filled with zeal and ambition.

Llewellyn, Richard, *How Green Was My Valley*. New York: The Macmillan Co., 1941.

A dramatic story of the struggles of a boy against the odds of poverty and class distinction in a mining area. His difficulties, privations, and thwartings are presented in an understanding manner.

Low, Elizabeth, *High Harvest*. New York: Harcourt, Brace & Co., Inc., 1948.

Life for fifteen-year-old Suzanne on a Vermont mountain farm was not easy. However, Suzanne found happiness and satisfaction from outlets available.

Maugham, Somerset W., *Of Human Bondage*. New York: Modern Library, Inc., 1940.

From a protected, pampered childhood Philip Cary is placed under the guardianship of his disciplinary uncle. Lack of understanding at home and humiliation caused by the ridicule of his club foot by his schoolmates and teachers makes him supersensitive and unhappy.

Maxwell, William, *The Folded Leaf*. New York: Harper and Brothers, 1945.

This is a story of the friendship begun by two normal boys at the age of fifteen. The conditions which draw them together, and the unfolding events of their friendships make the book an interesting as well as good portrayal of human relationships.

Miller, Sidney, *Roots in the Sky*. New York: The Macmillan Co., 1938.

Through poverty and nationality difficulties these children of Jewish-American stock had many chances, if not justifications, for going wrong. The unifying kinship and loyalty to race-standards upheld by the Jewish religion helped avert a tragedy.

Mitchell, William Ormond, *Who Has Seen the Wind*. Boston: Little, Brown & Co., 1947.

The story of a boy, Brian O'Connal, from his fourth year to his twelfth in a small prairie town.

Morris, Hilda, *The Long View*. New York: G. P. Putnam's Sons, 1937.

This author is interested in showing how Asher Allen was influenced by his sober, Quaker environment and forced pride in his family name.

O'Moran, Mabel, *Red Eagle, Buffalo Bill's Adopted Son*. Philadelphia: J. B. Lippincott Co., 1948.

The story of Red Eagle, a young Choctaw Indian boy, and his pioneer adventures is told in a fascinating manner.

Parrish, Anne, *Poor Child*. New York: Harper and Brothers, 1945.

A tragic story of a twelve-year-old boy in need of security and affection, established in a household where he was neither loved nor understood.

Powell, E. Alexander, *Gone Are the Days*. Boston: Little, Brown and Co., 1938.

The author visualizes objectively the conventional childhood experiences of the eighties and nineties in Syracuse, New York. The artificiality and restraint of the period are emphasized.

Pratt, Theodore, *Valley Boy*. New York: Duell, Sloan, and Pearce, Inc., 1946.

"Valley Boy" is a series of character sketches as seen through the eyes of Johnny Birch, a ten-year-old boy. Johnny is a sensitive, lonely lad seeking affection and security outside his home.

Raphaelson, Dorshka W., *Morning Song*. New York: Random House, 1948.

The story of a fifteen-year-old girl's efforts and problems as she attempts to support her neurotic mother and younger brother presents an interesting and touching picture of adolescence and family life.

Rawlings, Marjorie Kinnan, *The Yearling*. New York: E. P. Dutton and Co., Inc., 1938.

Jody's solitary environment leads him instinctively to find companionship in nature. His father's influence helps the boy to develop a strong character.

Robertson, Eileen A., *Summer's Lease*. Boston: Houghton Mifflin Co., 1940.

This is a story and a psychological study of a sensitive boy who was further handicapped by weak eyesight. There are the trials and pains of the boy as he tries to cope with problems accentuated by refined but unsuccessful home conditions.

Ricks, Peirson, *Hunter's Horn*. New York: Charles Scribner's Sons, 1947.

Life and values in post-bellum North Carolina are presented in this novel. Uncle Benjamin's grand-nephew falls in love with a girl from a poorer class status. This presents problems for all concerned.

Rölvaag, O. E., *Peder Victorious*. New York: A. L. Burt Co., Inc., 1938.

This sensitive, inquisitive Dakota boy is too eager to grow up, is old for his age, and lives through emotional periods approaching maladjustment. A contrast between two teachers is especially well done. Like a number of other authors, Rölvaag criticizes the effects of certain religious experiences upon children.

Sale, Elizabeth, *Recitation from Memory*. New York: Dodd, Mead and Co., Inc., 1943.

This is a story of the growing up of Fenella in the home of a happy letter carrier of Tacoma, Washington. It is written from the viewpoint of an older person looking back on happenings during the earlier years of life, and especially the years from the tenth to the fourteenth. The story is one filled with much action and adventure, with the introduction of many characters to enliven and make more realistic the events and happenings affecting Fenella.

Sessions, Ruth Huntington, *Sixty Odd: A Personal History*. Brattleboro, Vt.: Stephen Daye Press, 1936.

In this, another story of childhood in Cambridge at the time of the great intellectuals, the need for individual treatment in school, and the value of teaching skills for success, as well as many other modern practices, are mentioned.

Shanks, Edward, *Tom Tiddler's Ground*. Indianapolis: Bobbs-Merrill Co., 1934.

Tom Florey's high intelligence helps him to advance quickly in school in spite of himself. The antagonistic attitude of Tom's father toward his progress does not extinguish the boy's determination to escape from the confines of his home town.

Shaw, Lau, *Rickshaw Boy*. New York: Reynal and Hitchcock, Inc., 1945.
An adventure story of a Chinese boy whose dream of happiness was to own a rickshaw.

Skouen, Arne, *Stoker's Mess* (Translated from the Norwegian by Jordan Birkeland). New York: Alfred A. Knopf, Inc., 1948.
A symbolic novel of life at sea in which a mess boy becomes the hero. This is a tale showing the dreams and courage of an adolescent boy midst the lives of tough seamen.

Sperry, Armstrong, *The Little Eagle—A Navajo Boy*. New York: Winston, 1938.
An adventure story of a fourteen-year-old Navajo boy in the setting of the Arizona canyons is well presented. The warm relations of his home life helps him adjust to changed values and cultures encountered in the modern government school.

Smith, Betty, *A Tree Grows in Brooklyn*. New York: Harper & Brothers, 1943.
A colorful presentation of the childhood and youth of Francie Nolan, her family, and her friends. There is beauty and wholesomeness intermingled with plain realism in the problems faced by Francie as she strives to find for herself a place of usefulness in a larger social environment.

Spring, Howard, *My Son, My Son!* New York: Viking Press, Inc., 1938.
Parental attempts to shape the lives of two boys result in tragedy. One-sided personalities fail to fit the sons for adult life. Lack of understanding by the parents seems to have an effect upon the characters of all the children.

Strong, L. A. G., *The Seven Arms*. New York: Alfred A. Knopf, Inc., 1933.
This novel concerning the life of a large family of children reveals characters ranging from one extreme to the other. Included are the extrovert, bossy, thoughtful, daydreaming, weak, dependent, independent, and delicate types of children whose characters react upon one another.

Sullivan, Mark, *The Education of an American*. New York: Doubleday, Doran and Co., Inc., 1938.
Although the author probably overemphasizes his earliest recollections, we obtain a picture of his simple, hardworking childhood, his satisfying school experiences, and his relations with his brothers and sisters.

Summers, H. S., *City Limit*. Boston: Houghton Mifflin Co., 1948.
The story of the dilemma of a high school boy and girl who are driven into marriage by social criticisms and pressures. The idealism of youth is clearly revealed in contrast to the rigidity of the elders of a Kentucky community.

Tarkington, Booth, *The Fighting Littles*. New York: Doubleday, Doran and Co., Inc., 1941.
The author gives an interesting and oftentimes funny presentation of

the conflicts between youth and its parents.   The teen-age children appear as real characters in their revolt against parental controls.

Tarkington, Booth, *Little Orvie*.   New York: Grosset and Dunlap, Inc., 1933.
This is a commendable handbook of how to get results opposite to what you want in dealing with children.   Popularized through clever handling are the problems of nagging, thwarting, comparison with other children, discussion of child in his presence, shaming, and so on.

Tracy, E. B., *Great Husky*.   New York: Dodd, Mead and Co., Inc., 1949.
Vivid descriptions are presented of a boy and his dog on a journey into the Far North.   Their comradeship and heroism are stirring features of the story.

Van Etten, Winifred, *I Am the Fox*.   Boston: Little, Brown and Co., 1936.
Through lack of understanding and individual treatment by her elders, Selma Temple developed many misconceptions and fears.

Wells, H. G., *Experiment in Autobiography*.   New York: The Macmillan Co., 1934.
In this book about himself H. G. Wells attempts to explain all his reactions to situations and conditions in a psychological way.   He manages, however, to give a fairly good impression of child-life in the nineteenth century.

Wouk, H., *City Boy: The Adventures of Herbie Bookbinder and His Cousin, Cliff*.   New York: Simon and Schuster, 1948.
A refreshing and readable novel of an eleven-year-old Bronx boy.   The adventure of Herbie and his cousin at home, school, and elsewhere are vividly recounted.

APPENDIX D

The Vineland Social Maturity Scale[1]

A general application of this scale by those not schooled in this field will yield suggestively valuable facts; however, those who desire more reliable results should consult the Manual and carefully adhere to the standard method of administering and scoring the various items.

VINELAND SOCIAL MATURITY SCALE

Name ....Age ....M. A. ....Date ....
Descent ....Sex......Grade .........I.Q. ....Born ....
Occupation ....Yrs. Exp. ....Class ....Res. ....
Father's occupation ....Class ....Schooling ....
Mother's occupation ....Class ....Schooling ....
Informant ....Relationship ....Recorder ....
Remarks: Basal Score ....
                                         Additional pts. ....
                                         Total score ....
                                         Age equivalent ....
                                         Social quotient ....
                                         Informant's est. ....

CATEGORIES[2] ITEMS

0–I

C	1	"Crows"; laughs
SHG	2	Balances head

[1] The *Vineland Social Maturity Scale* is included here by courtesy of The Training School at Vineland, New Jersey, Department of Research. This scale is copyrighted and should not be reproduced without permission.

[2] Key to categorical arrangement of items:

SHG—Self-help general   SD—Self-direction   L—Locomotion
SHD—Self-help dressing   C—Communication   O—Occupation
SHE—Self-help eating   S—Socialization

SHG	3	Grasps objects within reach
S	4	Reaches for familiar persons
SHG	5	Rolls over
SHG	6	Reaches for near-by objects
O	7	Occupies self unattended
SHG	8	Sits unsupported
SHG	9	Pulls self upright
C	10	"Talks"; imitates sounds
SHE	11	Drinks from cup or glass assisted
L	12	Moves about on floor
SHG	13	Grasps with thumb and finger
S	14	Demands personal attention
SHG	15	Stands alone
SHE	16	Does not drool
C	17	Follows simple instructions

I–II

L	18	Walks about room unattended
O	19	Marks with pencil or crayon
SHE	20	Masticates food
SHD	21	Pulls off socks
O	22	Transfers objects
SHG	23	Overcomes simple obstacles
O	24	Fetches or carries familiar objects
SHE	25	Drinks from cup or glass unassisted
SHG	26	Gives up baby carriage
S	27	Plays with other children
SHE	28	Eats with spoon
L	29	Goes about house or yard
SHE	30	Discriminates edible substances
C	31	Uses names of familiar objects
L	32	Walks upstairs unassisted
SHE	33	Unwraps candy
C	34	Talks in short sentences

II–III

SHG	35	Asks to go to toilet
O	36	Initiates own play activities
SHD	37	Removes coat or dress
SHE	38	Eats with fork
SHE	39	Gets drink unassisted
SHD	40	Dries own hands
SHG	41	Avoids simple hazards
SHD	42	Puts on coat or dress unassisted
O	43	Cuts with scissors
C	44	Relates experiences

III–IV

L	45	Walks downstairs one step per tread
S	46	Plays co-operatively at kindergarten level
SHD	47	Buttons coat or dress
O	48	Helps at little household tasks
S	49	"Performs" for others
SHD	50	Washes hands unaided

IV–V

SHG	51	Cares for self at toilet
SHD	52	Washes face unassisted
L	53	Goes about neighborhood unattended
SHD	54	Dresses self except tying
O	55	Uses pencil or crayon for drawing
S	56	Plays competitive exercise games

V–VI

O	57	Uses skates, sled, wagon
C	58	Prints simple words
S	59	Plays simple table games
SD	60	Is trusted with money
L	61	Goes to school unattended

VI–VII

SHE	62	Uses table knife for spreading
C	63	Uses pencil for writing
SHD	64	Bathes self assisted
SHD	65	Goes to bed unassisted

VII–VIII

SHG	66	Tells time to quarter hour
SHE	67	Uses table knife for cutting
S	68	Disavows literal Santa Claus
S	69	Participates in preadolescent play
SHD	70	Combs or brushes hair

VIII–IX

O	71	Uses tools or utensils
O	72	Does routine household tasks
C	73	Reads on own initiative
SHD	74	Bathes self unaided

IX–X

SHE	75	Cares for self at table
SD	76	Makes minor purchases
L	77	Goes about home town freely

X–XI

C	78	Writes occasional short letters
C	79	Makes telephone calls
O	80	Does small remunerative work
C	81	Answers ads; purchases by mail

XI–XII

O	82	Does simple creative work
SD	83	Is left to care for self or others
C	84	Enjoys books, newspapers, magazines

XII–XV

S	85	Plays difficult games
SHD	86	Exercises complete care of dress
SD	87	Buys own clothing accessories
S	88	Engages in adolescent group activities
O	89	Performs responsible routine chores

XV–XVIII

C	90	Communicates by letter
C	91	Follows current events
L	92	Goes to near-by places alone
SD	93	Goes out unsupervised daytime
SD	94	Has own spending money
SD	95	Buys all own clothing

XVIII–XX

L	96	Goes to distant points alone
SD	97	Looks after own health
O	98	Has a job or continues schooling
SD	99	Goes out nights unrestricted
SD	100	Controls own major expenditures
SD	101	Assumes personal responsibility

XX–XXV

SD	102	Uses money providently
S	103	Assumes responsibilities beyond own needs
S	104	Contributes to social welfare
SD	105	Provides for future

XXV+

O	106	Performs skilled work
O	107	Engages in beneficial recreation
O	108	Systematizes own work
S	109	Inspires confidence
S	110	Promotes civic progress
O	111	Supervises occupational pursuits
SD	112	Purchases for others
O	113	Directs or manages affairs of others
O	114	Performs expert or professional work
S	115	Shares community responsibility
O	116	Creates own opportunities
S	117	Advances general welfare

INDEXES

Subject Index

A roman number in parentheses after an item is the number of the chapter that constitutes the major reference to that item.

499

Author Index

A

Abelson, H. H., 27
Abernathy, E. M., 123
Albert, E., 227
Allen, L., 118
Anderson, J. E., 120, 168, 176, 197, 329
Anderson, J. P., 230, 231
Anderson, R. G., 312
Appy, N., 271
Arlitt, A. H., 131, 250, 309
Arnold, A., 296, 395
Arnold, M., 188, 190, 193
Atkinson, R. K., 49, 50
Aub, J. C., 71
Averill, L. A., 67, 155, 188, 400, 453

B

Baker, H. J., 329, 400
Baker, H. N., 221
Baldwin, B. T., 43, 48, 117, 125
Barker, R. G., 179
Barker, R. S., 123
Baruch, D. W., 329
Baxter, B., 271
Bayard, B., 152
Beach, E. V., 233
Beck, A. G., 255
Beck, L. F., 227
Bell, H. M., 186, 241, 311, 319, 449, 453
Beneke, F. W., 73
Bent, R. K., 446
Bernard, W., 355
Biglow, M. A., 461
Blatz, W. E., 90, 100
Block, V. L., 238, 240, 248
Blose, D. I., 13
Bloss, P., 35, 112, 309, 366
Bluemel, C. S., 249, 250
Boder, D. B., 233
Bogardus, E. S., 309
Bohn, D. C., 286
Bolles, M. M., 329
Bonney, M. E., 321

Books, E., 221
Boorman, W. R., 133
Boswell, F. P., 40
Bowman, I., 469
Boynton, J. C., 58, 67, 131
Boynton, P., 10, 58, 67, 131, 151, 152
Bradley, W. A., 126, 127, 434, 435
Bresby, L. M., 48
Bridges, K. M. B., 90
Briggs, T. H., 20, 155, 424
Brill, J. G., 400
Britten, F., 223, 227
Bromley, D. B., 223, 227
Bronner, A. F., 218, 303, 342, 343, 345, 380, 401
Brooks F. D., 112
Brown, F., 243
Brown, F. J., 145, 158, 250
Burgess, E. W., 231
Butler, G. D., 288
Butterfield, O. M., 227, 240

C

Cabot, P. S. deQ., 355
Cabot, R. C., 382
Campbell, E. H., 227
Campbell, H. L., 61
Campbell, M. W., 347
Canivet, N., 460
Carmichael, L., 460
Carr, L. J., 400
Carter, H. D., 443, 453
Catchpole, H. R., 80, 86
Cattell, P., 48
Centers, R., 288
Chaffey, J., 170, 179, 180
Chamberlain, L. M., 453
Chambers, M. M., 472
Chant, S. N. F., 100
Chapman, P. W., 422
Chave, E. J., 393, 400
Clark, W. R., 146
Clem, O. M., 178
Cluver, E. H., 50

505